Compressible Fluid Dynamics and Shock Waves

Akihiro Sasoh

Compressible Fluid Dynamics and Shock Waves

Springer

Akihiro Sasoh
Department of Aerospace Engineering
Nagoya University
Nagoya, Japan

ISBN 978-981-15-0503-4 ISBN 978-981-15-0504-1 (eBook)
https://doi.org/10.1007/978-981-15-0504-1

Original Japanese edition: "Assyuku sei ryuutai rikigaku Shougeki ha" by Akihiro Sasoh. Copyright © 2017 by Akihiro Sasoh. Published by CORONA PUBLISHING CO.,LTD., 4-46-10, Sengoku, Bunkyo, Tokyo, Japan

This Springer imprint is published by the registered company Springer Nature Singapore Pte Ltd.
The registered company address is: 152 Beach Road, #21-01/04 Gateway East, Singapore 189721, Singapore

To Michiyo, Asako, Tomoko, Go and Hiromichi

Preface

Compressible fluid dynamics captures interactions between pressure waves and flow, which are accompanied by beautiful physical nature. It is based on the conservation relations of mass, momentum and energy over a control volume. The readers are believed to master its basic principle in a short period by using only simple mathematics. In many problems, we should consider how the flow reacts to the variations of the flow passage area, and to force and heat exchanges. The readers of this book can produce various flow functions in many applications, thereby generating important values to the society.

Most of the contents of this book is the translation from a book *Compressible Fluid Dynamics and Shock Waves* published in 2017 in Japanese from Corona Publishing Co., Ltd., Tokyo, Japan, who kindly permitted me to publish this English version. I am grateful for many illustrations provided by many colleagues and institutions acknowledged in the respective parts. Also, I appreciate the excellent editing support by Ms. Akiko Matsuda and many alumni of Shock Wave and Space Propulsion Research Group, Department of Aerospace Engineering, Nagoya University, Japan.

Nagoya, Japan Akihiro Sasoh

Contents

1 Propagation of Pressure Waves 1
- 1.1 Propagation of Sound 1
- 1.2 Sound Waves from Flying Object 2
- 1.3 Motion of One-Dimensional Beads and Wave Propagation 5
  - 1.3.1 Piston–Bead Collision 5
  - 1.3.2 Bead–Bead Collision 6
  - 1.3.3 Motions of Piston and Beads 6
  - 1.3.4 Characteristic Velocity 7
  - 1.3.5 Mean Bead Velocity 7
  - 1.3.6 Mean Kinetic Energy 7
  - 1.3.7 Compression Ratio 8
  - 1.3.8 Force on the Piston 9
- 1.4 Pressure-Wave Propagation After Solid–Solid Collision 10
- Reference 11

2 Motion of Gas Particles and Thermodynamics 13
- 2.1 Basics of Thermodynamics 13
- 2.2 Thermal Speed and Flow Velocity 17
- 2.3 Pressure 18
  - 2.3.1 Column: Thrust of a Rocket Engine 20
- 2.4 Internal Energy and Temperature 21
  - 2.4.1 Column: Velocity Distribution Function and Thermodynamic Properties in LTE 24
- 2.5 Equation of State of Ideal Gas 26
  - 2.5.1 Column: Mean Free Path 28
  - 2.5.2 Column: Real Gas 31
- 2.6 Isentropic Processes 32
- 2.7 Enthalpy, Total Temperature, and Total Pressure 33
- 2.8 Multicomponent Gas Mixture 37
- References 39

3 Basic Equations . . . 41
3.1 Conservation Equations . . . 41
3.1.1 Conservation of Mass . . . 41
3.1.2 Conservation of Momentum . . . 42
3.1.3 Conservation of Energy . . . 46
3.1.4 Other Relations . . . 48
3.1.5 Similarity in Inviscid Flow . . . 48
3.2 Galilean Transformation . . . 49
3.2.1 Inertial Frame of Reference . . . 49
3.2.2 Galilean Transformation . . . 50

4 Discontinuity . . . 55
4.1 Condition and Classification of Discontinuity . . . 55
4.1.1 Rankine–Hugoniot Relation . . . 55
4.1.2 Classification of Discontinuity . . . 58
4.2 Normal Shock Wave . . . 61
4.2.1 General Characteristics . . . 61
4.2.2 Equations for Calorically Perfect Gas . . . 66
4.2.3 Glancing Incidence . . . 77
4.2.4 Stability of Shock Wave Front . . . 78
4.2.5 Shock-Wave Propagation with Boundary Layer . . . 78
4.3 Oblique Shock Wave . . . 79
4.3.1 Oblique Shock Relations . . . 79
4.3.2 Mach Wave . . . 80
4.3.3 Two Solutions and Their Post-shock Mach Numbers . . . 82
4.3.4 Attached and Detached Shock Waves . . . 85
4.4 Interface and Its Stability . . . 87
4.5 Rayleigh–Taylor Instability . . . 88
4.6 Richtmyer–Meshkov (R–M) Instability . . . 88
4.6.1 Kelvin–Helmholtz (K–H) Instability . . . 92
References . . . 92

5 Quasi-One-Dimensional Flows . . . 93
5.1 Control Volume and Basic Equations . . . 93
5.1.1 Control Volume and Associated Equations . . . 93
5.1.2 Equations in Derivative Form . . . 97
5.2 Flow Characteristics . . . 99
5.2.1 Influence Coefficients . . . 99
5.2.2 Effects of Variation in Cross-Sectional Area . . . 99
5.2.3 Effects of Heating/Cooling . . . 101

5.2.4 Effects of Friction 101
5.2.5 Effects of Body Force 102
5.2.6 Choking Condition 102
5.3 Duct Flow with Friction 103

6 Systems with Source Terms 107
6.1 Generalized Rankine–Hugoniot Relations 107
6.2 Detonation/Deflagration 111
6.2.1 Solution Regime 112
6.2.2 Detonation 113
6.2.3 Deflagration 115
6.2.4 Entropy Variation 116
6.2.5 Energy Variation 117
6.2.6 ZND Model 119
6.2.7 Cellular Structure 120
6.3 Ram Accelerator 122
6.3.1 Operation Principle and Characteristics 122
6.3.2 Derivation of Thrust 123
6.3.3 Thermally Choked Operation 125
6.3.4 Experiments on the Ram Accelerator 127
6.4 Thrust by Exhaust Jet 128
6.5 Air-Breathing Engine 130
References 133

7 Two-Dimensional Flows 135
7.1 Compression/Expansion Waves 135
7.2 Prandtl–Meyer Expansion 143
7.3 Supersonic Flow Around a Cone 144
7.4 Reflection of Shock Waves 149
7.4.1 Shock Reflection Patterns in Steady Flow 149
7.4.2 Shock Polar 150
7.4.3 Two Shock Theory 151
7.4.4 Three-Shock Theory 152
7.4.5 Transition Criteria 154
7.4.6 Shock Wave Reflection in Pseudo-Steady Flows 155
7.5 Shock Wave—Boundary Layer Interaction 157
7.6 Practice: Supersonic Flow Incident on an Inverted Triangle Wing 158
References 162

8 Unsteady, One-Dimensional Flows 163
8.1 Sound Wave 163
8.2 Characteristic Velocity and Invariants 165
8.3 Compression Wave 171

8.4 Expansion Wave 176
8.4.1 Exercise: Piston Falling in Tube 177
8.5 Pressure-Wave Propagation Around Normal Shock Wave 181
8.6 Shock-Wave Propagation in Variable Area Duct 183
8.7 Blast Wave 185
References 195

9 Riemann Problem 197
9.1 Definition and Solution 197
9.2 Shock Tube 204
9.3 Reflection of Normal Shock Wave 209
9.4 Reflection of Expansion Fan 213
9.5 Shock–Shock Interactions 217
9.5.1 Head-on Collision 217
9.5.2 Shock Overtaking Another One 218
9.6 Shock Interaction with Contact Surface 220
References 227

10 Method of Characteristics 229
10.1 Design of Supersonic Nozzle 229
10.1.1 Characteristics and Flow Variation 229
10.1.2 Design Procedure of Laval Nozzle 235
10.2 Wave Diagram of Shock Tube Operation 236

11 Generation and Utilization of Compressible Flows 241
11.1 Nozzle and Orifice 241
11.1.1 Isentropic Flow with Varying Cross Section 242
11.1.2 Mass Flow Rate 244
11.1.3 Thrust 247
11.1.4 Nozzle Flow Patterns with Various Nozzle Pressure Ratios 250
11.2 Supersonic Diffuser 253
11.2.1 Quasi-One-Dimensional Operation 254
11.2.2 Multidimensional Effects 258
11.2.3 Pseudo-Shock 259
11.3 Supersonic Test Facilities 260
11.3.1 Supersonic Wind Tunnel 260
11.3.2 Supersonic Free Flight 262
11.4 Unsteady Operation Driver 262
11.5 Shock Tunnel 263
11.6 Expansion Tube 267
11.7 Ballistic Range 272
References 276

Chapter 1
Propagation of Pressure Waves

Gas and liquid are termed *fluid* because they can flow with changes to their shape. Most flows occurring in our daily life are not significantly influenced by *compressibility*. However, in high-speed flows and/or when velocity or pressure rapidly varies, compressibility is significant. The *compression* and *expansion* of a fluid, which correspond to variations of its density, lead to the propagation of *pressure waves*. In compressible fluid dynamics, we study the relationship between pressure-wave propagation and its impact on flow properties. In this chapter, we will learn the basic concepts of pressure-wave propagation through several examples.

1.1 Propagation of Sound

Sound waves are the weakest pressure waves. When sound waves propagate through a quiescent gas, the gas particles oscillate around a fixed location. A sound wave is characterized by its amplitude and frequency spectra. Moreover, sound waves are not accompanied by flow, which is the movement of the center of gravity, but by the propagation of "information."

Let us consider the example shown in Fig. 1.1. When a rubber balloon bursts behind Mr. A, its information has not yet propagated to him in Fig. 1.1a at $t = t_0$. Later, at $t = t_1$ (Fig. 1.1b), the information has still not reached Mr. A. Mr. A hears the "bun" sound only at $t = t_2$ (Fig. 1.1c), thereby getting to know about the event.

In the *wave diagram* of Fig. 1.1d, the regions in which the sound wave arrives are bounded by the *ray* of the leading sound. Here, x is defined as the horizontal coordinate, and its positive direction is from the balloon location toward Mr. A. x_A and x_B are the locations of Mr. A and the balloon, respectively. Sound waves propagate at the *speed of sound*. The gray, inverse rectangular region with an apex at (x_B, t_0) is the *domain of influence* of the burst. In the region outside of it, the information has not arrived yet. The domain of influence is bounded by the right- and left-running rays. The magnitude of the speed of sound is equal to the reciprocal

A. Sasoh, *Compressible Fluid Dynamics and Shock Waves*,
https://doi.org/10.1007/978-981-15-0504-1_1

Fig. 1.1 A balloon bursts behind Mr. A

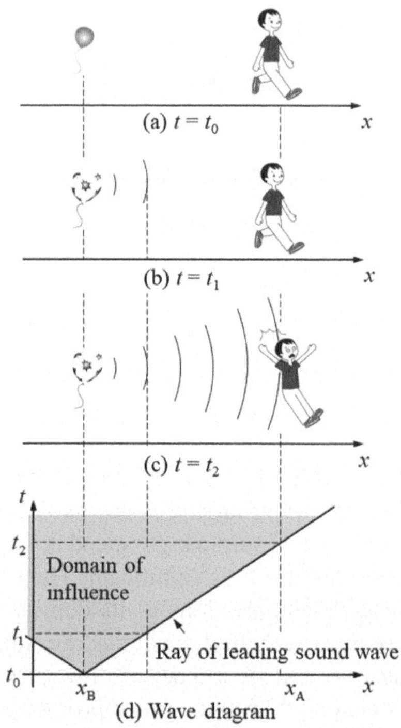

of the slope of the rays, or the reciprocal of $\mathrm{d}t/\mathrm{d}x$. A smaller slope means that the sound propagates faster.

If the speed of sound is constant, a sound cannot catch up with preceding ones. In Fig. 1.2a, Mr. A says, "The answer is XX." However, he found the answer wrong shortly after, and in Fig. 1.2b, he says, "I cancel it!" Unfortunately, he cannot cancel the sound of his first answer because a propagating sound cannot catch up with preceding ones.

In compressible fluid dynamics, pressure waves other than sound waves occur. A "strong" pressure wave influences the flow behind it. The stronger the wave is, the more the flow changes. Usually, the strength of a pressure wave is evaluated from the increment of pressure. In high-speed flows and/or when the flow velocity suddenly varies, a *shock wave* is generated. Pressure waves behind a shock wave can catch up with it, as shown in Fig. 1.3. This is an important characteristic in compressible fluid dynamics, the details of which we will learn step-by-step hereafter.

1.2 Sound Waves from Flying Object

Let us consider sound propagation from a flying object. The topology of sound propagation depends on whether the flight speed is higher than the speed of sound.

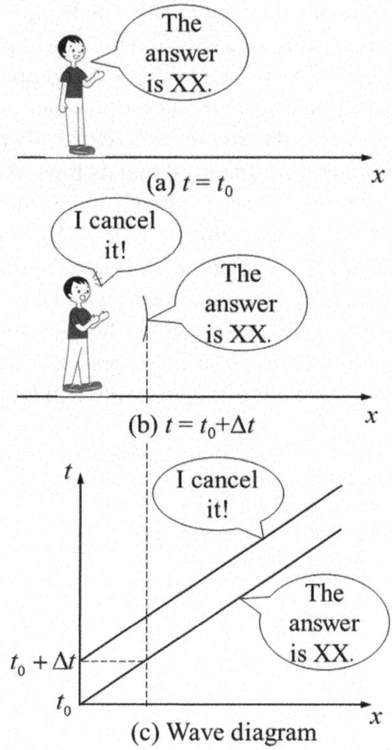

Fig. 1.2 A sound wave cannot be canceled

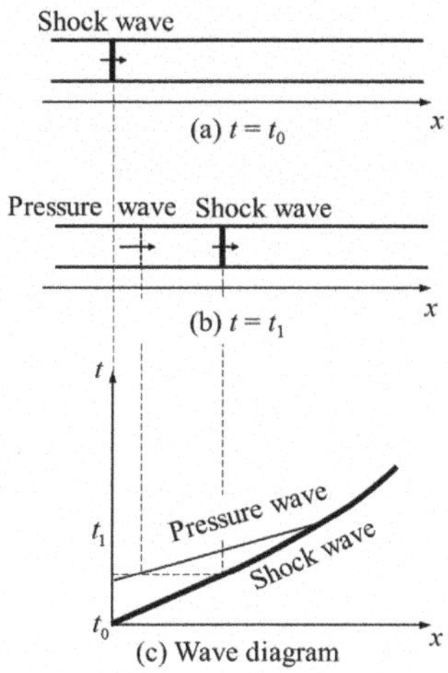

Fig. 1.3 A pressure wave can catch up with a shock wave

In *subsonic flight*, in which the object flies at a speed lower than the speed of sound, sound waves from the object propagate around it in all directions, as shown in Fig. 1.4. As shown in Fig. 1.2, a propagating wave does not catch up with a preceding wave; therefore, the latter propagates ahead of the former. Although few waves are shown in the figure, a large number of waves occur in reality; consequently, the domain of influence expands toward a wide region.

In *supersonic flight* in which the flight speed exceeds the speed of sound, the sound waves from the flying object form an envelope, as shown in Fig. 1.5. Such an envelope of sound waves is termed a "*Mach wave*." Neglecting the size of the object, the envelope forms a *Mach cone*. The domain of influence is the inside of the Mach cone, as shown by the gray region in Fig. 1.5. If the size of the object is not negligible, even stronger pressure waves, compression waves, or shock waves will be generated. Further details will be presented in Chap. 4 and thereafter.

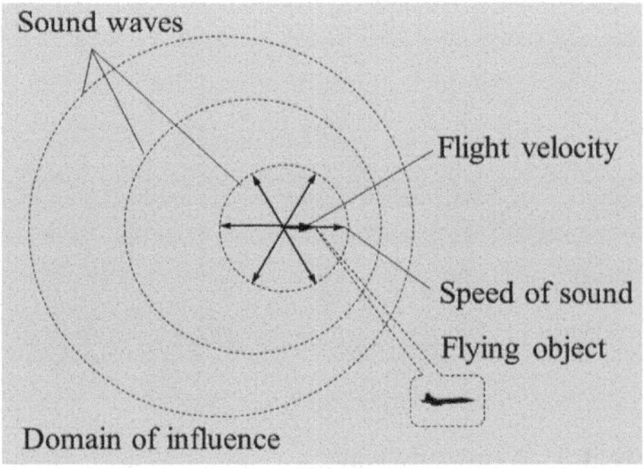

Fig. 1.4 Propagation of sound waves in subsonic flight. The gray region is the domain of influence

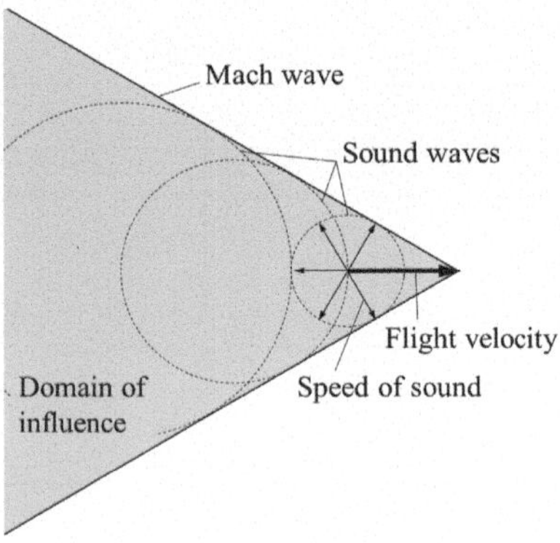

Fig. 1.5 Propagation of sound waves in supersonic flight. The gray region is the domain of influence

1.3 Motion of One-Dimensional Beads and Wave Propagation

Let us analyze wave propagation and the associated flow in a simple, one-dimensional model [1]. Consider the beads on the wire model shown in Fig. 1.6. The beads are aligned with a separation of l along the x-axis. Each bead has a mass of m. A piston with a mass of m_p collides against the leftmost bead at a constant speed of u_p. m_p is assumed to be much larger than m. Even in such a simple system, important characteristics of compressible fluid dynamics are evident.

1.3.1 Piston–Bead Collision

When the piston makes an "elastic" collision against a bead at rest, the following relations hold.

Conservation of momentum:

$$m_\mathrm{p}u_\mathrm{p} = m_\mathrm{p}u'_\mathrm{p} + mu', \tag{1.1}$$

Conservation of energy:

$$\frac{1}{2}m_\mathrm{p}u_\mathrm{p}^2 = \frac{1}{2}m_\mathrm{p}u_\mathrm{p}'^2 + \frac{1}{2}mu'^2, \tag{1.2}$$

where quantities after the collision are denoted by a prime. From (1.1) and (1.2), and using the assumption that $m/m_\mathrm{p} << 1$,

$$u' = \frac{2u_\mathrm{p}}{1 + m/m_\mathrm{p}} \simeq 2u_\mathrm{p}, \tag{1.3}$$

$$u'_\mathrm{p} = \frac{1 - m/m_\mathrm{p}}{1 + m/m_\mathrm{p}}u_\mathrm{p} \simeq u_\mathrm{p}. \tag{1.4}$$

After the piston collides against the bead at rest, the bead recoils at a speed of $2u_\mathrm{p}$, while the speed of the piston remains unchanged. This important collision dynamics, that is, the fact that a light object recoils from a heavy object after collision with double the speed of the heavy object, is applied to high-speed launch.

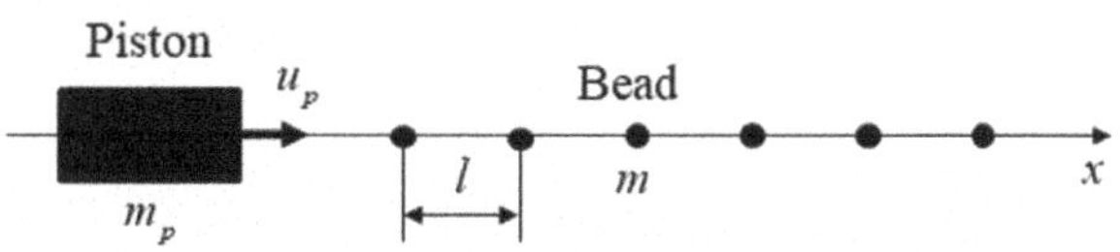

Fig. 1.6 A model of beads on a wire

1.3.2 Bead–Bead Collision

Consider that Bead 1 with a speed of $u_1 = 2u_\mathrm{p}$ collides from left to right against Bead 2 with a speed of $u_2 = 0$. By making the substitutions $u_\mathrm{p} \to u_1$, $u \to u_2$, and $m_\mathrm{p} \to m$ in (1.3) and (1.4),

$$u_1' = 0, \tag{1.5}$$

$$u_2' = u_1. \tag{1.6}$$

That is, the velocities are swapped after the collision such that the bead motion continues to the right.

1.3.3 Motions of Piston and Beads

The motion of this system is expressed as shown in Fig. 1.7. Because the piston has a constant velocity of u_p, its trajectory is a straight line. Let τ denote the time interval between successive piston–bead collisions:

$$\tau = \frac{l}{u_\mathrm{p}}. \tag{1.7}$$

Each bead repeats cycles in which it moves with a velocity of $2u_\mathrm{p}$ for a period of $\tau/2$ and stays still for the same period.

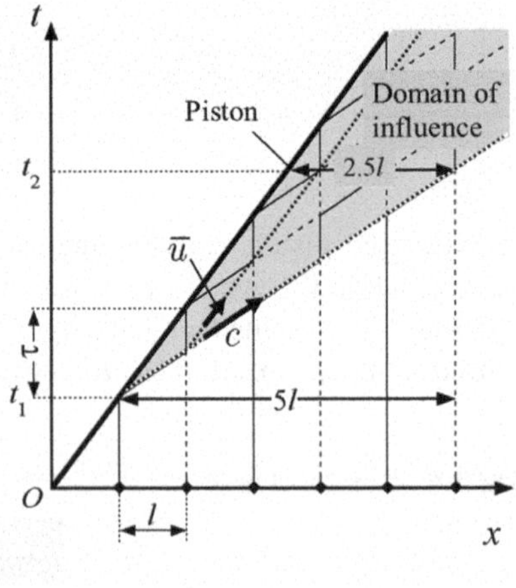

Fig. 1.7 Piston and bead trajectories. The gray region is the domain of influence of the piston motion

1.3.4 Characteristic Velocity

Up to where does the signal of the piston propagate? The velocity of the signal is termed the *characteristic velocity* and is denoted by c. This signal is delivered by the rightmost moving bead. Therefore,

$$c = 2u_{\mathrm{p}}. \tag{1.8}$$

The domain of influence, depicted as the gray region in Fig. 1.7, is formed behind it.

1.3.5 Mean Bead Velocity

The time-averaged bead velocity, $\bar{u}$, is expressed as

$$\bar{u} = \frac{l}{\frac{\tau}{2} + \frac{\tau}{2}} = u_{\mathrm{p}}. \tag{1.9}$$

1.3.6 Mean Kinetic Energy

Let us obtain the mean kinetic energy of a bead after a long period when many beads have started to move.

Mean kinetic energy obtained from bead motions, K_1.

At any moment, half of the beads propagate at a velocity of $2u_{\mathrm{p}}$, and the other half stay still. Therefore,

$$K_1 = \frac{1}{2}m(2u_{\mathrm{p}})^2 \times \frac{1}{2} + \frac{1}{2}m \cdot 0^2 \times \frac{1}{2} = mu_{\mathrm{p}}^2. \tag{1.10}$$

Mean kinetic energy obtained from input energy from piston, K_2.

At the moment after a period τ' as the piston first collided against the leftmost bead,

$$(\text{Number of beads in the domain of influence}) = \frac{c\tau'}{l} = \frac{2u_{\mathrm{p}}\tau'}{l}, \tag{1.11}$$

$$(\text{Number of collisions against the piston}) = \frac{u_{\mathrm{p}}\tau'}{l}, \tag{1.12}$$

$$\text{(Kinetic energy acquired by the leftmost bead after each collision)} = \frac{1}{2}m(2u_\mathrm{p})^2 = 2mu_\mathrm{p}^2. \tag{1.13}$$

Therefore, the mean kinetic energy is

$$K_2 = \frac{2mu_\mathrm{p}^2 \frac{u_\mathrm{p}\tau'}{l}}{\frac{2u_\mathrm{p}\tau'}{l}} = mu_\mathrm{p}^2 = K_1. \tag{1.14}$$

Kinetic energy of center of gravity, K_3.

When the mean kinetic energy is estimated from the mean velocity, $\bar{u}$,

$$K_3 = \frac{1}{2}m\bar{u}^2 = \frac{1}{2}mu_\mathrm{p}^2. \tag{1.15}$$

Therefore,

$$K_1 = K_2 = 2K_3. \tag{1.16}$$

What is the physical meaning of these relations? Because the mean velocity $\bar{u}$ is equal to the velocity of the center of gravity, K_3 is equal to the kinetic energy of the system divided by the number of beads. Yet, in the frame of reference fixed to the center of gravity, no bead stays still: they exhibit "isotropic" motions; half of the beads move at a velocity of u_p to the right, while the other half move at $-u_\mathrm{p}$ to the left. These motions are equivalent to *thermal* motion. This thermal energy, which is equivalent to the *translational* energy of a gas,[1] yields

$$K_4 = \frac{1}{2}\cdot\frac{1}{2}m(-u_\mathrm{p})^2 + \frac{1}{2}\cdot\frac{1}{2}m(+u_\mathrm{p})^2 = \frac{1}{2}mu_p^2. \tag{1.17}$$

It follows from these results that

$$K_1 = K_2 = K_3 + K_4\,(K_3 = K_4). \tag{1.18}$$

1.3.7 Compression Ratio

Let ρ denote the "density," which is the mass of beads in a unit length in this case. The length of the system decreases after piston–bead and bead–bead collisions.

[1]See Chap. 2.

That is, the system becomes "compressed." Let us obtain the compression *ratio*, that is, the ratio of densities before and after the compression[2]:

Before the compression,

$$\rho_0 = \frac{m}{l}. \tag{1.19}$$

After the compression,

$$\rho_1 = \frac{2u_p\tau'}{2u_p\tau' - u_p\tau'}\frac{m}{l} = \frac{2m}{l}. \tag{1.20}$$

Therefore,

$$\frac{\rho_1}{\rho_0} = 2. \tag{1.21}$$

1.3.8 Force on the Piston

The force exerted on the piston is obtained as the time average of impulse input onto the leftmost bead:

$$\begin{aligned} F_1 &= (\text{impulse by single collision}) \times (\text{collision frequency}) \\ &= m \cdot 2u_p \cdot \frac{u_p}{l} = \frac{m}{l} \cdot u_p \cdot 2\,u_p = \rho_0 u_p c. \end{aligned} \tag{1.22}$$

F is obtained from another consideration. The length of the domain of influence at $\tau'(\gg \tau)$ after the first piston–bead collision equals $c\tau'$. The velocity of the center of gravity of this domain equals $\bar{u}$. The force is equivalent to the momentum that the system acquires in unit time; that is,

$$F_2 = \frac{m\frac{c\tau'}{l}\bar{u}}{\tau'} = \rho_0\bar{u}c. \tag{1.23}$$

As $\bar{u} = u_p$, $F_1 = F_2 \equiv F$. Therefore, the same physical quantity is obtained from different evaluation processes. Equation (1.23) gives an *impulsive force* during collision, F, which is the product of the intact density, particle velocity, and wave-propagation speed.

In the above analyses, we have obtained the following fluid-dynamics quantities in this "elastic" system:

---

[2]Usually, in fluid dynamics, "density" refers to the mass per unit volume. Here, the unit of density is different from the usual one.

- Speed of the wave (characteristic speed): c
- Speed of the center of mass (flow speed): $\bar{u}$
- Average energy per bead in the laboratory frame: $K_1(=K_3+K_4)$
- Average energy per bead of the center of gravity: K_3
- Average energy per bead around the center of gravity (internal energy): K_4
- Compression ratio: $\frac{\rho_1}{\rho_0}$
- Impulsive force: $F=\rho_0\bar{u}c$

If the collisions are inelastic, the energy is distributed not only to kinetic energies but also to "thermal" energy, which changes the temperature of the piston and beads. In a real fluid, the energy is distributed even to the rotational, vibrational, and electronic excitation energies of atoms and/or molecules.

1.4 Pressure-Wave Propagation After Solid–Solid Collision

Usually, a solid is not regarded as a fluid. However, the processes and behavior of their mutual collision are equivalent to the processes of compressible-flow dynamics. When we push a book on a desk or hit a golf ball with a club, the object starts moving. Even in these cases, "waves" should propagate so that the "propulsive" force is transferred to other sections. In real life, the propagation period is so short that we cannot sense this propagation; the object seems to start moving instantaneously. However, in a physical sense, it should take a finite time for the waves to propagate through the object before it starts moving. Here, let us consider such processes.

Figure 1.8a shows a wave diagram of solid–solid collision. The two solid objects, A and B, which have the same thickness and are made of the same material, experience a normal collision. Here, we neglect shearing forces. Object A moves at a speed of u_1 to the right (state A1) and collides against Object B, which is at rest. Here, the

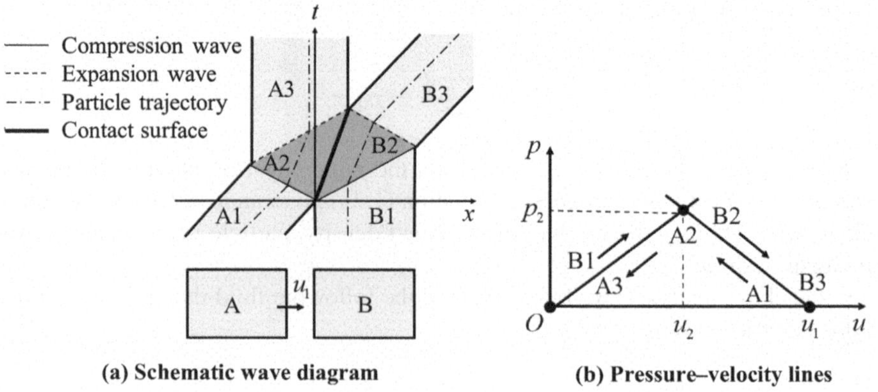

(a) Schematic wave diagram (b) Pressure–velocity lines

Fig. 1.8 Collision process between two solid objects

collision and associated wave propagation are assumed to be one-dimensional. A wave resulting in a density increase is termed as a *compression wave*, whereas a wave resulting in a density decrease is termed as an *expansion wave* or a *rarefaction wave*. The pressure also increases behind a compression wave and decreases behind an expansion wave. Immediately after A collides against B, the *contact surface*, that is, the leftmost surface of B contacting A, first starts to move because of the compression stress, which is equivalent to a pressure p_2. Subsequently, a compression wave propagates to the right. Behind the compression wave, in B2, a *particle velocity* u_2 is induced to the right with a pressure increase to p_2, as shown in Fig. 1.8b.[3] Up to the moment when the wave completes a round trip across Object B, this particle velocity and pressure remain constant. When the compression wave arrives at the right free surface of B, an expansion wave propagates as the reflected wave.[4] Behind the expansion wave (B3), the particle velocity to the right is further increased to u_1, whereas the pressure becomes restored to the ambient one, which vanishes in the figure. By contrast, a left-propagating compression wave in A propagates with a pressure increase to p_2 and a velocity decrease to u_2. Note here that, across the contact surface, both the particle velocity and pressure are continuous. At the free surface of A on the left, the compression wave is reflected as an expansion wave. Behind the expansion wave (A3), the particle velocity is further decreased, and the pressure is restored to the ambient one. In the case of Fig. 1.8, in state A3, the particle velocity vanishes because the objects are made of the same material.

When A collides with a velocity of u_1 against B, after all the wave-propagation processes, A becomes still and B starts moving at a velocity of u_1. These results are consistent with the macroscopic behavior of the elastic collision.

As presented in this chapter, the characteristics of compressible flow are closely related to those of pressure-wave propagation. In this book, we will study compressible fluid dynamics while capturing wave generation induced by object motions, its impact on flow, and the mutual interactions of waves.

Reference

1. Asay JR, Shahinpoor M (eds) (1993) Beads on a wire model: high-pressure shock compression of solids. Springer, New York, pp 12–14

[3]In fluid dynamics, a pressure usually has a positive value, whereas in solid mechanics, a tensile stress is defined to be positive.

[4]For simplicity, the expansion wave is drawn as a single ray.

Chapter 2
Motion of Gas Particles and Thermodynamics

The static characteristics of a gas are described by *thermodynamics*. Thermodynamics starts with the formulation of experimentally observed phenomena, following which the formulae are related as results of the behavior of a group of gas particles. Compressible fluid dynamics is also based on thermodynamics. The local motion of gas particles is decomposed into that of the center of gravity and that around it. *Flow* is the motion of the center of gravity, while *flow velocity* is the velocity of such motion. Pressure, temperature, and other thermodynamic properties are determined from the motion around the center of gravity, that is, the thermal motion. The flow changes when it experiences a force and/or exchange energy. In this chapter, we will derive the relations between gas-particle motions and thermodynamic properties.

2.1 Basics of Thermodynamics

In thermodynamics, we study how the behavior of a gas changes with force and/or heat. A quantity that expresses a thermodynamic condition is termed a *property* or *state variable*. Examples are the temperature T, pressure p, density ρ, internal energy e, enthalpy h, and entropy s. A dependent property is given as a function of two independent properties. For example, e is expressed as a function of T and ρ:

$$e = e(T, \rho). \tag{2.1}$$

This relation becomes particularly important for differentiation

$$\mathrm{d}e = \left(\frac{\partial e}{\partial T}\right)_\rho \mathrm{d}T + \left(\frac{\partial e}{\partial \rho}\right)_T \mathrm{d}\rho. \tag{2.2}$$

An *equation of state* is a representative equation that relates properties. A gas with negligible molecule sizes and intermolecular forces is termed an *ideal gas*. The

A. Sasoh, *Compressible Fluid Dynamics and Shock Waves*,
https://doi.org/10.1007/978-981-15-0504-1_2

pressure of an ideal gas is determined only from thermal molecular motions, and it is given by the following equation of state:

$$p = \rho RT \text{ or } pv = RT. \tag{2.3}$$

We must be mindful of which unit each property is evaluated in. The unit may be different depending on the field of study. In fluid dynamics, many quantities are evaluated for a unit mass. Such a quantity is called *specific*.[1] For example, a volume per unit mass, v, is termed as *specific volume*, which has an international system of units (SI unit) of [m^3/kg]. Furthermore, a gas constant, R, is a quantity per unit mass. By using a molecular mass, m_{mole}, R is related to the *universal gas constant*, $\Re$, by

$$R\left[\frac{\text{J}}{\text{kg}\cdot\text{K}}\right] = \frac{\Re\left[\frac{\text{J}}{\text{mol}\cdot\text{K}}\right]}{m_{\text{mole}}\left[\frac{\text{kg}}{\text{mol}}\right]}, \quad \Re = 8.3144. \tag{2.4}$$

Let us consider a gas element that expands on heating. In Fig. 2.1a, the gas expands at the same location without flow. The expanding gas element pushes the surrounding gas against the ambient pressure, thereby exerting work on the surroundings. Furthermore, the energy of the gas element is decreased by the amount of work. If heat is extracted from the gas element, that is, if it is cooled down, the element is shrunk, thereby receiving work from the surroundings. Such processes follow the *first law of thermodynamics*, which is equivalent to the *conservation of energy*:

$$\delta q = \text{d}e + \delta w. \tag{2.5}$$

Here, δq, de, and δw represent the heat input per unit mass, increment of internal energy, and work done by the gas per unit mass, respectively. Because the internal energy is a thermodynamic property, its variation is expressed by "d." The variation of other quantities is expressed by "δ" because the variation depends on processes.

Such heating/cooling and expansion/shrinkage can occur even with the flow, as shown in Fig. 2.1b. In this case, the processes are observed by tracing a gas element with a flow velocity of $\mathbf{u}$.

Here, it is necessary to note the difference between a reversible process and an irreversible process. If a gas element is slowly heated while it gradually expands and then slowly cooled down to be restored to its initial volume, each thermodynamic property is also restored to the initial value. Such a process is termed a *reversible process*. On the other hand, if the element is rapidly heated up and cooled down to its initial volume, other properties would not be restored to their initial values. Such processes are accompanied by the so-called *dissipative* processes, such as diffusion, friction, and heat conduction, and are thereby called *irreversible processes*. As will be shown later, if this element is made to expand rapidly, the pressure variation becomes different from that during a slow expansion.

---

[1]In this book, if understood from context, "specific" is omitted for simplicity.

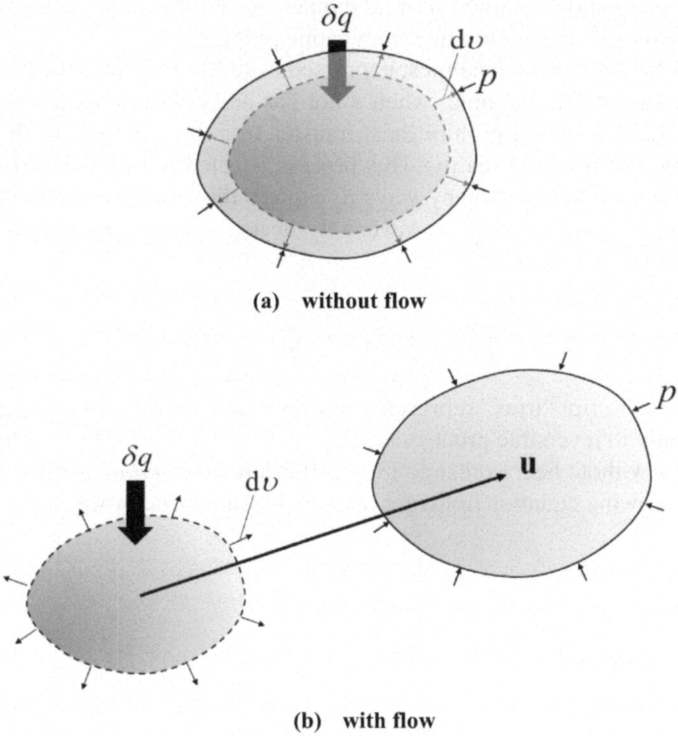

Fig. 2.1 Expansion of a gas element. The gas element is inside the broken line before expansion and inside the solid line after expansion

An (specific) *entropy*, s, is an important thermodynamic property that is indicative of how thermodynamic processes proceed. It is defined by

$$\mathrm{d}s = \frac{\delta q_{\mathrm{rev}}}{T}. \tag{2.6}$$

Here, the subscript "rev" indicates that the heat is input through a reversible process. The SI unit of s is $[\mathrm{J}/(\mathrm{kg}\cdot\mathrm{K})]$. In a reversible process, when a gas element slowly expands by $\mathrm{d}\upsilon$, it exerts work equal to $p\mathrm{d}\upsilon$ toward the surroundings. Therefore, for reversible processes,

$$\delta w_{\mathrm{rev}} = p\mathrm{d}\upsilon. \tag{2.7}$$

By applying (2.5)–(2.7) to reversible processes,

$$T\mathrm{d}s = \mathrm{d}e + p\mathrm{d}\upsilon. \tag{2.8}$$

This is an equation common in fluid dynamics, and it expresses the first law of thermodynamics by using only thermodynamic properties.

The first law of thermodynamics, expressed as (2.8), does not give the direction of energy transfer. For example, when a hot gas and cold gas are in contact with each other, (2.8) does not prohibit heat transfer from the cold gas to the hot one, which would not occur in reality. This process is ruled out by the *second law of thermodynamics*. There are many ways to express this law. A typical statement is "the entropy of a system that is irreversibly heated becomes greater than that of relevant, reversible processes."

$$\mathrm{d}s = \frac{\delta q_{\mathrm{irrev}}}{T} + \mathrm{d}s_{\mathrm{irrev}} \geq \frac{\delta q_{\mathrm{irrev}}}{T}, \quad \mathrm{d}s_{\mathrm{irrev}} \geq 0. \tag{2.9}$$

Here, the subscript "irrev" represents an irreversible process. In (2.9), the quality is applied only to reversible processes.

A process without heat exchange ($\delta q = 0$) is termed an *adiabatic process*. From (2.6), the following equation holds for a reversible, adiabatic process:

$$\mathrm{d}s = 0. \tag{2.10}$$

Therefore, such a process is *isentropic*.[2]

Enthalpy is defined as

$$h \equiv e + \frac{p}{\rho} = e + pv. \tag{2.11}$$

The second term on the rightmost side of (2.11), pv, corresponds to the energy "saved" in a volume v at a pressure p. In order to let an infinitesimally small element expand slowly from zero volume to v, the element needs to exert a work of pv on the surroundings. On the other hand, if an element with volume v slowly shrinks to an infinitesimally small volume, it experiences a work of pv from the surroundings. In fluid dynamics, it is often convenient to use enthalpy for modeling flows involving such energy exchanges.

The thermodynamic properties of a gaseous mixture of various chemically reacting species depend on the mixture ratios of the components, which in turn depend on the temperature and pressure. However, if a gas is composed only of a single species and the intermolecular force is negligible, the internal energy, e, and enthalpy, h, can be modeled as a function of T alone:

$$e = e(T), \tag{2.12}$$

$$h = h(T). \tag{2.13}$$

[2]Note the differences between reversible, adiabatic, and isentropic processes.

Such a gas is termed a *thermally perfect gas*. Without volume variation, the internal energy increases by the amount of input heat, as expressed in (2.8). Furthermore, from (2.8) and (2.11),

$$T\mathrm{d}s = \mathrm{d}(h - pv) + p\mathrm{d}v = \mathrm{d}h - v\mathrm{d}p. \quad (2.14)$$

Therefore, at a constant pressure, the enthalpy increases by the amount of input heat. Differentiating (2.12) and (2.13) yields

$$\mathrm{d}e = C_v(T)\mathrm{d}T, \quad (2.15)$$

$$\mathrm{d}h = C_p(T)\mathrm{d}T. \quad (2.16)$$

Here, C_v and C_p are the *specific heat at constant volume* and *specific heat at constant pressure*, respectively. For a thermally perfect gas, they are functions of T alone. Moreover, if they are constant,

$$e = C_vT \quad (C_v = \text{const.}), \quad (2.17)$$

$$h = C_pT \quad (C_p = \text{const.}). \quad (2.18)$$

A gas satisfying (2.17) and (2.18) is termed a *calorically perfect gas*. For air with temperature and pressure around their standard values, this simplified treatment is effective to evaluate flow properties. In this case, the *specific heat ratio,*

$$\gamma \equiv \frac{C_p}{C_v}, \quad (2.19)$$

is also constant. Consequently, variables such as flow properties and shock-wave relations can be expressed in explicit forms.

2.2 Thermal Speed and Flow Velocity

Particles in a gas move in a random manner with collisions at a finite rate. Here, we will consider particle motions and the resulting macroscopic quantities, such as flow velocity and thermodynamic properties.

Consider a group of gas particles with a mass of m with its center of gravity moving at a velocity of $\mathbf{u}$, as shown in Fig. 2.2. The velocity vector of each particle, $\mathbf{V}$, is equal to the sum of $\mathbf{u}$ and its relative velocity vector, $\mathbf{v}$:

$$\mathbf{V} = \mathbf{u} + \mathbf{v}. \quad (2.20)$$

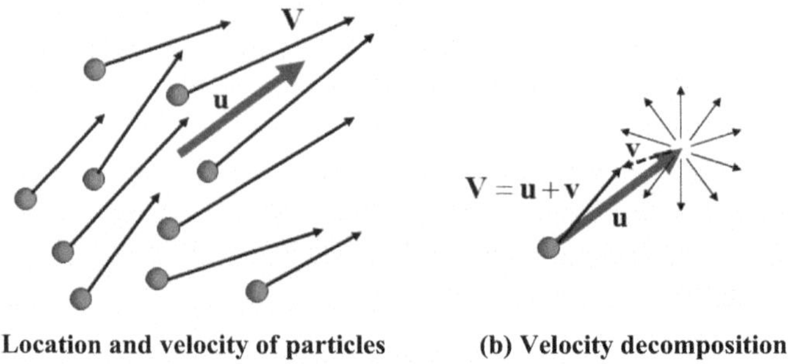

Fig. 2.2 Particle motions with a flow velocity of **u**

Let an upper bar $^{-}$ denote an ensemble average in the group at a moment. Generally, for functions f and g, the following relations hold:

$$\left.\begin{array}{l}\overline{f\cdot g}\neq\overline{f}\cdot\overline{g}\\ \overline{\overline{f}}=\overline{f}\\ \overline{f+g}=\overline{f}+\overline{g}\\ \overline{\overline{f}\cdot g}=\overline{\overline{f}}\cdot\overline{g}=\overline{f}\cdot\overline{g}\end{array}\right\}. \tag{2.21}$$

By applying (2.21) to (2.20),

$$\overline{\mathbf{V}}=\overline{\mathbf{u}}+\overline{\mathbf{v}}=\overline{\mathbf{u}}=\mathbf{u},\quad \overline{\mathbf{v}}=\mathbf{0}, \tag{2.22}$$

where $\mathbf{v}$ is equivalent to the thermal velocity. Here, $\mathbf{v}$ is isotropically distributed because the velocity of the center of mass (hereafter flow velocity), $\mathbf{u}$, is constant.

2.3 Pressure

A pressure p is a compressive force exerted in the direction normal to a unit surface. As per

$$(\text{force})=(\text{mass})\times(\text{acceleration})=\frac{(\text{mass})\times(\text{velocity})}{(\text{time})}=\frac{(\text{momentum})}{(\text{time})}, \tag{2.23}$$

a force in a physical sense is equal to the momentum produced per unit time. Based on these relations, we will derive the pressure due to the particle motions.

Now, we set a *control volume*, which is the virtual domain enclosed by the cube of broken lines in Fig. 2.3a, and analyze particle motions across the *control surface* A. Let us evaluate the number of particles and the value of the corresponding momentum

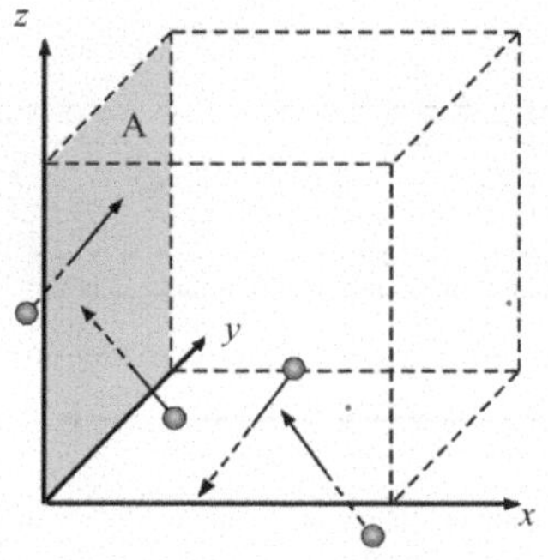

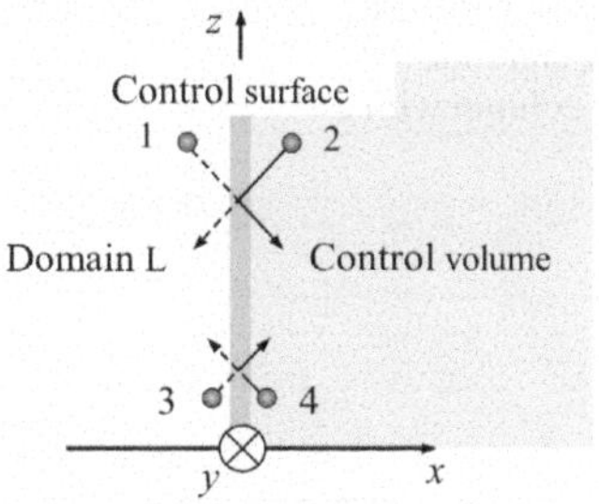

Fig. 2.3 Particle motions across a control surface A

across A. In the side view of Fig. 2.3b, for particles 1 and 3, $V_x > 0$. That is, when particles 1 and 3 enter the control volume through the control surface A, the momentum in the x -direction of the control volume is increased if they enter the control volume from the left. Particles 2 and 4 have a negative component of the velocity, that is, $V_x < 0$. If they exit from the control volume to the surroundings through the control surface A, the "negative" momentum decreases, that is, the momentum in the x-direction increases here as well. It follows from these results that the momentum in the x-direction is increased irrespective of the direction of particle motions.

A *flux* is a physical quantity that quantifies penetration through a control surface per unit time and area. A particle with $V_x > 0$ at the control surface A enters the control volume from the left domain L. The particles passing through the control surface A stayed in a column with a unit cross-sectional area and a length of V_x a unit time ago. The *mass flux* across the control surface A is equal to the mass in the column,

$$\rho V_x \,. \tag{2.24}$$

Since the momentum of a unit mass is equal to V_x, the *momentum flux* in the x-direction, I_x, is

$$I_x = \rho V_x^2, \tag{2.25}$$

which has a positive value for $V_x \neq 0$ (Fig. 2.4).

In the same manner, particles with $V_x < 0$ on the control surface A only leave the control volume to the region L. Their mass flux is also given by (2.24), but they have a negative value of mass flux. The momentum flux is obtained by multiplying the mass flux and V_x, yielding (2.25). In conclusion, irrespective to the sign of V_x, the momentum flux of (2.25) has a positive value. By using (2.21) and (2.22), we take the ensemble average of the momentum flux:

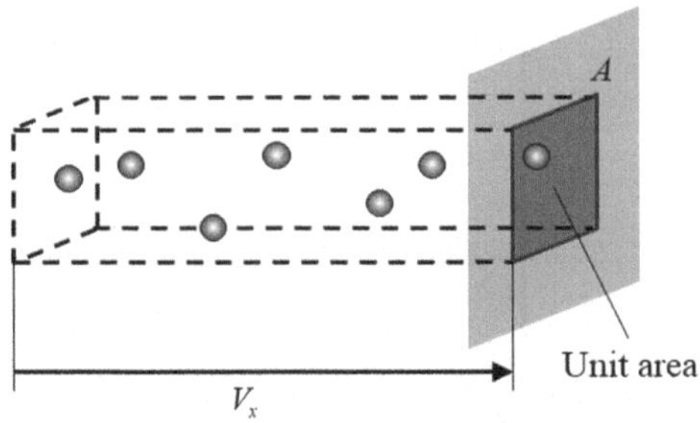

Fig. 2.4 Particles passing through a control surface A with a unit cross-sectional area during a unit time stayed in the column with a length of V_x

$$\overline{I_x} = \rho\overline{V_x^2} = \rho\overline{(u_x + v_x)^2} = \rho\left(u_x^2 + 2u_x\overline{v_x} + \overline{v_x^2}\right) = \rho\left(u_x^2 + \overline{v_x^2}\right). \quad (2.26)$$

The first term on the rightmost side of (2.26) corresponds to the contribution of the motion of the center of gravity,[3] and the second one corresponds to that of thermal motions. For isotropic, three-dimensional particle motions,

$$\overline{|\mathbf{v}|^2} = \overline{v^2} = \overline{v_x^2} + \overline{v_y^2} + \overline{v_z^2}, \quad (2.27)$$

$$\overline{v_x^2} = \overline{v_y^2} = \overline{v_z^2} = \frac{1}{3}\overline{v^2}. \quad (2.28)$$

By substituting (2.26) in (2.27) and (2.28),

$$p = \frac{1}{3}\rho\overline{v^2}, \quad (2.29)$$

where p corresponds to a force due to the thermal motion of the particles, that is, a *pressure*.[4]

2.3.1 Column: Thrust of a Rocket Engine[5]

By applying the results in Sect. 2.3, we can derive the *thrust* of a rocket engine, F. When the engine is not in operation, the pressure inside and outside of the engine is uniform at the ambient value of p_0, and $F = 0$. When the engine is in operation as shown in Fig. 2.5, exhaust gas is ejected from the exit at a speed of u_e. F is equal to the difference between the momentum ejected per unit time and the force due to the pressure on the front projection surface. By adding a subscript "e" to quantities

[3]In this book, we do not use the term "dynamic pressure" because it may cause confusion with incompressible fluid dynamics.

[4]Depending on necessity, it is also termed as *static pressure*, to be distinguished from *total pressure*, which will appear later.

[5]See also Sect. 11.1.3.

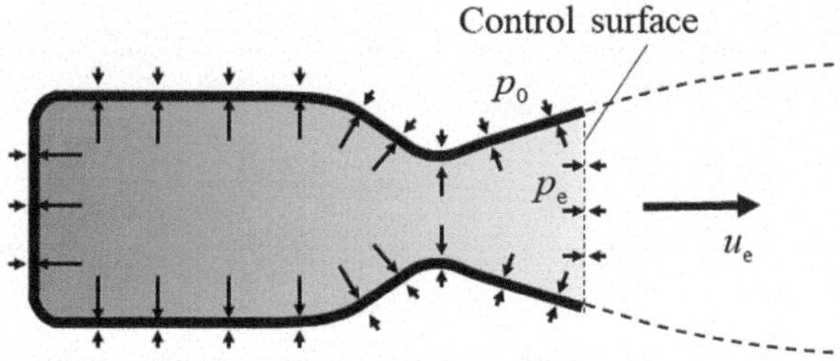

Fig. 2.5 Pressure distribution of a rocket engine in operation

at the engine exit, we obtain

$$F = \left(\rho_e u_e^2 + p_e\right) A_e - p_0 A_e. \tag{2.30}$$

Here, A_e is the cross-sectional area at the exit. The mass flow rate of the exhaust gas, $\dot{m}$, is given by

$$\dot{m} = \rho_e u_e A_e, \tag{2.31}$$

$$F = \dot{m} u_e + (p_e - p_0) A_e. \tag{2.32}$$

The first term on the right-hand side of (2.32) is the *momentum thrust*, which is caused by the momentum of the center of gravity, and the second one is the *pressure thrust* caused by the pressure difference between the engine exit and the surroundings. Usually, in a rocket for space transportation, the engine exit pressure is designed to be balanced with the ambient pressure at an appropriate altitude. This implies that, in the beginning of launch near the ground, the pressure thrust has a negative value, corresponding to *pressure loss* in rocket engineering.

2.4 Internal Energy and Temperature

Let us take the ensemble average of kinetic energy of a particle group, e_k.

$$e_k = \frac{1}{2}\overline{|\mathbf{V}|^2} = \frac{1}{2}\overline{|\mathbf{u}+\mathbf{v}|^2} = \frac{1}{2}\left(\overline{|\mathbf{u}|^2} + 2\mathbf{u}\cdot\overline{\mathbf{v}} + \overline{|\mathbf{v}|^2}\right) = \frac{1}{2}|\mathbf{u}|^2 + \frac{1}{2}\overline{v^2}. \tag{2.33}$$

The first term on the rightmost side of (2.33) corresponds to the kinetic energy of the center of gravity, and the second one corresponds to the kinetic energy due to thermal motion, which is equivalent to the *translational energy*, e_{tr}, and is a part of the *internal energy* of molecules, as shown in Fig. 2.6.

$$e_{tr} = \frac{1}{2}\overline{v^2}. \tag{2.34}$$

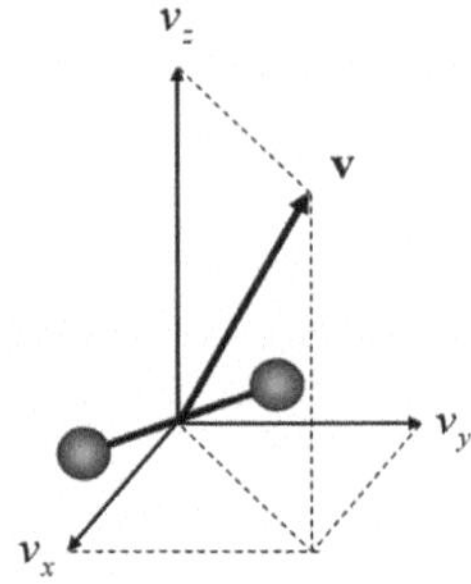

Fig. 2.6 Translational motion of a diatomic molecule

From (2.29) and (2.34),

$$e_{\mathrm{tr}} = \frac{3}{2}\frac{p}{\rho}. \tag{2.35}$$

Temperature is a physical quantity defined based on the proportion of energy of a molecule. In this book, we introduce the following results obtained from statistical physics [1]:

An energy of $\frac{1}{2}kT$ is distributed to a single degree of freedom in a molecule.

Here, $k = 1.38 \times 10^{-23}$[J/K] is the *Boltzmann constant*. In a single degree of freedom, the energy is expressed by

$$\frac{1}{2}\alpha X^2 \quad (\alpha = \text{const.}). \tag{2.36}$$

The total energy of a molecule having ϕ degrees of freedom is equal to

$$\frac{\phi}{2}kT. \tag{2.37}$$

By quantifying it for a unit mass,

$$e = \frac{\phi}{2}RT, \tag{2.38}$$

$$R \equiv \frac{k}{m}. \tag{2.39}$$

Here, R and m are the gas constant and molecular mass, respectively, as described in Sect. 2.1.

The translational energy, e_{tr}, has three degrees of freedom ($\phi = 3$):

$$\frac{1}{2}m\overline{v_x^2} = \frac{1}{2}m\overline{v_y^2} = \frac{1}{2}m\overline{v_z^2} = \frac{1}{2}kT. \tag{2.40}$$

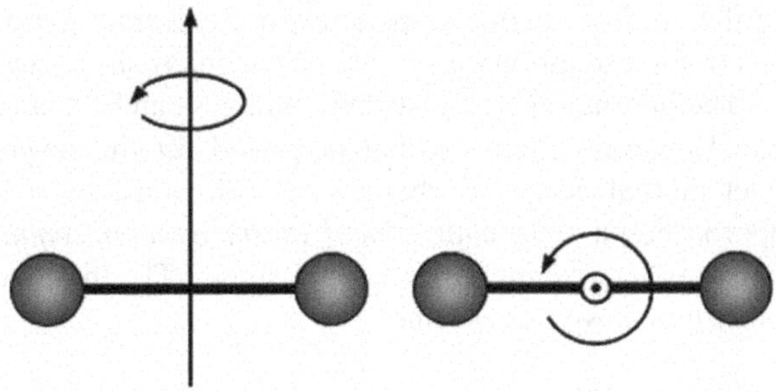

Fig. 2.7 Rotation of a diatomic molecule

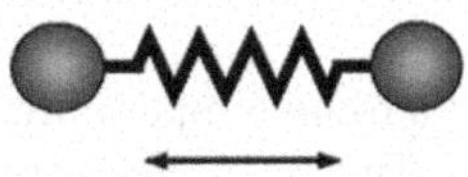

Fig. 2.8 Vibration of a diatomic molecule

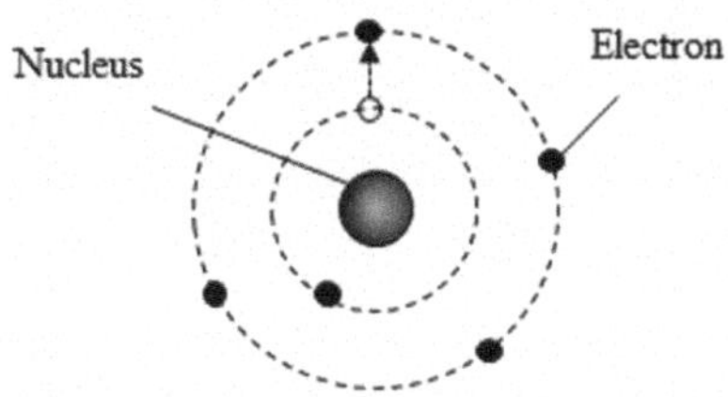

Fig. 2.9 Electronic excitation

From (2.28), (2.34), and (2.40),

$$e_{\text{tr}} = \frac{3}{2}\frac{k}{m}T = \frac{3}{2}RT. \tag{2.41}$$

The internal energy is the sum of the translational energy (Fig. 2.6), rotational energy (Fig. 2.7), vibrational energy (Fig. 2.8), and electronic excitation energy (Fig. 2.9):

$$e = e_{\text{tr}} + e_{\text{rot}} + e_{\text{vib}} + e_{\text{el}}. \tag{2.42}$$

The molecular internal modes are excited by collisions. The higher the translational energy, the higher is the rate of excitation. This implies that the higher the temperature, the larger is the degree of freedom. A diatomic molecule has two axes of rotation perpendicular to the internuclear axis, thereby having two degrees of freedom, as shown in Fig. 2.7. In molecule, the rotational energy, e_{rot}, is excited at several tens of kelvins and higher. At several hundreds of kelvins and higher, the effects of vibrational energy, e_{vib}, gradually increases with increasing temperature. The threshold value of temperature depends on the chemical species. In air, it is of the order of 600 K. In diatomic molecules, as shown in Fig. 2.8, the vibrational energy has two degrees of freedom, corresponding to the kinetic and potential energy. With further increase in the temperature up to several thousands of kelvins and higher, the electronic excitation energy, e_{el}, is added, as shown in Fig. 2.9.

The internal-energy distribution in gas can be changed by the flow, energy input, etc. However, after a sufficient period for the respective energy modes to conduct mutual energy exchanges by collisions, the local condition asymptotically approaches a state called *local thermodynamic equilibrium (LTE)*. In this book, unless otherwise mentioned, we assume LTE. In monoatomic gases, air, and many other gases, we can assume

$$\phi = \begin{cases} 3\,(\text{monoatimic gas}) = 3(\text{translational}) \\ 5\,(\text{diatomic gas}) = 3\,(\text{translational}) + 2\,(\text{rotational}) \end{cases}, \tag{2.43}$$

at a room temperature up to approximately 1,000 K.

The *total energy* of a fluid, e_t, is the sum of the kinetic energy of the center of mass and the internal energy:

$$e_t = \frac{1}{2}|\mathbf{u}|^2 + e. \tag{2.44}$$

2.4.1 Column: Velocity Distribution Function and Thermodynamic Properties in LTE

According to gas kinetics, the distribution of thermal velocity, $\mathbf{v}$, follows a Maxwell distribution, as shown in Fig. 2.10 [2]:

$$f(v_x, v_y, v_z) = \left(\frac{m}{2\pi kT}\right)^{3/2} \exp\left\{-\frac{m}{2kT}\left(v_x^2 + v_y^2 + v_z^2\right)\right\}. \tag{2.45}$$

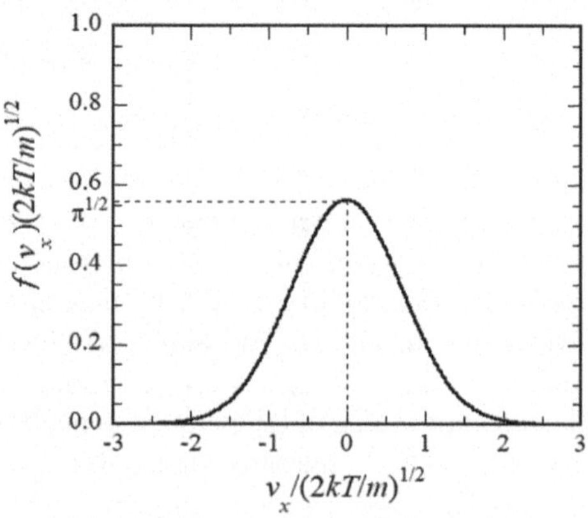

Fig. 2.10 Maxwell distribution of thermal velocity

Here, f is a velocity distribution function, which is defined such that the probability of velocity components in the range of $v_x \sim v_x + dv_x$, $v_y \sim v_y + dv_y$, and $v_z \sim v_z + dv_z$ is given by

$$f(v_x, v_y, v_z)dv_x dv_y dv_z = f(v_x)dv_x f(v_y)dv_y f(v_z)dv_z. \tag{2.46}$$

From statistical physics,

$$f(v_x) = \left(\frac{m}{2\pi kT}\right)^{1/2} \exp\left(-\frac{m}{2kT}v_x^2\right), \tag{2.47}$$

with the normalized condition of

$$\int_{v_x=-\infty}^{\infty} f(v_x)dv_x = 1, \tag{2.48}$$

$$\int_{-\infty}^{\infty} \exp(-a\xi^2)d\xi = \sqrt{\frac{\pi}{a}}, \tag{2.49}$$

$$\int_{-\infty}^{\infty} \xi^2 \exp(-a\xi^2)d\xi = \frac{\sqrt{\pi}}{2a^{3/2}}. \tag{2.50}$$

Using these equations, the following relations are confirmed:

$$p = \int_{-\infty}^{\infty}\int_{-\infty}^{\infty}\int_{-\infty}^{\infty} \rho v_x^2 f(v_x, v_y, v_z)dv_x dv_y dv_z = \rho RT, \tag{2.51}$$

$$e_{\text{tr}} = \int_{-\infty}^{\infty}\int_{-\infty}^{\infty}\int_{-\infty}^{\infty} \frac{1}{2}(v_x^2 + v_y^2 + v_z^2) f(v_x, v_y, v_z)dv_x dv_y dv_z = \frac{3}{2}\frac{kT}{m} = \frac{3}{2}RT. \tag{2.52}$$

Let us obtain the ensemble average of thermal speed. The result depends on the type of averaging. The ensemble average of the absolute value of thermal speed, $\bar{v}$, is given by

$$\bar{v} \equiv \int_{-\infty}^{\infty}\int_{-\infty}^{\infty}\int_{-\infty}^{\infty} |v| f(v_x, v_y, v_z)dv_x dv_y dv_z, \tag{2.53}$$

with the spherical symmetry,

$$dv_x dv_y dv_z = 4\pi v^2 dv, \tag{2.54}$$

$$v^2 = v_x^2 + v_y^2 + v_z^2. \tag{2.55}$$

From (2.45), (2.53) to (2.55),

$$\bar{v} \equiv 4\pi\left(\frac{m}{2\pi kT}\right)^{3/2}\int_0^\infty v^3 \exp\left(-\frac{m}{2kT}v^2\right)dv = \sqrt{\frac{8kT}{\pi m}}. \tag{2.56}$$

From (2.50), the root mean square, $\sqrt{\overline{v^2}}$, is obtained as

$$\sqrt{\overline{v^2}} = \sqrt{\frac{3kT}{m}}. \tag{2.57}$$

The value of v that maximizes

$$f(v) = 4\pi\left(\frac{m}{2\pi kT}\right)^{3/2} v^2 \exp\left(-\frac{m}{2kT}v^2\right) \tag{2.58}$$

is obtained when its derivative is 0:

$$v_{\mathrm{m}} = \sqrt{\frac{2kT}{m}}. \tag{2.59}$$

These three averages have the following relation:

$$v_{\mathrm{m}} < \bar{v} < \sqrt{\overline{v^2}}. \tag{2.60}$$

2.5 Equation of State of Ideal Gas

Up to this point, we have ignored the size of molecules and intermolecular forces. Let us confirm how these assumptions are applicable to reality.

Let us denote the *number density* by n. The value of n under the standard condition of $p_0 = 1.013 \times 10^5$[Pa] and $T_0 = 273.15$[K] is known as the *Loschmidt constant*, which is $n_0 \cong 2.687 \times 10^{25} \simeq 2.7 \times 10^{25}[\mathrm{m}^{-3}]$. It follows that the average intermolecular distance is approximately equal to $n_0^{-1/3} \cong (1/3) \times 10^{-8} \cong 3$[nm]. The frequency and travel distance for molecular collisions are dependent on the *collision cross section*, σ. The effective diameter of nitrogen, which is the largest constituent of air, equals $d \cong 3.8 \times 10^{-10}$[m] $= 0.38$[nm]; for oxygen, it is 0.36 [nm]. By using this value, we obtain $\sigma = \pi d^2 \cong 4.5 \times 10^{-19}[\mathrm{m}^2]$. The mean free path, λ, is the travel distance between successive collisions: $\lambda = 1/(\sqrt{2}\sigma n_0) \cong 6 \times 10^{-8}$[m] $= 60$[nm]. In summary, in air under standard conditions, the intermolecular distance and mean free path are one and two orders of magnitude larger than the molecular size, respectively. The volume occupied by the molecules is three orders of magnitude larger than the volume of molecules. Under such conditions, we can reasonably adopt an

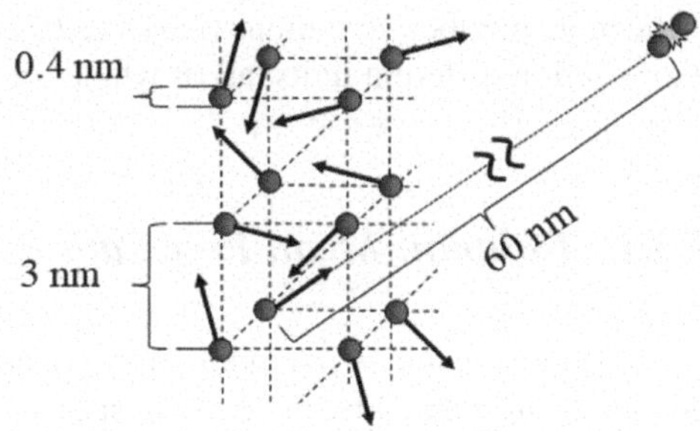

Fig. 2.11 Typical dimensions in air under standard conditions

ideal gas model in which the effects of molecular size and intermolecular forces are neglected (Fig. 2.11).

The equation of state of an ideal gas is obtained by combining (2.35) and (2.41):

$$p = \rho RT. \tag{2.61}$$

By substituting

$$\rho = mn \tag{2.62}$$

and (2.39) in (2.61), we obtain

$$p = nkT. \tag{2.63}$$

Since

$$n = \frac{N}{V}, \; N, \text{ number of molecules}; V, \text{ volume}, \tag{2.64}$$

$$pV = NkT. \tag{2.65}$$

By using the *universal gas constant,* $\Re$[J/(mol × K)], and *Avogadro's number,* $N_A = 6.022 \times 10^{23}$[mol^{-1}], we obtain

$$pV = \frac{N}{N_A} kN_A T = \widehat{M}\Re T, \tag{2.66}$$

$$\Re = kN_A, \tag{2.67}$$

where

$$\widehat{M} = \frac{N}{N_A} \tag{2.68}$$

is the mole number. Equations (2.61), (2.63), (2.65), and (2.66) are equations of state expressed in different units, all of which are equivalent to each other.

2.5.1 Column: Mean Free Path

Let us derive the *mean free path*, λ, by considering the *mean free time*, τ. These two quantities are related by the particle speed, v, as follows:

$$\lambda = v\tau. \tag{2.69}$$

Let us consider the motions of two particles, particle 1 with velocity $\mathbf{v}_1$ and particle 2 with velocity $\mathbf{v}_2$, as shown in Fig. 2.12a. The relative velocity vector, as shown in Fig. 2.12b, is given by

$$\mathbf{v} = \mathbf{v}_1 - \mathbf{v}_2. \tag{2.70}$$

By using this relative velocity, the motion of particle 1 is observed from particle 2 in the manner shown in Fig. 2.12c. Assume that they are rigid spheres with radii of r_1 and r_2, respectively. For these particles to collide, the distance, R, between the two lines in Fig. 2.12c should satisfy the following condition:

$$R \leq r_1 + r_2. \tag{2.71}$$

If we regard each particle as a dot, particle 1 has an effective collision cross section against particle 2 with a radius of R_m:

$$R_\mathrm{m} = r_1 + r_2. \tag{2.72}$$

If the particles are of the same type, R_m is equal to the diameter of the particle. In any case, the collision cross section, σ, is given by

$$\sigma = \pi R_\mathrm{m}^2. \tag{2.73}$$

The condition for a certain particle to collide with another one is equivalent to the condition that one particle exists in the volume swept by that particle; that is,

$$\sigma\lambda n = 1, \tag{2.74}$$

or

$$\lambda = \frac{1}{\sigma n}. \tag{2.75}$$

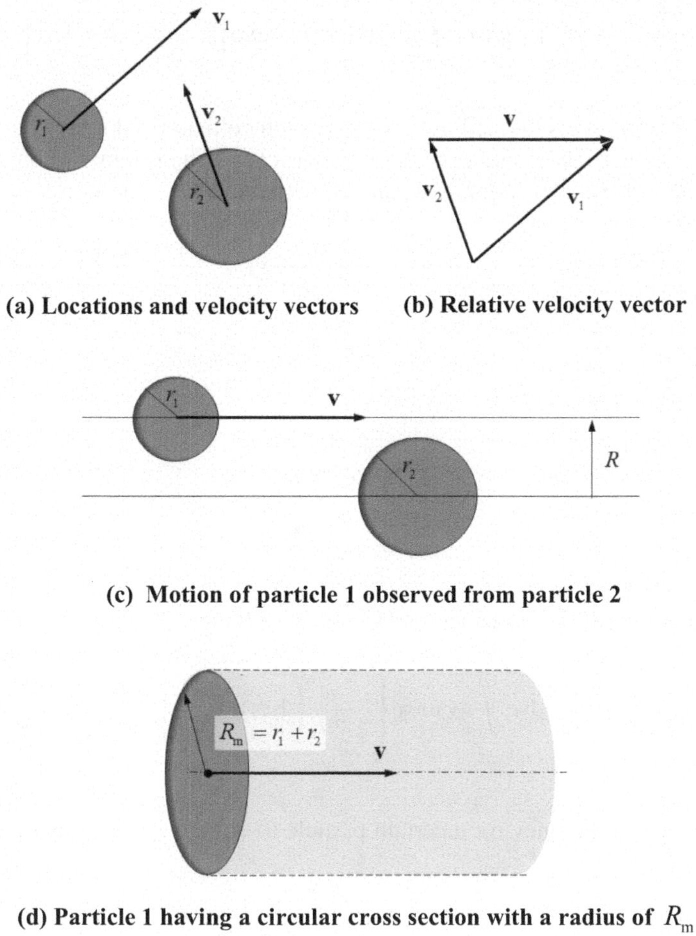

Fig. 2.12 Motions of two particles, particle 1 (left) and 2 (right)

Equation (2.75) is valid if all particles have the same speed. However, in reality, as shown in the previous column, the particles have a velocity distribution determined from statistics physics. In this case, the value of λ needs to be corrected.

In order to obtain an accurate value of λ, let us determine the collision frequency with the effect of velocity distribution taken into account. Here, we assume gas particles of the same type. Let a colliding particle be labeled 1 and the counterpart particle be labeled 1'. Of course, in practice, we cannot discriminate between particles. The particle collision frequency, Z_{11}, per unit volume and unit time is obtained by multiplying n^2 and the integration of the reciprocal of the free flight time before collision, $|\mathbf{v}_1 - \mathbf{v}_{1'}|\sigma$, with respect to the respective velocity distribution functions. By using (2.45),

$$Z_{11} = \frac{n^2\sigma}{2}\left(\frac{m}{2\pi kT}\right)^3 \int \mathrm{d}\mathbf{v}_1 \int \mathrm{d}\mathbf{v}_{1'}|\mathbf{v}_1 - \mathbf{v}_{1'}| \exp\left\{-\frac{m}{2kT}\left(v_1^2 + v_{1'}^2\right)\right\}, \tag{2.76}$$

where the integration is divided by 2 because each collision is doubly counted such as particle 1 to particle 1' and particle 1' to particle 1. The velocity of the center of mass, $\mathbf{v}_G$, and the relative velocity, $\mathbf{v}$, are expressed using $\mathbf{v}_1$ and $\mathbf{v}_{1'}$:

$$\mathbf{v}_G = \frac{m_1}{m_1 + m_{1'}}\mathbf{v}_1 + \frac{m_{1'}}{m_1 + m_{1'}}\mathbf{v}_{1'} = \frac{\mathbf{v}_1 + \mathbf{v}_{1'}}{2}, \tag{2.77}$$

$$\mathbf{v} = \mathbf{v}_1 - \mathbf{v}_{1'}. \tag{2.78}$$

Therefore,

$$\mathbf{v}_1 = \mathbf{v}_G + \frac{\mathbf{v}}{2}, \tag{2.79}$$

$$\mathbf{v}_{1'} = \mathbf{v}_G - \frac{\mathbf{v}}{2}. \tag{2.80}$$

By substituting (2.76) in (2.79) and (2.80), we obtain

$$Z_{11} = \frac{n^2\sigma}{2}\left(\frac{m}{2\pi kT}\right)^3 \int \mathrm{d}\mathbf{v}_G \int \mathrm{d}\mathbf{v}\, v \exp\left\{-\frac{m}{2kT}\left(2v_G^2 + \frac{v^2}{2}\right)\right\} = \frac{n^2\sigma}{\sqrt{2}}\sqrt{\frac{8kT}{\pi m}} = \frac{n^2\sigma\bar{v}}{\sqrt{2}}. \tag{2.81}$$

The collision frequency for a certain particle to collide against other particles is given as the reciprocal of τ:

$$\frac{1}{\tau} = \frac{2Z_{11}}{n} = \sqrt{2}n\sigma\bar{v}. \tag{2.82}$$

Here, Z_{11} is multiplied by 2 because we specify the colliding particle. From (2.69) and (2.82) with $v = \bar{v}$,

$$\lambda = \frac{\bar{v}}{\sqrt{2}n\sigma\bar{v}} = \frac{1}{\sqrt{2}n\sigma}. \tag{2.83}$$

Note here that (2.83) differs from (2.75) by a factor of $1/\sqrt{2}$ because only (2.83) is obtained by taking the effects of velocity distribution into account.

2.5.2 Column: Real Gas

When the density of the gas is such that the effects of molecular size and intermolecular forces cannot be neglected, the equation of state does not follow (2.61). Such a gas is termed a *real gas*. If we apply the order estimation in Sect. 2.5 to a gas at room temperature with a pressure 1,000 times as high as the atmospheric pressure, from simple calculus, the intermolecular distance would become of the same order as the molecular dimensions. With such a high density, *real gas effects* become significant. A typical equation of state for a real gas includes the compressibility, z:

$$p = z\rho RT. \tag{2.84}$$

Here, $z = 1$ leads to the ideal gas equation. As shown in Fig. 2.13, at $T = 300$ [K] and $p < 15.5$ [MPa], $z = 1.00 \pm 0.01$. Therefore, air is regarded as an ideal gas in this range.

Another form of the real gas equation is the *van der Waals equation of state* [3]:

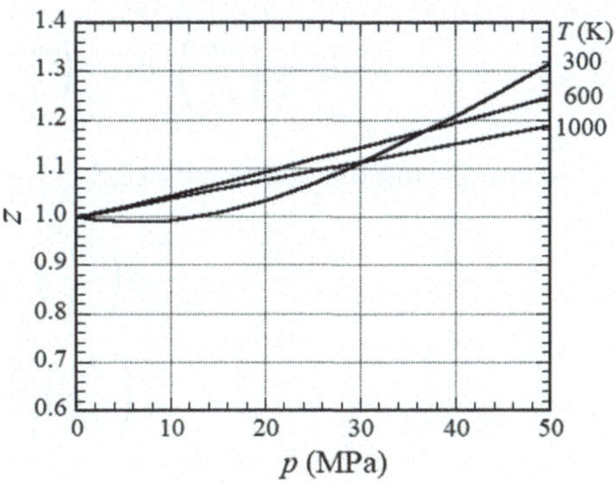

Fig. 2.13 Pressure dependence of compressibility of air

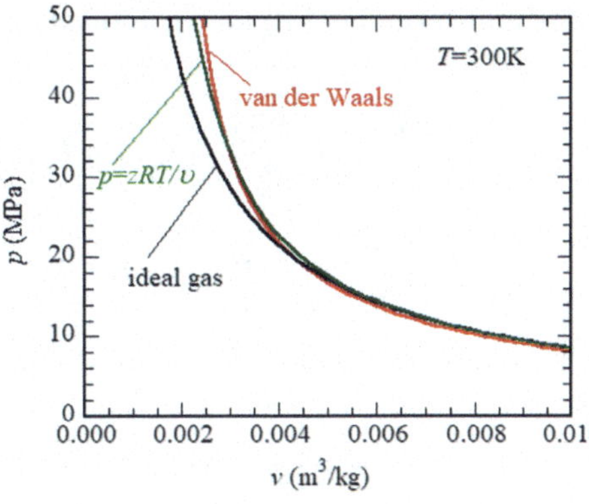

Fig. 2.14 Comparison of pressure–volume curves of real gases and an ideal gas

$$\left(p + \frac{a}{\upsilon^2}\right)(\upsilon - b) = (p + a\rho^2)\left(\frac{1}{\rho} - b\right) = RT, \tag{2.85}$$

where a and b are constants. A resultant force exerted on a molecule is proportional to the density. The number of molecules experiencing the intermolecular force is also proportional to the density. Consequently, the effect of the intermolecular force scales as $\rho^2 = 1/\upsilon^2$. Therefore, the term a/υ^2 corresponds to a decrease in pressure due to the intermolecular force. The term $-b$ reflects a decrease in the effective volume due to the volume occupied by the molecules.[6]

2.6 Isentropic Processes

Generally, flow is accompanied by loss due to heat, friction, vortices, shock waves, etc. Entropy, s, is a thermodynamic property that quantifies such losses. Unlike other thermodynamic properties, the variation in s is usually more important than its absolute value. Let us formulate e as a function of s and υ.

$$\mathrm{d}e = \left(\frac{\partial e}{\partial s}\right)_\upsilon \mathrm{d}s + \left(\frac{\partial e}{\partial \upsilon}\right)_s \mathrm{d}\upsilon, \quad e = e(s, \upsilon). \tag{2.86}$$

On comparing the above with (2.8),

$$\left(\frac{\partial e}{\partial s}\right)_\upsilon = T, \tag{2.87}$$

$$\left(\frac{\partial e}{\partial \upsilon}\right)_s = -p. \tag{2.88}$$

Even without heating, entropy increases with the generation of friction, vortices, shock waves, etc.

By transforming (2.8) for a calorically perfect gas,

$$\mathrm{d}s = C_\upsilon \frac{dT}{T} + R\frac{\mathrm{d}\left(\frac{1}{\rho}\right)}{\frac{1}{\rho}} = C_\upsilon \mathrm{d}\ln T + R\mathrm{d}\ln\left(\frac{1}{\rho}\right) = C_\upsilon \mathrm{d}\ln(T\rho^{1-\gamma}), \tag{2.89}$$

$$\frac{\mathrm{d}s}{C_\upsilon} = \mathrm{d}\ln(T\rho^{1-\gamma}) = \mathrm{d}\ln\left(p^{1-\gamma}T^\gamma\right) = \mathrm{d}\ln(p\rho^{-\gamma}). \tag{2.90}$$

Therefore, for isentropic processes with $\mathrm{d}s = 0$, the following relations hold:

$$\frac{T}{\rho^{\gamma-1}} = \text{const.}, \quad \frac{p}{T^{\frac{\gamma}{\gamma-1}}} = \text{const.}, \quad \frac{p}{\rho^\gamma} = \text{const.} \tag{2.91}$$

[6]For air, $a = 1.6 \times 10^2 [\mathrm{J} \cdot \mathrm{m}^3/\mathrm{kg}^2]$ and $b = 1.3 \times 10^{-3} [\mathrm{m}^3/\mathrm{kg}]$ (Fig. 2.14).

2.7 Enthalpy, Total Temperature, and Total Pressure

In steady flow, a stream line is drawn by connecting the tangents to the velocity vector of a flow element. For inviscid flow, the stream line is equivalent to the wall that separates adjoining flow elements. Let us consider a control volume bounded by two stream lines, as shown by the gray region in Fig. 2.15. Here, we neglect gravity. The fluid flows in through control surface 1 and flows out through control surface 2. How much energy does this fluid gain after passing through the control volume? The fluid element in the control volume experiences a force of $p_1 A_1$ in the flow direction on control surface 1. The fluid on the left-hand side of the flow element pushes it in the same manner as of a "piston" with a velocity of u_1, thereby inputting a power of $p_1 u_1 A_1$ to the element. On the other hand, on control surface 2, the flow element pushes the fluid on the right-hand side, consuming a power of $p_2 u_2 A_2$. By using a constant mass flow rate of $\rho_1 u_1 A_1 = \rho_2 u_2 A_2 (\neq 0)$, this power balance can be expressed as follows:

$$\rho_2 u_2 A_2 e_{t,2} - \rho_1 u_1 A_1 e_{t,1} = p_1 u_1 A_1 - p_2 u_2 A_2$$

$$\rho_1 u_1 A_1 \left(e_{t,1} + \frac{p_1}{\rho_1} \right) = \rho_2 u_2 A_2 \left(e_{t,2} + \frac{p_2}{\rho_2} \right),$$

$$e_{t,1} + \frac{p_1}{\rho_1} = e_{t,2} + \frac{p_2}{\rho_2}. \tag{2.92}$$

Here,

$$e_t = e + \frac{1}{2} u^2 \tag{2.93}$$

is the total energy. By using the enthalpy, as expressed in (2.11), (2.92) is expressed in a simpler form:

$$h_t \equiv h + \frac{1}{2} u^2 = \text{const.}, \tag{2.94}$$

where h_t is the *total enthalpy* or *stagnation enthalpy*. Equation (2.94) implies that the total enthalpy is conserved along a stream line. When a flow becomes stagnated at a

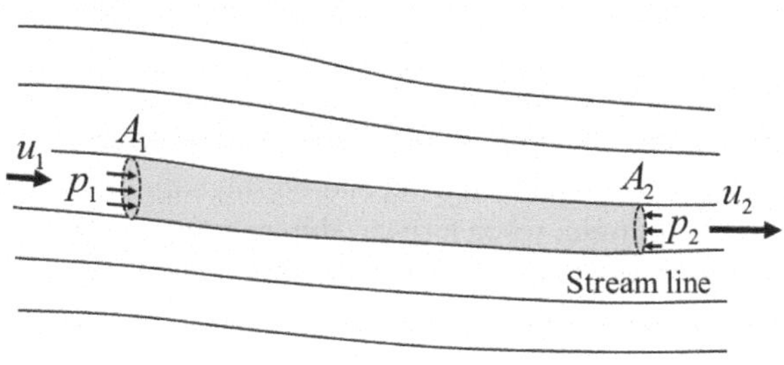

Fig. 2.15 Work exerted on a flow element bounded by two stream lines

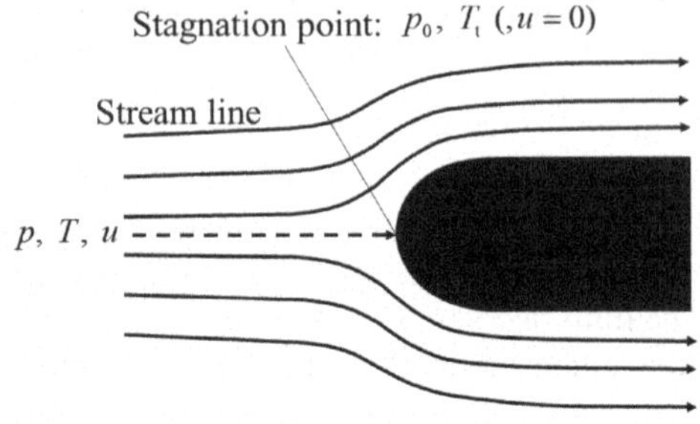

Fig. 2.16 Subsonic flow over an immersed body

point on an immersed body, as shown in Fig. 2.16, h_t is equal to the *static enthalpy* at the stagnation point.

From (2.11), (2.38), and (2.61),

$$h \equiv e + pv = e + \frac{p}{\rho} = e + RT = \left(\frac{\phi}{2} + 1\right)RT. \tag{2.95}$$

Equation (2.95) implies that h is larger than e by two degrees of freedom. This difference corresponds to an energy of pv, which is necessary to create a vacant volume of v at a pressure of p. For a calorically perfect gas, they are given by

$$e = C_v T = \frac{\phi}{2}RT = \frac{1}{\gamma - 1}RT = \frac{1}{\gamma - 1}\frac{p}{\rho}, \tag{2.96}$$

$$h = C_p T = \frac{\phi + 2}{2}RT = \frac{\gamma}{\gamma - 1}RT = \frac{\gamma}{\gamma - 1}\frac{p}{\rho}, \tag{2.97}$$

$$\gamma = \frac{\frac{\phi+2}{2}}{\frac{\phi}{2}} = \frac{\phi + 2}{\phi} = \begin{cases} \frac{5}{3}(\text{monoatomic gas at room temperature}, \phi = 3) \\ \frac{7}{5}(\text{diatomic gas at room temperature}, \phi = 5) \end{cases}. \tag{2.98}$$

When the flow of a calorically perfect gas with u, p, and T stagnates to yield a pressure of p_0 and a temperature of T_t, from (2.94),

$$\frac{\gamma R}{\gamma - 1}T_t = \frac{\gamma R}{\gamma - 1}T + \frac{1}{2}u^2, \tag{2.99}$$

$$T_t = T + \frac{\gamma - 1}{2\gamma R}u^2. \tag{2.100}$$

The temperature when the flow stagnated, T_t, is termed the *total temperature* or *stagnation temperature*. In order to distinguish T from T_t, the former is termed *static temperature*. T_t does not depend on stagnation processes or on entropy variation.

For example, when a space ship reenters the Earth's atmosphere, the entry speed is as high as 8 km/s. For a calorically perfect gas with $T = 200$ [K], the stagnation temperature is as high as

$$T_t = 200 + \frac{1.4 - 1}{2 \times 1.4 \times \frac{8.31}{29 \times 10^{-3}}} \times (8 \times 10^3)^2 = 3.2 \times 10^4 \text{ [K]}.$$

Note here that, in reality, the temperature would not increase so much because the values of specific heat would increase with the increasing temperature.

Let us see how the pressure varies with stagnation. From (2.90) and (2.100),

$$\frac{p_0}{p} = \left(\frac{T_t}{T}\right)^{\frac{\gamma}{\gamma-1}} \exp\left(-\frac{\Delta s}{R}\right) = \left(1 + \frac{\gamma - 1}{2} M^2\right)^{\frac{\gamma}{\gamma-1}} \exp\left(-\frac{\Delta s}{R}\right), \tag{2.101}$$

$$a = \sqrt{\gamma RT} = \sqrt{\frac{\gamma p}{\rho}}, \tag{2.102}$$

$$M \equiv \frac{u}{a}. \tag{2.103}$$

where a and M are the *speed of sound* and *Mach number* which is the ratio of the flow speed to the speed of sound, respectively. In particular, the pressure, p_t, resulting from isentropic stagnation, $\Delta s = 0$, is termed as a *stagnation pressure*.

$$\frac{p_t}{p} = \left(\frac{T_t}{T}\right)^{\frac{\gamma}{\gamma-1}} = \left(1 + \frac{\gamma - 1}{2} M^2\right)^{\frac{\gamma}{\gamma-1}} \tag{2.104}$$

Comparing (2.101) and (2.104), we see that the pressure is decreased by a factor of $\exp(-\frac{\Delta s}{R})$ which is caused by the entropy increase. This is termed as a *pressure loss*.

Pitot tube, Fig. 2.17, is a device to measure the Mach number of flow. A *Pitot pressure*, p_{Pitot}, is measured with its tip being directed against the flow direction. Also, the static pressure, p_1, on the side wall is measured.[7] From (2.94),

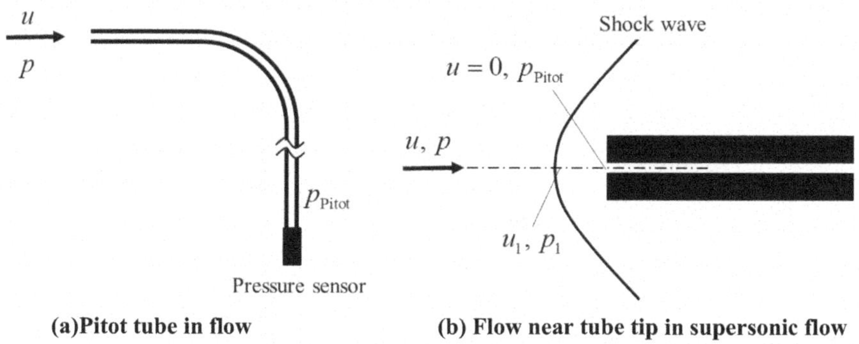

Fig. 2.17 Measurement of flow Mach number using Pitot tube

[7] For subsonic flow, the static pressure is measured using through the outer passage of a double tube.

$$\mathrm{d}h_t = \mathrm{d}\left(h + \frac{1}{2}u^2\right) = 0. \tag{2.105}$$

Letting $\mathrm{d}s = 0$ in (2.14),

$$\mathrm{d}h - \frac{\mathrm{d}p}{\rho} = 0, \tag{2.106}$$

$$\frac{\mathrm{d}p}{\rho} + \mathrm{d}\frac{u^2}{2} = 0. \tag{2.107}$$

The integration of (2.107),

$$\int \frac{\mathrm{d}p}{\rho} + \frac{u^2}{2} = \text{const.} \tag{2.108}$$

is *Bernoulli's equation*.[8] Note here that this equation is applicable only for isentropic flows.

As shown in Fig. 2.17b, if we put a Pitot tube in supersonic flow, a shock wave is generated ahead of it, being accompanied with an entropy increase. On the center axis of the Pitot tube, the shock wave becomes normal to the flow. Behind the normal shock wave, the flow becomes subsonic, then is isentropically decelerated to the stagnation point. From these relations and using results of Chap. 4, *Rayleigh's Pitot tube formula* is obtained.

$$\frac{p_1}{p} = 1 + \frac{2\gamma}{\gamma+1}(M^2 - 1) \tag{2.109}$$

$$\frac{\rho_1}{\rho} = \frac{(\gamma+1)M^2}{(\gamma-1)M^2+2} \tag{2.110}$$

$$M_1 = \left(\frac{p_1}{p}\right)^{-1/2}\left(\frac{\rho_1}{\rho}\right)^{-1/2} M \tag{2.111}$$

$$\frac{p_{\text{Pitot}}}{p_1} = \left(1 + \frac{\gamma-1}{2}M_1^2\right)^{\frac{\gamma}{\gamma-1}} \tag{2.112}$$

$$\frac{p_{\text{Pitot}}}{p} = \frac{p_1}{p}\frac{p_{\text{Pitot}}}{p_1} = \frac{\left(\frac{\gamma+1}{2}M^2\right)^{\frac{\gamma}{\gamma-1}}}{\left(\frac{2\gamma M^2-\gamma+1}{\gamma+1}\right)^{\frac{1}{\gamma-1}}} \tag{2.113}$$

As seen in Fig. 2.18, the Pitot pressure is approximately equal to the total pressure for $M \cong 1$, yet becomes lower for high Mach numbers. This is caused by the entropy

[8] Equation (2.108) differs from that for incompressible flow.

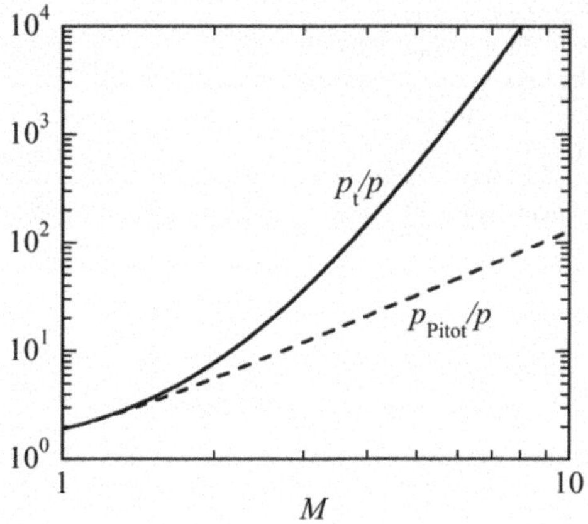

Fig. 2.18 Comparison between Pitot pressure and total pressure

increase due to the normal shock wave. With $M = 4.4$, p_{Pitot} becomes one order in magnitude smaller than p_{t}.

2.8 Multicomponent Gas Mixture

Many gases including air are mixtures of plural chemical species. Let us derive relations about thermodynamics properties of multicomponent gas. Here, we assume that each component behaves as ideal and thermally perfect gas at a temperature of T. According to *Dalton's law*, the pressure, p, is equal to the sum of the partial pressure of the components, p_i $(i = 1 \cdots N)$.

$$p = \sum_{i=1}^{N} p_i = \sum_{i=1}^{N} \rho_i R_i T = \sum_{i=1}^{N} Y_i \rho \frac{\Re}{W_i} T = \rho \overline{R} T, \tag{2.114}$$

$$\overline{R} = \frac{\Re}{\overline{W}} = \Re \sum_{i=1}^{N} \frac{Y_i}{W_i}, \tag{2.115}$$

where Y_i and W_i are the mass fraction and molecular mass, respectively, of a chemical species i. A quantity with - is a mass-averaged value. $\overline{R}$ is the gas constant of the mixture, and has a unit of $[\text{J/kg} \cdot \text{K}]$. In the same way, the enthalpy of the mixture is given by

$$h = \sum_{i=1}^{N} Y_i h_i, \tag{2.116}$$

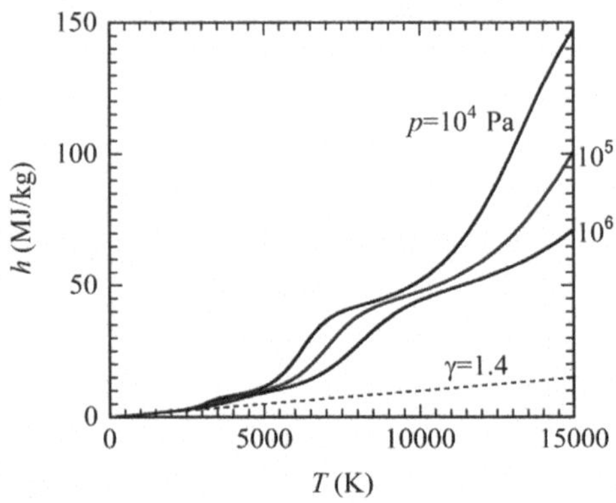

Fig. 2.19 Enthalpy of air in equilibrium, $T_{\mathrm{ref}} = 0\,(\mathrm{K})$, the broken line corresponds to calorically perfect gas with $\gamma = 1.4$ courtesy to Prof. T. Sakai (Tottori University)

$$h_i = h_{\mathrm{f},i}(T_{\mathrm{ref}}) + \int_{T_{\mathrm{ref}}}^{T} C_{p,i} dT. \tag{2.117}$$

$h_{\mathrm{f},i}(T_{\mathrm{ref}})$ is the *standard enthalpy of formation* of a species i at a reference temperature, T_{ref}, which is a necessary energy to produce the chemical species from fundamental elements such as nitrogen gas, oxygen gas, solid carbon, etc. Although h_i is a function only of T, h depends not only on T but also on p because Y_i depends on both.

Figure 2.19 shows the equilibrium air enthalpy at various temperatures and pressures. Up to around 1,000 K, the assumption of calorically perfect gas with $\gamma = 1.4$ is reasonably applied. Up to about 2,000 K, the pressure dependence is small.

In order to obtain the speed of sound of the mixture,[9] it is necessary to relate variation of species mass fraction with pressure. If the chemical reactions are so fast that the mass fractions follow to an equilibrium state, we can obtain an *equilibrium" speed of sound*.

$$a_{\mathrm{e}} = \left(\frac{\partial p}{\partial \rho}\right)_{s, Y_i = Y_{i,\mathrm{e}}(s,\rho)}, \tag{2.118}$$

where $Y_{i,\mathrm{e}}(s, \rho)$ is a mass fraction in equilibrium at an entropy s and density ρ.

Yet, in many practical conditions as is dealt with in this book, pressure fluctuation is so weak that we can neglect variation in mole fractions. Under this condition, we use a *frozen speed of sound*, which can be expressed in an explicit form. For isentropic processes,

$$dh - \frac{dp}{\rho} = 0.$$

[9]We will learn the definition and derivation of speed of sound in Chap. 8. Here, we only use the result.

Assuming calorically perfect gas,

$$dh_i = C_{p,i} dT. \tag{2.119}$$

$$h = \sum_{i=1}^{N} Y_i h_i. \tag{2.120}$$

Under the condition that Y_i of each species is kept constant,

$$dh = \sum_{i=1}^{N} (Y_i dh_i + h_i dY_i) = \sum_{i=1}^{N} Y_i dh_i = \sum_{i=1}^{N} Y_i C_{p,i} dT. \tag{2.121}$$

Substituting (2.118) with (2.121) with

$$\overline{C}_p = \sum_{i=1}^{N} Y_i C_{p,i}. \tag{2.122}$$

$$\overline{C}_p dT - \frac{dp}{\rho} = \frac{\overline{C}_p}{\overline{R}} d\left(\frac{p}{\rho}\right) - \frac{dp}{\rho} = \frac{\overline{C}_p - \overline{R}}{\rho \overline{R}} dp - \frac{p\overline{C}_p}{\rho^2 \overline{R}} d\rho = 0. \tag{2.123}$$

where (2.114) is applied. Transforming this to substituting (8.6), the frozen speed of sound is obtained such that

$$a_{\mathrm{f}} \equiv \left(\frac{\partial p}{\partial \rho}\right)_{s, Y_i (i=1 \cdots N)} = \sqrt{\overline{\gamma} \frac{p}{\rho}} = \sqrt{\overline{\gamma} \overline{R} T}, \tag{2.124}$$

$$\overline{\gamma} = \frac{\overline{C}_p}{\overline{C}_p - \overline{R}} = \frac{\overline{C}_p}{\overline{C}_v}. \tag{2.125}$$

Here, the specific heats and the specific heat ratio do no need to be constant, but should depend only on the temperature. Equation (2.124) is in the same form as of a single component gas.

References

1. Smith NO (1982) Elementary statistical thermodynamics. a problems approach. Springer
2. Vincenti WG, Kruger CH Jr (1965) Introduction to physical gas dynamics, Krieger Publishing Co, Malabar, Florida, USA (Chaps. II-IV)
3. Perry RH, Green DW (1984) Perry's chemical engineers' handbook, 6th edn. McGraw-Hill

Chapter 3
Basic Equations

When analyzing flow, it is often sufficient to observe the spatiotemporal variation of macroscopic quantities of a *continuum fluid*, such as the flow velocity and thermodynamic properties, rather than to trace all particles. In this chapter, we will derive the basic equations of compressible fluid dynamics. Although diffusion, viscosity, and heat conduction are important transport phenomena, their influence on pressure-wave propagation is quite limited. Therefore, unless otherwise stated, we will deal with *inviscid flow* in this book.

3.1 Conservation Equations

Let us consider a *control volume* with a volume V, as shown in Fig. 3.1. This is enclosed by a *control surface* A, which is a virtual surface constructed for the analysis. Let $\mathbf{u}$ be the flow velocity, which is a vector quantity. We shall formulate the conservation relations of mass, momentum, and energy.

First, let us learn an important quantity, *flux*, which quantifies a physical quantity through a control surface per unit area and during a unit time. We have "mass," "momentum," and "energy" fluxes, depending on the type of quantity.

3.1.1 Conservation of Mass

A mass flux passing through a control surface with a cross-sectional area of $\mathrm{d}A$ and an outward normal vector (unit vector) of $\mathrm{d}\mathbf{n}$ is equal to

$$\rho\mathbf{u}\cdot\mathbf{n}\mathrm{d}A. \tag{3.1}$$

Here, the flux exiting through the control surface is defined as positive. The time variation of mass in the control volume is equal to the net mass that enters and exits

A. Sasoh, *Compressible Fluid Dynamics and Shock Waves*,
https://doi.org/10.1007/978-981-15-0504-1_3

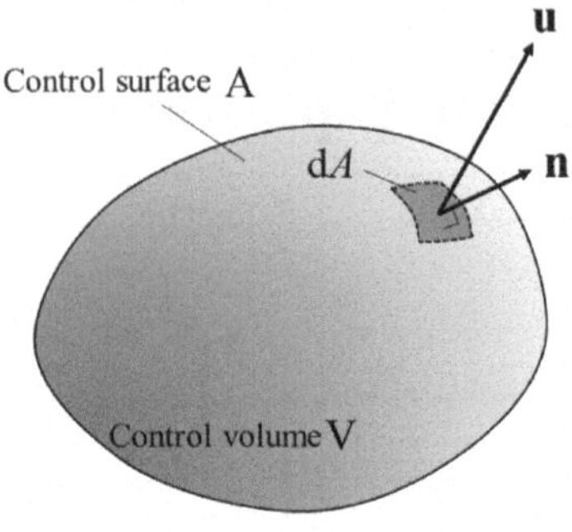

Fig. 3.1 Control volume

through the control surface:

$$\frac{\partial}{\partial t}\int_V \rho \mathrm{d}V = -\int_A \rho(\mathbf{u}\cdot\mathbf{n})dA. \tag{3.2}$$

Assume that the control volume does not move or change its shape. By applying Gauss's divergence theorem, we obtain

$$\int_V \frac{\partial \rho}{\partial t}\mathrm{d}V + \int_V \nabla\cdot(\rho\mathbf{u})dV = 0,$$
$$\int_V \left\{\frac{\partial \rho}{\partial t} + \nabla\cdot(\rho\mathbf{u})\right\}\mathrm{d}V = 0. \tag{3.3}$$

Since the above equation is applicable to any control volume,

$$\frac{\partial \rho}{\partial t} + \nabla\cdot(\rho\mathbf{u}) = 0. \tag{3.4}$$

The above is the *equation of mass conservation*.

3.1.2 Conservation of Momentum

For the variation of momentum in the control volume during a unit time, the following equation holds:

[momentum increment in control volume] = [momentum through control surface] + [force on control surface]+[force in control volume]. (3.5)

Since momentum is a vector quantity, its conservation equation should be formulated for the respective components. The momentum per unit volume is $\rho\mathbf{u}$, and the momentum flux is $-\rho\mathbf{u}(\mathbf{u}\cdot\mathbf{n})$. Several types of forces are directly exerted on each molecule, such as the force of gravity and electromagnetic forces. We name such a force as a *volume force*. While the force due to pressure over a control volume is evaluated only by accounting for the pressure on the control surface, the force due to the volume force should be evaluated by integrating over the whole volume. Let a volume force be designated by **f**. Thus, (3.5) yields

$$\frac{\partial}{\partial t}\int_V \rho\mathbf{u}dV = -\int_A \rho\mathbf{u}(\mathbf{u}\cdot\mathbf{n})dA + \int_A \boldsymbol{\sigma}\cdot\mathbf{n}dA + \int_V \rho\mathbf{f}dV. \tag{3.6}$$

By applying Gauss's divergence theorem to the i-th component of the first term on the right-hand side, we obtain

$$\left[\int_A \rho\mathbf{u}(\mathbf{u}\cdot\mathbf{n})dA\right]_i = \int_A \rho u_i(\mathbf{u}\cdot\mathbf{n})dA = \int_V \nabla\cdot(\rho u_i\mathbf{u})dV = \int_V \sum_{j=1}^{3}\frac{\partial\rho u_i u_j}{\partial x_j}dV. \tag{3.7}$$

$\boldsymbol{\sigma}$ in the second term on the right-hand side of (3.6) denotes the stress tensor. Assuming an inviscid fluid, it only has diagonal pressure components:

$$\sigma_{ij} = -p\delta_{ij}, \tag{3.8}$$

where i and j correspond to the direction of force and the direction of the normal vector, respectively. δ_{ij} is the Kronecker delta function.

$$\delta_{ij} = \begin{cases} 1\,(i=j) \\ 0\,(i\neq j) \end{cases},$$

$$\left[\int_A \boldsymbol{\sigma}\cdot\mathbf{n}dA\right]_i = \sum_{j=1}^{3}\int_A \sigma_{ij}n_j dA = \int_V \sum_{j=1}^{3}\frac{\partial\sigma_{ij}}{\partial x_j}dV = -\int_V \frac{\partial p}{\partial x_i}\mathrm{d}V. \tag{3.9}$$

Therefore, in the same manner as for the mass conservation, the i-th component of the momentum conservation equation, (3.6), is written as

$$\int_V \left\{\frac{\partial}{\partial t}(\rho u_i) + \sum_{j=1}^{3}\frac{\partial\rho u_i u_j}{\partial x_j} + \frac{\partial p}{\partial x_i} - \rho f_i\right\}\mathrm{d}V = 0,$$

$$\frac{\partial}{\partial t}(\rho u_i) + \sum_{j=1}^{3}\frac{\partial\rho u_i u_j}{\partial x_j} = -\frac{\partial p}{\partial x_i} + \rho f_i. \tag{3.10}$$

By designating the flow velocity components of **u** by (u, v, w) on Cartesian coordinates (x, y, z), the x component of (3.10) is expressed as (Fig. 3.2)

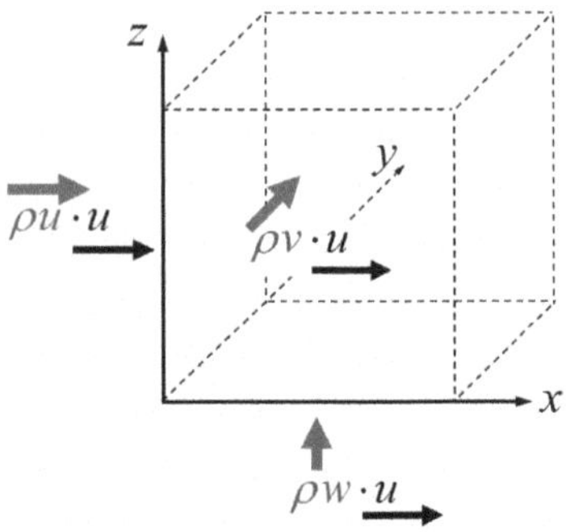

Fig. 3.2 The x-component of momentum fluxes

$$x\text{ - component: } \frac{\partial \rho u}{\partial t} + \frac{\partial \rho u^2}{\partial x} + \frac{\partial \rho u v}{\partial y} + \frac{\partial \rho u w}{\partial z} = -\frac{\partial p}{\partial x} + \rho f_x, \tag{3.11}$$

where the three momentum flux terms on the left-hand side correspond to the momentum flux past a control surface normal to the x-, y-, and z-axes, respectively:
$\frac{\partial \rho u^2}{\partial x} = \frac{\partial \rho u \cdot u}{\partial x}$: x component of momentum difference through control surfaces normal to the x-axis,
$\frac{\partial \rho u v}{\partial y} = \frac{\partial \rho v \cdot u}{\partial y}$: x component of momentum difference through control surfaces normal to the y-axis,
$\frac{\partial \rho u w}{\partial z} = \frac{\partial \rho w \cdot u}{\partial z}$: x component of momentum difference through control surfaces normal to the z-axis,

$$y\text{ - component: } \frac{\partial \rho v}{\partial t} + \frac{\partial \rho v u}{\partial x} + \frac{\partial \rho v^2}{\partial y} + \frac{\partial \rho v w}{\partial z} = -\frac{\partial p}{\partial y} + \rho f_y, \tag{3.12}$$

$$z\text{ - component: } \frac{\partial \rho w}{\partial t} + \frac{\partial \rho w u}{\partial x} + \frac{\partial \rho w v}{\partial y} + \frac{\partial \rho w^2}{\partial z} = -\frac{\partial p}{\partial z} + \rho f_z. \tag{3.13}$$

In vector form,

$$\frac{\partial \rho \mathbf{u}}{\partial t} + \nabla \cdot (\rho \mathbf{u} \otimes \mathbf{u}) = -\nabla p + \rho \mathbf{f}, \tag{3.14}$$

where $\mathbf{a} \otimes \mathbf{b}$ denotes a tensor product[1]:

$$\mathbf{a} \otimes \mathbf{b} = \begin{pmatrix} a_1 \\ a_2 \\ a_3 \end{pmatrix} \begin{pmatrix} b_1 & b_2 & b_3 \end{pmatrix} = \begin{pmatrix} a_1 b_1 & a_1 b_2 & a_1 b_3 \\ a_2 b_1 & a_2 b_2 & a_2 b_3 \\ a_3 b_1 & a_3 b_2 & a_3 b_3 \end{pmatrix},$$

[1]Here, $x_1 = x$, $x_2 = y$, and $x_3 = z$.

$$\nabla \cdot (\mathbf{a} \otimes \mathbf{b}) = \begin{pmatrix} \dfrac{\partial a_1 b_1}{\partial x_1} + \dfrac{\partial a_1 b_2}{\partial x_2} + \dfrac{\partial a_1 b_3}{\partial x_3} \\ \dfrac{\partial a_2 b_1}{\partial x_1} + \dfrac{\partial a_2 b_2}{\partial x_2} + \dfrac{\partial a_2 b_3}{\partial x_3} \\ \dfrac{\partial a_3 b_1}{\partial x_1} + \dfrac{\partial a_3 b_2}{\partial x_2} + \dfrac{\partial a_3 b_3}{\partial x_3} \end{pmatrix}.$$

Equation (3.14) is the momentum conservation equation without viscosity, named *Euler's equation of motion* in a *conserved* form, in which no differentiation has a variable coefficient. This form fits the instinctive understanding of a conservation relation and minimizes discretization error.

Euler's Eq. (3.14) is transformed into another form after decomposing it to a vector form and utilizing the mass conservation equation given by (3.4):

$$\rho \frac{\partial \mathbf{u}}{\partial t} + \underbrace{\mathbf{u} \left\{ \frac{\partial \rho}{\partial t} + \nabla \cdot (\rho \mathbf{u}) \right\}}_{=0} + \rho (\mathbf{u} \cdot \nabla) \mathbf{u} = -\nabla p + \rho \mathbf{f},$$

$$\rho \frac{\partial \mathbf{u}}{\partial t} + \rho (\mathbf{u} \cdot \nabla) \mathbf{u} = -\nabla p + \rho \mathbf{f}. \tag{3.15}$$

In this, non-conservative form of Euler's equation, the physical meaning of each component is clear. The *convective derivative*,

$$\mathbf{u} \cdot \nabla = u \frac{\partial}{\partial x} + v \frac{\partial}{\partial y} + w \frac{\partial}{\partial z}, \tag{3.16}$$

corresponds to a variation observed on a flow element. The second term on the left-hand side of (3.15) is named the *convective term*, which is the origin of flow nonlinearity.

A *substantial derivative*,

$$\frac{\mathrm{D}}{\mathrm{D}t} \equiv \frac{\partial}{\partial t} + \mathbf{u} \cdot \nabla, \tag{3.17}$$

is the total derivative with respect to $\frac{\mathrm{d}x}{\mathrm{d}t} = \mathbf{u}$, which quantifies flow variation observed on a flow element. By using the substantial derivative, (3.15) yields

$$\rho \frac{\mathrm{D}\mathbf{u}}{\mathrm{D}t} = -\nabla p + \rho \mathbf{f}. \tag{3.18}$$

3.1.3 Conservation of Energy

The *total energy*, e_t, of a fluid element is the sum of the internal energy and kinetic energy.

$$e_t = e + \frac{1}{2}|\mathbf{u}|^2 \tag{3.19}$$

The variations of the total energy in a control volume per unit time are formulated as follows:

[total energy increment] = [energy flux through control surface]
+ [work with pressure on control surface] + [heat transfer on control surface],
+ [heat input to control volume] + [work with volume force],

$$\frac{\partial}{\partial t}\int_V \rho e_t dV = -\int_A \rho e_t \mathbf{u}\cdot\mathbf{n}dA - \int_A p\mathbf{u}\cdot\mathbf{n}dA - \int_A \mathbf{q}\cdot\mathbf{n}dA + \int_V \rho\dot{Q}dV + \int_V \rho\mathbf{f}\cdot\mathbf{u}dV \tag{3.20}$$

where $\dot{Q}$ denotes the heat input per unit time to a unit volume. By using Gauss's divergence theorem,

$$\frac{\partial}{\partial t}\int_V \rho e_t dV = -\int_A \nabla\cdot(\rho e_t\mathbf{u})dV - \int_V \nabla\cdot(p\mathbf{u})dV - \int_V \nabla\cdot\mathbf{q}dV + \int_V \rho\dot{Q}dV + \int_V \rho\mathbf{f}\cdot\mathbf{u}dV. \tag{3.21}$$

For (3.21) to be applicable to any volume element,

$$\frac{\partial \rho e_t}{\partial t} + \nabla\cdot(\rho e_t + p)\mathbf{u} = -\nabla\cdot\mathbf{q} + \rho\dot{Q} + \rho\mathbf{f}\cdot\mathbf{u}, \tag{3.22}$$

$$\rho\frac{\partial e_t}{\partial t} + \rho\mathbf{u}\cdot\nabla e_t + \underbrace{e_t\left\{\frac{\partial\rho}{\partial t} + \nabla\cdot(\rho\mathbf{u})\right\}}_{[A]} + p\nabla\cdot\mathbf{u} + \mathbf{u}\cdot\underbrace{\nabla p}_{[B]} = -\nabla\cdot\mathbf{q} + \rho\dot{Q} + \rho\mathbf{f}\cdot\mathbf{u}.$$

By applying (3.4) and (3.15) to [A] and [B] in the above equation, respectively, and by using (3.19), we obtain

$$\rho\frac{\partial}{\partial t}\left(e + \frac{1}{2}|\mathbf{u}|^2\right) + \rho\mathbf{u}\cdot\nabla\left(e + \frac{1}{2}|\mathbf{u}|^2\right) + p\nabla\cdot\mathbf{u} + \mathbf{u}\cdot\left\{-\rho\frac{\partial\mathbf{u}}{\partial t} - \rho(\mathbf{u}\cdot\nabla)\mathbf{u} + \rho\mathbf{f}\right\}$$
$$= -\nabla\cdot\mathbf{q} + \rho\dot{Q} + \rho\mathbf{f}\cdot\mathbf{u}$$

$$\rho\frac{\partial e}{\partial t} + \rho\mathbf{u}\cdot\nabla e + p\nabla\cdot\mathbf{u} + \rho\mathbf{u}\cdot\nabla\frac{|\mathbf{u}|^2}{2} - \rho\mathbf{u}\cdot\{(\mathbf{u}\cdot\nabla)\mathbf{u}\} = -\nabla\cdot\mathbf{q} + \rho\dot{Q}.$$

Since

$$\mathbf{u} \cdot \{(\mathbf{u} \cdot \nabla)\mathbf{u}\} = \begin{pmatrix} u \\ v \\ w \end{pmatrix} \cdot \begin{pmatrix} (\mathbf{u} \cdot \nabla)u \\ (\mathbf{u} \cdot \nabla)v \\ (\mathbf{u} \cdot \nabla)w \end{pmatrix} = \frac{1}{2}(\mathbf{u} \cdot \nabla)\, |\mathbf{u}|^2,$$

$$\rho\frac{\mathrm{D}e}{\mathrm{D}t} + p\nabla \cdot \mathbf{u} = -\nabla \cdot \mathbf{q} + \rho\dot{Q}. \tag{3.23}$$

By using (3.4) and (3.17), we obtain

$$\nabla \cdot \mathbf{u} = -\frac{1}{\rho}\frac{\mathrm{D}\rho}{\mathrm{D}t}. \tag{3.24}$$

Combing with (3.23), we obtain

$$\frac{\mathrm{D}e}{\mathrm{D}t} = \frac{p}{\rho^2}\frac{\mathrm{D}\rho}{\mathrm{D}t} - \frac{1}{\rho}\nabla \cdot \mathbf{q} + \dot{Q} = -p\frac{\mathrm{D}\upsilon}{\mathrm{D}t} - \frac{1}{\rho}\nabla \cdot \mathbf{q} + \dot{Q},, \tag{3.25}$$

$$\upsilon \equiv \frac{1}{\rho}. \tag{3.26}$$

Equation (3.25) gives the variation in internal energy due to a variation in volume, heat transfer, and volumetric heat input. By substituting the second and third terms on the rightmost side of (3.25) with the heat input δQ and the substantial derivative of X with $\mathrm{d}X$, we obtain

$$\delta Q = \mathrm{d}e + p\mathrm{d}\upsilon. \tag{3.27}$$

The above is the *first law of thermodynamics* applied to a flow element. By using the enthalpy, h:

$$h = e + \frac{p}{\rho} = e + p\upsilon, \tag{3.28}$$

$$\delta Q = \mathrm{d}h - \upsilon\mathrm{d}p. \tag{3.29}$$

Note here that, in (3.27) and (3.29), the effect of volume force does not appear, because it affects only the motion of the center of gravity and not the compression and expansion of the element.

Next, let us consider the variation of kinetic energy of the center of mass. By taking the inner product of (3.15) and $\mathbf{u}$, we obtain

$$\mathbf{u} \cdot \frac{\partial \mathbf{u}}{\partial t} + \mathbf{u} \cdot (\mathbf{u} \cdot \nabla)\mathbf{u} = \frac{\mathrm{D}}{\mathrm{D}t}\left(\frac{1}{2}|\mathbf{u}|^2\right) = \left(-\frac{\nabla p}{\rho} + \mathbf{f}\right) \cdot \mathbf{u}. \tag{3.30}$$

This equation implies that the flow kinetic energy varies with the pressure gradient and volume force. Because the pressure varies with the internal energy, the flow kinetic energy and internal energy are coupled with each other.

To summarize these results, in the total energy, the internal energy varies based on the first law of thermodynamics, as expressed by (3.27), and the flow kinetic energy varies based on the equation of motion, as expressed by (3.30).

3.1.4 Other Relations

In three-dimensional flows, a flow condition is defined by five independent parameters—two thermodynamic properties and three velocity components—each of which is a function of position (x, y, z) and time t. In this chapter, we have derived five conservation equations: one for mass, three for momentum components, and one for energy. Therefore, the number of governing equations is equal to the number of unknown parameters. However, the conservation equations involve more than two thermodynamic parameters: e, p, ρ, and, after a transformation, a, which is expressed by thermodynamic properties such as h and T. A thermodynamic property can be given as a function of two other properties. For example, the internal energy is given as a function of pressure and density:

$$e = e(p, \rho). \tag{3.31}$$

Therefore, the number of governing equations is equal to that of independent unknown parameters. Such a flow problem can be solved with appropriate initial and/or boundary conditions.

3.1.5 Similarity in Inviscid Flow

In this book, we deal with inviscid flows, in which we implicitly neglect diffusion, viscosity, and heat conduction. In such flows, no characteristic length exists, that is, the flow is self-similar. Let us show this similarity in Euler's equation. Let the variables appearing in (3.15) be normalized using three characteristic parameters p_∞, L, and U_∞:

$$\tilde{\mathbf{f}} \equiv \frac{\mathbf{f}}{U_\infty^2/L}, \tilde{p} \equiv \frac{p}{p_\infty}, \tilde{\mathbf{u}} \equiv \frac{\mathbf{u}}{U_\infty}, \tilde{t} \equiv \frac{t}{L/U_\infty}, \tilde{\nabla} \equiv L\nabla, \tilde{\rho} \equiv \frac{\rho}{p_\infty/U_\infty^2}, \tag{3.32}$$

$$\frac{\partial \tilde{\mathbf{u}}}{\partial \tilde{t}} + (\tilde{\mathbf{u}} \cdot \tilde{\nabla})\tilde{\mathbf{u}} = -\frac{\tilde{\nabla}\tilde{p}}{\tilde{\rho}} + \tilde{\mathbf{f}}. \tag{3.33}$$

The Euler's equation in a dimensionless form, given by (3.33), holds irrespective of the value of the length scale, L. Therefore, under the same flow conditions, flows with similar shapes are similar.

3.2 Galilean Transformation

A flow field around an airplane flying at a constant velocity observed from the ground is different from that observed from the airplane with respect to the flow velocity. In the sense of dynamics, the flow is observed on different *inertial frames of reference* that move at a constant relative velocity. However, the distribution of thermodynamic properties such as pressure and temperature should be uniquely determined irrespective of the inertial frame of reference. For the flow to be consistent, the form of governing equations must remain unchanged on any inertial frame of reference. In this section, we will confirm this.

3.2.1 *Inertial Frame of Reference*

A frame of reference fixed on the ground is termed a *laboratory frame*. A laboratory frame on Earth is approximately regarded as an inertial frame of reference.[2] Assume that an airplane is flying with a constant velocity vector of $\mathbf{U}$ in a quiescent atmosphere, as shown in Fig. 3.3. When observed from the airplane, as shown in Fig. 3.4, the flow comes toward the airplane with a velocity vector of $-\mathbf{U}$. For transformation between these frames, we will perform the *Galilean transformation* as follows.

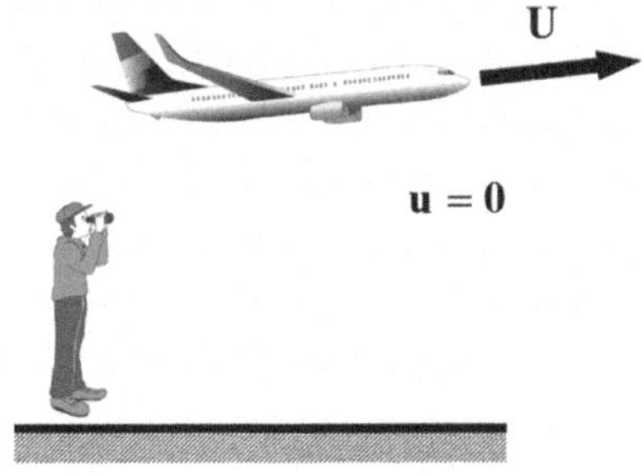

Fig. 3.3 Motion of an airplane flying in a quiescent atmosphere at a constant velocity of **U** observed from the ground

[2]Strictly speaking, a frame fixed on Earth is not an inertial frame of reference, because an object experiences an inertial force due to Earth's spin. Yet, for example, a centrifugal force due to Earth's spin is equal to $0.03\cos\phi[\mathrm{m/s^2}]$ (φ, latitude) of Earth's gravity, which is only 0.3% at most.

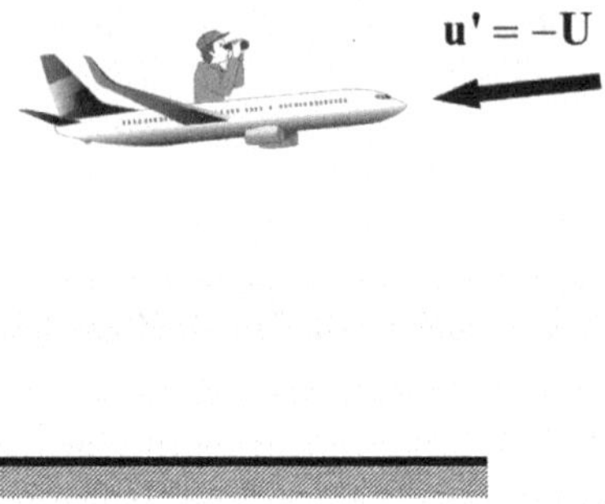

Fig. 3.4 Incoming flow with a velocity vector of $-\mathbf{U}$ observed from the airplane

3.2.2 Galilean Transformation

Let us consider an inertial frame of reference A with the origin O, as shown in Fig. 3.5a. Let the pointing vector of a particle, $\mathbf{x}$, be defined by

$$\mathbf{x} = \begin{pmatrix} x \\ y \\ z \end{pmatrix}. \tag{3.34}$$

Next, consider another inertial frame of reference B with the origin O', which moves at a constant translational velocity vector $\mathbf{U}$ on Frame A, as shown in Fig. 3.5b. There is no relative rotation between these frames. The pointing vector of the same particle on Frame B, as shown in Fig. 3.5b, is defined as

(a) Frame A **(b) Frame B, moving at a constant velocity of U with respect to Frame A**

Fig. 3.5 Two inertial frames

$$\mathbf{x}' = \begin{pmatrix} x' \\ y' \\ z' \end{pmatrix}. \tag{3.35}$$

As shown in Fig. 3.6, these two pointing vectors are related to each other by

$$\mathbf{x}' \equiv \mathbf{x} - \mathbf{x}_{O'}, \tag{3.36}$$

$$\begin{pmatrix} x' \\ y' \\ z' \end{pmatrix} = \begin{pmatrix} x - x_{O'}(t) \\ y - y_{O'}(t) \\ z - z_{O'}(t) \end{pmatrix},, \tag{3.37}$$

$$t' = t, \tag{3.38}$$

where t and t' are the time in Frame A and B, respectively. Here, we assume that **U** is constant. Then, the position of O', $\mathbf{x}_{O'}$, is a function of t alone.

$$\mathbf{U} = \frac{\mathrm{d}\mathbf{x}_{O'}(t)}{\mathrm{d}t} \equiv \begin{pmatrix} U_x \\ U_y \\ U_z \end{pmatrix}. \tag{3.39}$$

By using the above relations, let us conduct a transformation from Frame A to B. The derivatives between these frames are related by a *Jacobian matrix* **Y**:

$$\mathbf{Y} = \frac{|\partial(t', \mathbf{x}')|}{|\partial(t, \mathbf{x})|} = \begin{pmatrix} \frac{\partial t'}{\partial t} & \frac{\partial x'}{\partial t} & \frac{\partial y'}{\partial t} & \frac{\partial z'}{\partial t} \\ \frac{\partial t'}{\partial x} & \frac{\partial x'}{\partial x} & \frac{\partial y'}{\partial x} & \frac{\partial z'}{\partial x} \\ \frac{\partial t'}{\partial y} & \frac{\partial x'}{\partial y} & \frac{\partial y'}{\partial y} & \frac{\partial z'}{\partial y} \\ \frac{\partial t'}{\partial z} & \frac{\partial x'}{\partial z} & \frac{\partial y'}{\partial z} & \frac{\partial z'}{\partial z} \end{pmatrix}. \tag{3.40}$$

Therefore,

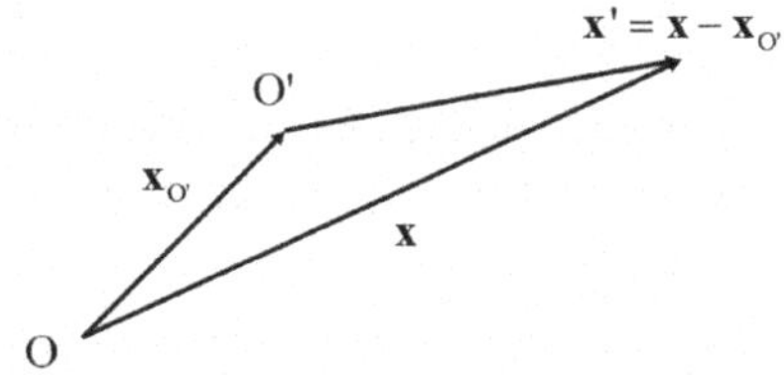

Fig. 3.6 Relation between two pointing vectors referring to the same particle

$$\begin{pmatrix} \dfrac{\partial}{\partial t} \\ \dfrac{\partial}{\partial x} \\ \dfrac{\partial}{\partial y} \\ \dfrac{\partial}{\partial z} \end{pmatrix} = Y \begin{pmatrix} \dfrac{\partial}{\partial t'} \\ \dfrac{\partial}{\partial x'} \\ \dfrac{\partial}{\partial y'} \\ \dfrac{\partial}{\partial z'} \end{pmatrix} = \begin{pmatrix} \frac{\partial}{\partial t'} & -U_x \frac{\partial}{\partial x'} & -U_y \frac{\partial}{\partial y'} & -U_z \frac{\partial}{\partial z'} \\ & \frac{\partial}{\partial x'} & & \\ & & \frac{\partial}{\partial y'} & \\ & & & \frac{\partial}{\partial z'} \end{pmatrix}. \tag{3.41}$$

In a vector form,

$$\frac{\partial}{\partial t} = \frac{\partial}{\partial t'} - \mathbf{U} \cdot \nabla', \tag{3.42}$$

$$\nabla = \nabla' = \begin{pmatrix} \dfrac{\partial}{\partial x'} \\ \dfrac{\partial}{\partial y'} \\ \dfrac{\partial}{\partial z'} \end{pmatrix}. \tag{3.43}$$

Equations (3.42) and (3.43) are the equations for Galilean transformation. The spatial derivatives are expressed in the same form. However, note that the term $-\mathbf{U}\cdot\nabla'$ should be added to correct for the relative translational motion between the frames.

3.2.2.1 Galilean Transformation of Conservation Equations

Let flow velocities $\mathbf{u}$ and $\mathbf{u}'$ on Frame A and B, respectively, be defined by

$$\mathbf{u} = \frac{\mathrm{d}\mathbf{x}}{\mathrm{d}t}, \tag{3.44}$$

$$\mathbf{u}' = \frac{\mathrm{d}\mathbf{x}'}{\mathrm{d}t'}. \tag{3.45}$$

From (3.36), (3.38), and (3.39),

$$\mathbf{u}' \equiv \mathbf{u} - \mathbf{U}. \tag{3.46}$$

Let us transform the conservation equations, given by. (3.4), (3.15), and (3.23), by using (3.42), (3.43), and (3.46).

(1) Conservation of mass
By applying Galilean transformation to (3.4), we obtain

$$\frac{\partial \rho}{\partial t'} - \mathbf{U} \cdot \nabla' \rho + \nabla' \cdot \rho(\mathbf{U} + \mathbf{u}') = 0. \tag{3.47}$$

U is a spatiotemporal invariant, that is,

$$\nabla' \cdot \mathbf{U} = \mathbf{0}, , \tag{3.48}$$

$$\frac{\partial \mathbf{U}}{\partial t'} = \mathbf{0}. \tag{3.49}$$

Thus, Eq. (3.47) becomes

$$\frac{\partial \rho}{\partial t'} + \nabla' \cdot (\rho \mathbf{u}') = 0. \tag{3.50}$$

The above equation has the same form as (3.4).

(2) Conservation of momentum

In the same manner as for the conservation of mass, Galilean transformation is applied to (3.15):

$$\begin{aligned} &\rho\left(\frac{\partial}{\partial t'} - \mathbf{U} \cdot \nabla'\right)(\mathbf{U} + \mathbf{u}') + \rho(\mathbf{U} + \mathbf{u}') \cdot \nabla'(\mathbf{U} + \mathbf{u}') = -\nabla' p + \rho \mathbf{f}, \\ &\rho \frac{\partial \mathbf{u}'}{\partial t'} + \rho \mathbf{u}' \cdot \nabla' \mathbf{u}' = -\nabla' p + \rho \mathbf{f}. \end{aligned} \quad , \tag{3.51}$$

which has the same form as (3.15).

(3) Conservation of energy

By applying Galilean transformation to (3.23),

$$\begin{aligned} &\rho\left(\frac{\partial}{\partial t'} - \mathbf{U} \cdot \nabla'\right)e + \rho\{(\mathbf{U} + \mathbf{u}') \cdot \nabla'\}e + p\nabla' \cdot (\mathbf{U} + \mathbf{u}') = -\nabla' \cdot \mathbf{q} + \rho \delta \dot{Q}, \\ &\rho \frac{\partial e}{\partial t'} + \rho(\mathbf{u}' \cdot \nabla')e + p\nabla' \cdot \mathbf{u}' = -\nabla' \cdot \mathbf{q} + \rho \delta \dot{Q}. \end{aligned} \tag{3.52}$$

From (3.50) to (3.52), each conservation equation is expressed in the same form on $(t', \mathbf{x}')$ coordinates. Moreover, a substantial derivative does not change its form by Galilean transformation.

$$\frac{\mathrm{D}}{\mathrm{D}t} = \frac{\partial}{\partial t} + \mathbf{u} \cdot \nabla = \frac{\partial}{\partial t'} - \mathbf{U} \cdot \nabla' + (\mathbf{U} + \mathbf{u}') \cdot \nabla' = \frac{\partial}{\partial t'} + \mathbf{u}' \cdot \nabla' = \frac{\mathrm{D}\prime}{\mathrm{D}t'}. \tag{3.53}$$

Note here that Galilean transformation is applicable only under the condition that **U** does not vary; it is not applicable to frames with relative acceleration.

Chapter 4
Discontinuity

When we drive a car on a highway, we occasionally have to press the brake pedal at a sudden traffic jam, the cause of which we do not readily understand. From a bird's-eye view, we see a boundary past which the car density and speed sharply change, as shown in Fig. 4.1a. Moreover, Fig. 4.1b shows water and oil in contact with each other, while Fig. 4.1c shows a stone slipping on ice. In such cases, two media with different densities and/or velocities are in contact. Such a discontinuity can appear even in flows.

In this book, we deal with flows of *continuum fluids*; the dimensions of objects are much larger than the mean free path of the fluids. It seems contradictory that discontinuity exists in continuum fluid. Yet, the flow can have a significant spatial variation within a short distance of the order of tens of the mean free path (or, in a traffic flow, within a distance of tens of the vehicle-separation distance). We regard such a short-distance variation as a *discontinuity*. There are several types of discontinuities classified according to how the conservation relations in fluid dynamics are satisfied through them.

4.1 Condition and Classification of Discontinuity

4.1.1 Rankine–Hugoniot Relation

Let us assume a discontinuity in a flow and consider a control volume Ω (ABCDE-FGH) that encloses it by the control surface Φ, as shown in Fig. 4.2. The control volume Ω is assumed to be so thin that the flow does not temporally change in the period during which the flow passes through it. The flow is uniform along the discontinuity. By applying Galilean transformation to the conservation equations in a conserved form, given by (3.4), (3.14), and (3.22), we obtain

A. Sasoh, *Compressible Fluid Dynamics and Shock Waves*,
https://doi.org/10.1007/978-981-15-0504-1_4

(a) Traffic jam suddenly appearing on a highway, courtesy of Prof. K. Nishinari (The University of Tokyo)

(b) Contact between water and oil

(c) Slip surface

Fig. 4.1 Examples of discontinuity

$$\nabla \cdot (\rho \mathbf{u}) = 0, \tag{4.1}$$

$$\nabla \cdot (\rho \mathbf{u} \otimes \mathbf{u}) = -\nabla p + \rho \mathbf{f}, \tag{4.2}$$

$$\nabla \cdot \{(\rho e_{\mathrm{t}} + p)\mathbf{u}\} = \nabla \cdot (\rho h_{\mathrm{t}} \mathbf{u}) = \rho\left(\delta \dot{Q} + \mathbf{f} \cdot \mathbf{u}\right). \tag{4.3}$$

By using Gauss's divergence theorem with respect to the control volume Ω with a control surface Φ, we obtain

$$\int_{\Omega} \nabla \cdot (\rho \mathbf{u}) \mathrm{d}V = \int_{\Phi} \rho \mathbf{u} \cdot \mathbf{n} \mathrm{d}A = 0, \tag{4.4}$$

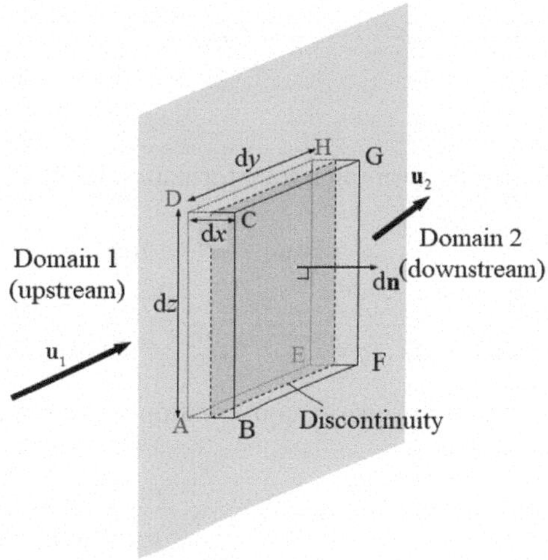

Fig. 4.2 A control volume (ABCDEFGH) with a control surface Φ enclosing a discontinuity

$$\int_\Omega \{\nabla \cdot (\rho\mathbf{u} \otimes \mathbf{u}) + \nabla p\} \mathrm{d}V = \int_\Omega \rho \mathbf{f} \mathrm{d}V,$$

$$\int_\Phi \{(\rho\mathbf{u} \otimes \mathbf{u})\} \cdot \mathbf{n} \mathrm{d}A + \int_\Omega \nabla p \mathrm{d}V = \int_\Omega \rho \mathbf{f} \mathrm{d}V, \tag{4.5}$$

$$\int_\Omega \nabla \cdot (\rho h_\mathrm{t} \mathbf{u}) \mathrm{d}V = \int_\Phi (\rho h_\mathrm{t} \mathbf{u}) \cdot \mathbf{n} \mathrm{d}A = \int_\Omega \rho(\delta \dot{Q} + \mathbf{f} \cdot \mathbf{u}) \mathrm{d}V. \tag{4.6}$$

We set the x coordinate to be normal to the discontinuity, with the control volume having a thickness of dx and widths of dy and dz. $\mathbf{n}$ is a unit vector normal to a control surface element toward the outward direction. With dx decreasing to an infinitesimally small value, d$x \to 0$, the volume of Ω also decreases to an infinitesimally small value as $\mathrm{d}V = \mathrm{d}x\mathrm{d}y\mathrm{d}z \to 0$, with $\mathrm{d}y\mathrm{d}z$ remaining finite. Under this ultimate condition, we can neglect fluxes through the side elements of the control surface and the source terms.

From (4.4),

$$(\rho\mathbf{u})_{\mathrm{ADHE}} \cdot \mathbf{n}_{\mathrm{ADHE}} dydz + (\rho\mathbf{u})_{\mathrm{BFGC}} \cdot \mathbf{n}_{\mathrm{BFGC}} dydz = 0,$$

$$\{(\rho\mathbf{u})_{\mathrm{ADHE}} - (\rho\mathbf{u})_{\mathrm{BFGC}}\} \cdot \mathbf{n}_{\mathrm{ADHE}} = 0. \tag{4.7}$$

By denoting the value on the upstream side on ADHE and that on the downstream side on BFGC by subscripts 1 and 2, respectively, we have

$$\rho_1 u_1 = \rho_2 u_2. \tag{4.8}$$

In the same manner, the right-hand side of the momentum conservation equation, expressed as (4.5), vanishes with $dx \to 0$:

$$\begin{aligned} \rho_1 u_1^2 + p_1 &= \rho_2 u_2^2 + p_2 \\ \rho_1 u_1 v_1 &= \rho_2 u_2 v_2, \\ \rho_1 u_1 w_1 &= \rho_2 u_2 w_2. \end{aligned} \tag{4.9}$$

By performing a transformation using (4.8),

$$\rho_1 u_1 (v_1 - v_2) = 0, \tag{4.10}$$

$$\rho_1 u_1 (w_1 - w_2) = 0. \tag{4.11}$$

From the energy conservation equation, expressed as (4.6), we obtain

$$\begin{aligned} \rho_1 u_1 h_{t,1} &= \rho_2 u_2 h_{t,2}, \\ \rho_1 u_1 \left(h_{t,1} - h_{t,2}\right) &= 0. \end{aligned} \tag{4.12}$$

Equations (4.8)–(4.12) are known as *Rankine–Hugoniot equations*. They describe flow variations even past a discontinuity. Note here that, except for the momentum equation along the normal to the discontinuity, expressed as (4.9), each equation has a form of the product with the mass flux of (4.8).

4.1.2 Classification of Discontinuity

The solutions of Rankine–Hugoniot equations are classified depending on whether the flow values change or on whether the mass flux has a non-null value past the control surface.

(1) *Continuous solution*: $X_1 = X_2$

This solution corresponds to the condition that the flow remains continuous past the surface, and it is applicable to any place except for the following discontinuities.

(2) *Shock wave*: $u_1 \neq 0$

Because we assume a continuous fluid with $\rho \neq 0$, we obtain the following equations for $u_1 \neq 0$:

$$\rho_1 u_1 = \rho_2 u_2 \neq 0, \tag{4.13}$$

$$\rho_1 u_1^2 + p_1 = \rho_2 u_2^2 + p_2, \tag{4.14}$$

$$v_1 = v_2, \tag{4.15}$$

$$w_1 = w_2, \tag{4.16}$$

$$h_{t,1} = h_{t,2}. \tag{4.17}$$

A discontinuity that satisfies (4.13)–(4.17) is a shock wave. In a shock wave, the flow velocity normal to the wave front is not zero and discontinuously varies across it, together with the density, pressure, and temperature. However, the total enthalpy and flow velocity along the wave front do not change. Important and interesting characteristics of compressible fluid dynamics are often attributed to shock waves. Depending on whether a tangential velocity component accompanies the shock wave, shock waves are classified into two: normal shock waves and oblique shock waves (Fig. 4.3).

(2.1) *Normal shock wave*: $v_1 = v_2 = w_1 = w_2 = 0$
This is a shock wave without a tangential velocity component. The wave front is normal to the flow velocity vector, and it is the strongest shock wave, with the largest pressure ratio for a certain Mach number.

(2.2) *Oblique shock wave*: $v_1 = v_2 \neq 0$ and/or $w_1 = w_2 \neq 0$
This is a shock wave accompanied by a tangential velocity component, as shown in Fig. 4.4. For the same upstream conditions, the post-shock pressure, density, and temperature are lower than those of a normal shock wave.

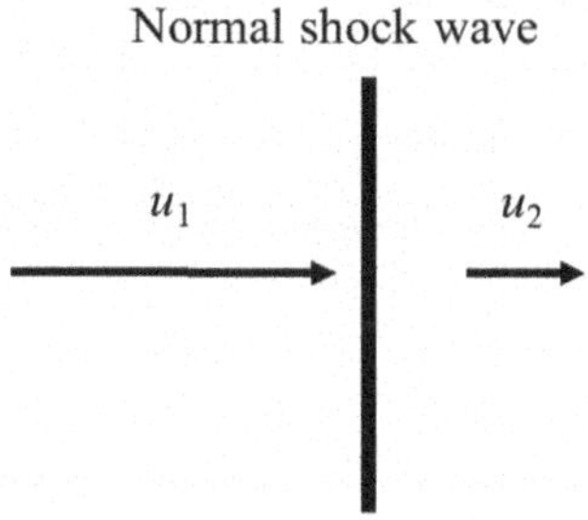

Fig. 4.3 Normal shock wave

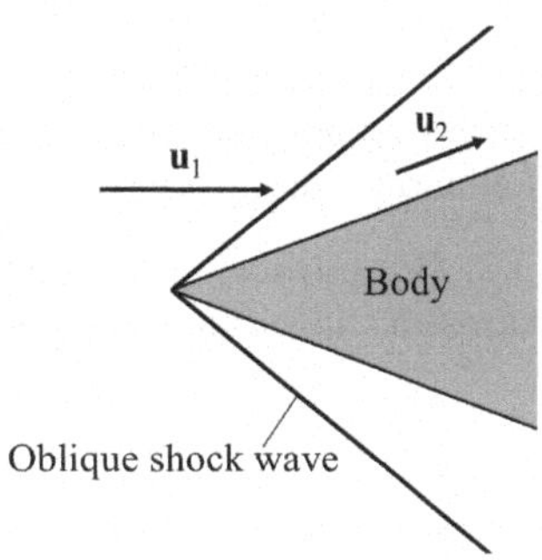

Fig. 4.4 Oblique shock wave

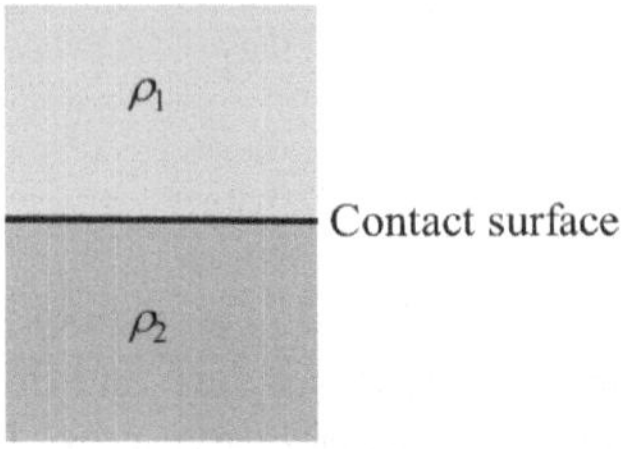

Fig. 4.5 Contact surface

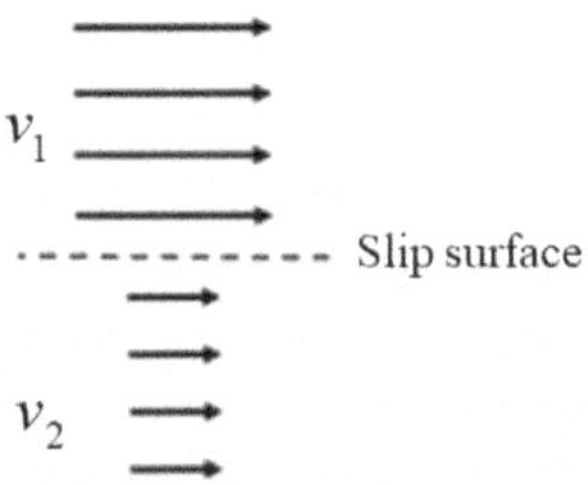

(a) Schematic illustration

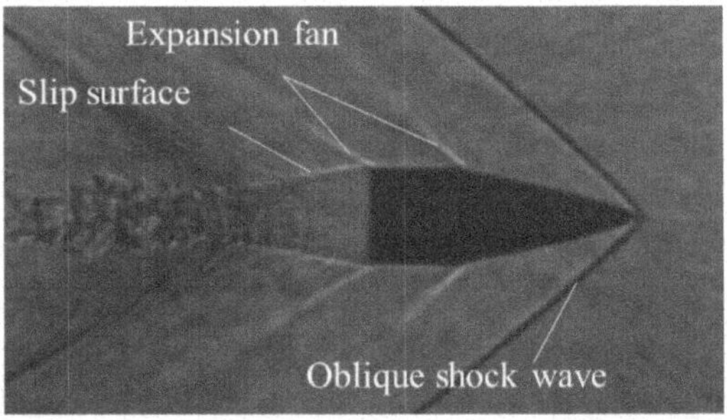

(b) Schlieren image around a supersonically flying body.

Fig. 4.6 Slip surface. We will learn about expansion waves in Chap. 7

(3) Interface: $u_1 = u_2 = 0$

Without flow across the discontinuity, by combining $u_1 = u_2 = 0$ and (4.9),

$$p_1 = p_2. \tag{4.18}$$

That is, the pressure is continuous. In this case, from (4.8) and (4.10)–(4.12), any of ρ, v, w, and h can be discontinuous. Such a surface is called an *interface*. Interfaces are subdivided into two depending on the discontinuous quantity.

(3.1) *Contact surface*: $v = w = 0$ ($\rho_1 \neq \rho_2$ and/or $h_1 \neq h_2$)
Across a contact surface, two media with different densities and/or temperatures (enthalpies) are in contact with each other, as shown in Fig. 4.5.

(3.2) *Slip surface*: $v_1 \neq v_2$ and/or $w_1 \neq w_2$
Across a slip surface, tangential velocities are discontinuous, as shown in Fig. 4.6.

4.2 Normal Shock Wave

A normal shock wave has fundamental characteristics applicable even to oblique shock waves. We will first learn general characteristics applicable to any kind of gas, following which we will explore useful relations for a calorically perfect gas.

4.2.1 General Characteristics

4.2.1.1 Derivation of Post-shock Conditions

Here, we consider a normal shock wave normal to the x-direction, as shown in Fig. 4.7a. Quantities on the upstream and downstream sides of the shock wave are labeled "1" and "2," respectively. The velocity relative to the wave front is designated by u. From the Rankine–Hugoniot relations, expressed by (4.8), (4.9), and (4.12), we have

$$\rho_1 u_1 = \rho_2 u_2 \equiv j, \tag{4.19}$$

$$\rho_1 u_1^2 + p_1 = \rho_2 u_2^2 + p_2, \tag{4.20}$$

$$h_1 + \frac{1}{2}u_1^2 = h_2 + \frac{1}{2}u_2^2. \tag{4.21}$$

From (4.19) and (4.20),

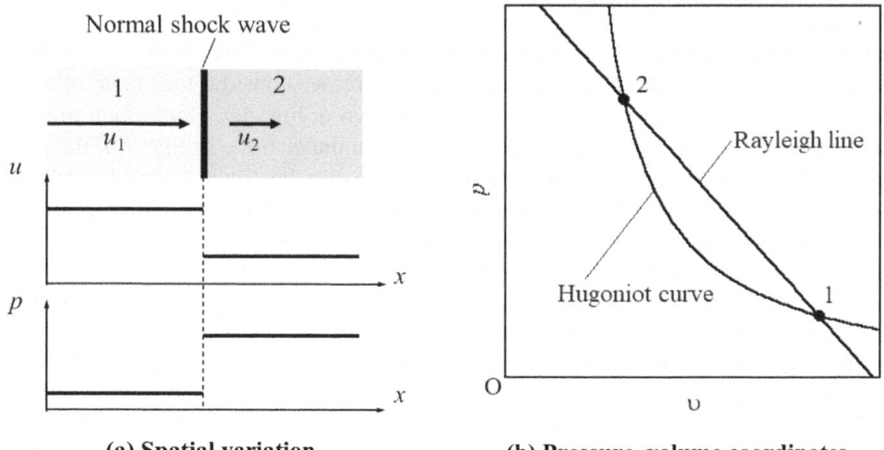

(a) Spatial variation **(b) Pressure–volume coordinates**

Fig. 4.7 Shock relation on the shock coordinates and pressure–volume coordinates

$$\frac{p_2 - p_1}{\upsilon_2 - \upsilon_1} = -j^2. \tag{4.22}$$

This equation gives the *Rayleigh line* on $\upsilon - p$ coordinates with a slope of $-j^2$.

Next, an equation involving only thermodynamic properties is obtained from (4.19)–(4.22):

$$h_2 - h_1 = \frac{p_2 - p_1}{2}\left(\frac{1}{\rho_2} + \frac{1}{\rho_1}\right) = \frac{p_2 - p_1}{2}(\upsilon_1 + \upsilon_2). \tag{4.23}$$

By using the internal energy, e,

$$e_2 - e_1 = \frac{p_1 + p_2}{2}\left(\frac{1}{\rho_1} - \frac{1}{\rho_2}\right) = \frac{p_1 + p_2}{2}(\upsilon_1 - \upsilon_2). \tag{4.24}$$

Equations (4.23) and (4.24) are applicable to any gas, irrespective of its equation of state, and are termed *Hugoniot equations*. This is a kind of equation of state that relates thermodynamic properties across a normal shock wave. Since enthalpy and internal energy can be expressed as functions of volume and pressure, that is, $e = e(\upsilon, p)$ and $h = h(\upsilon, p)$, (4.23) and (4.24) give a unique relation between υ and p, which corresponds to the *Hugoniot curve*, as shown in Fig. 4.7b.

As shown in Fig. 4.7b, the conditions upstream and downstream of a normal shock wave correspond to intersections between the Rayleigh line and Hugoniot curve. Intersection 4.1, which occurs at a lower pressure, corresponds to the upstream, while intersection 4.2 corresponds to the downstream. On the shock coordinate, the flow velocity is decreased across the shock wave.

4.2.1.2 Variation in Energy

Equation (4.24) gives the increment in internal energy across the shock wave, which is equal to the area of Trapezoid A in Fig. 4.8. Here, we consider the shock relation on a laboratory frame in the x-coordinate, as shown in Fig. 4.9. Here, quantities on the laboratory frame are discriminated by using an under bar $\underline{\ }$. In Fig. 4.9, the flow velocity and $\underline{x}$ to the right are defined to be positive, but the direction of x is opposite to that in Fig. 4.7a. By using the shock velocity on the laboratory frame, $\underline{U}_s$, u on the shock frame and $\underline{u}$ on the laboratory frame are related by

$$u = \underline{U}_s - \underline{u}. \tag{4.25}$$

By using (4.19), (4.20), and (4.25), the post-shock velocity is obtained as

$$\frac{1}{2}\left(\underline{u}_2 - \underline{u}_1\right)^2 = \frac{1}{2}(u_1 - u_2)^2 = \frac{1}{2}(p_2 - p_1)(\upsilon_1 - \upsilon_2). \tag{4.26}$$

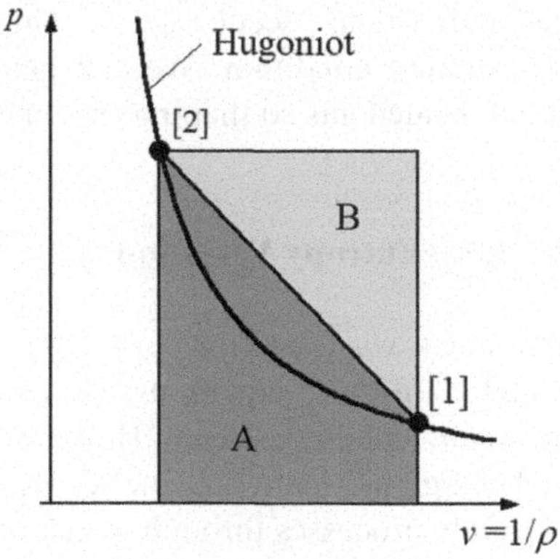

Fig. 4.8 Increase in energy across a shock wave

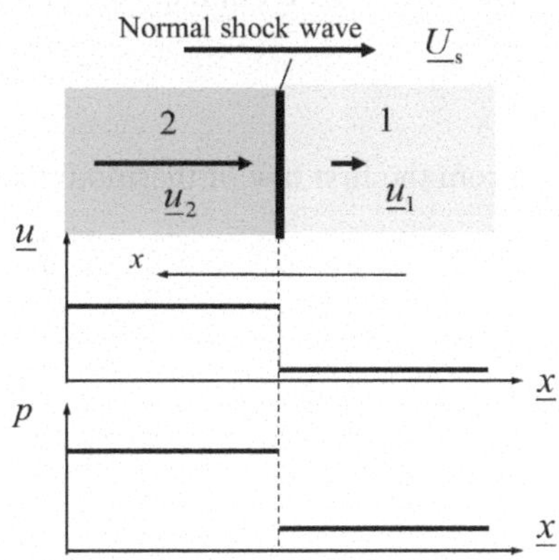

Fig. 4.9 Properties across a shock wave in laboratory coordinates

This is equal to the area of Triangle B. The increment in the total energy is

$$e_{t,2} - e_{t,1} = e_2 - e_1 + \frac{1}{2}\left(\underline{u}_2 - \underline{u}_1\right)^2 = p_2(\upsilon_1 - \upsilon_2), \tag{4.27}$$

which corresponds to the area of Rectangle A + B. From (4.24) and (4.26),

$$\underbrace{e_2 - e_1}_{\substack{\text{Internal energy increment} \\ \text{(Area A)}}} \quad > \quad \underbrace{\frac{1}{2}\left(\underline{u}_2 - \underline{u}_1\right)^2}_{\substack{\text{Kinetic energy increment} \\ \text{(Area B)}}} \quad . \tag{4.28}$$

As expressed in (4.28), past a shock wave, the increment in internal energy is greater than that in the kinetic energy. In other words, a shock wave "heats" the gas, rather than "accelerate" it. This characteristic is useful for generating a high-temperature condition. We can generate even high-speed flow by expanding the shock-heated gas so that internal energy is converted to kinetic energy.[1]

4.2.1.3 Entropy Variation

The shock-wave relations of (4.19)–(4.21) are applicable even after swapping states 1 and 2. In these equations, we cannot determine which state corresponds to the upstream and downstream. However, from a microscopic point of view, momentum and energy dissipation occur across the shock wave; these processes are irreversible [1]. Only processes through which the entropy increases are physically possible. Let us consider the variation of entropy across a shock wave.

Here, we regard enthalpy as a function of entropy and pressure:

$$h = h(s, p). \tag{4.29}$$

From the first law of thermodynamics,

$$\mathrm{d}h = T\mathrm{d}s + \upsilon \mathrm{d}p, \tag{4.30}$$

$$T = \left(\frac{\partial h}{\partial s}\right)_p, \tag{4.31}$$

$$\upsilon = \left(\frac{\partial h}{\partial p}\right)_s. \tag{4.32}$$

With these equations, the Taylor expansion of enthalpy yields

$$h_2 - h_1 = T_1(s_2 - s_1) + \upsilon_1(p_2 - p_1) + \tfrac{1}{2}\left(\tfrac{\partial \upsilon}{\partial p}\right)_{s,1}(p_2 - p_1)^2 + \tfrac{1}{6}\left(\tfrac{\partial^2 \upsilon}{\partial p^2}\right)_{s,1}(p_2 - p_1)^3$$
$$+O\left[(p_2 - p_1)^4\right] + O[(s_2 - s_1)(p_2 - p_1)] + O\left[(s_2 - s_1)^2\right]$$
$$\upsilon_2 - \upsilon_1 = \left(\tfrac{\partial \upsilon}{\partial p}\right)_{s,1}(p_2 - p_1) + \tfrac{1}{2}\left(\tfrac{\partial^2 \upsilon}{\partial p^2}\right)_{s,1}(p_2 - p_1)^2 + O\left[(p_2 - p_1)^3\right] + O[s_2 - s_1].$$

Substituting the above into the equation of the Hugoniot curve, (4.23), yields

$$T_1(s_2 - s_1) + \upsilon_1(p_2 - p_1) + \frac{1}{2}\left(\frac{\partial \upsilon}{\partial p}\right)_{s,1}(p_2 - p_1)^2 + \frac{1}{6}\left(\frac{\partial^2 \upsilon}{\partial p^2}\right)_{s,1}(p_2 - p_1)^3 + [\text{higher order}]$$
$$= \frac{p_2 - p_1}{2}\left[2\upsilon_1 + \left(\frac{\partial \upsilon}{\partial p}\right)_{s,1}(p_2 - p_1) + \frac{1}{2}\left(\frac{\partial^2 \upsilon}{\partial p^2}\right)_{s,1}(p_2 - p_1)^2 + O[(p_2 - p_1)^3] + O[s_2 - s_1]\right]$$

$$s_2 - s_1 = \frac{1}{12T_1}\left(\frac{\partial^2 \upsilon}{\partial p^2}\right)_{s,1}(p_2 - p_1)^3 + [\text{higher order}]. \tag{4.33}$$

[1] Detailed methodologies will be presented in Sects. 11.5 and 11.6.

Therefore, the entropy variation is of the third order of the pressure variation. On $\upsilon - p$ coordinates, the Hugoniot curve and the entrope are tangential to the second order. We call a shock wave with a small pressure increment a *weak shock wave*. A weak shock wave is isentropic up to the second order. With a *strong shock wave*, which is accompanied by a large pressure increment, the entropy increases to a large extent.

Typical gases including air satisfy[2]

$$\left(\frac{\partial^2 \upsilon}{\partial p^2}\right)_s > 0. \tag{4.34}$$

Because shock waves act as irreversible processes, the entropy increases behind it. Therefore, from (4.33) and Inequality (4.34), $p_2 - p_1 > 0$. Following Fig. 4.8, the density increases, and the flow speed relative to the shock wave decreases, as expressed by (4.19).

4.2.1.4 Variation in Mach Number

The slope of the Rayleigh line has an important meaning related to the flow Mach number. Let us obtain the slope $-j_t^2$ of the tangent to the Hugoniot curve at state 1, upstream of the shock wave. For $\upsilon_2 \to \upsilon_1$, we can assume an isentropic process.

$$j_{t,1}^2 = \rho_1^2 u_1^2 = -\left(\frac{\partial p}{\partial \upsilon}\right)_{s,1} = \rho_1^2 \left(\frac{\partial p}{\partial \rho}\right)_{s,1} = \rho_1^2 a_1^2, \tag{4.35}$$

where we utilize the equation for the speed of sound.[3]

$$a = \sqrt{\left(\frac{\partial p}{\partial \rho}\right)_s}. \tag{4.36}$$

From (4.35), $u_1 = a_1$; the tangent to the Hugoniot curve corresponds to a sonic flow. Since $\left(\frac{\partial^2 \upsilon}{\partial p^2}\right)_s > 0$, the Hugoniot curve, as shown in Fig. 4.10, has a downward convex shape.

For the Rayleigh line,

$$\begin{aligned} j^2 \upsilon_1^2 = u_1^2 = -\frac{p_2 - p_1}{\upsilon_2 - \upsilon_1} \upsilon_1^2 > j_{t,1}^2 \upsilon_1^2 = -\upsilon_1^2 \left(\frac{\partial p}{\partial \upsilon}\right)_{s,1} = a_1^2, \\ u_1 > a_1, \end{aligned} \tag{4.37}$$

---

[2] In special cases where the differential has a negative value, an *expansion shock wave* is generated.

[3] See Chap. 8.

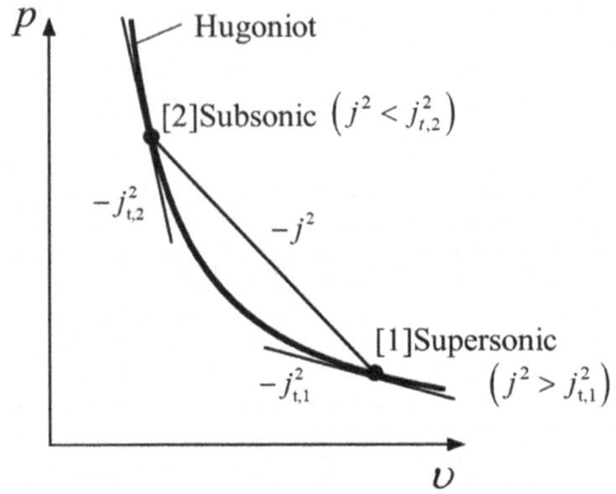

Fig. 4.10 Slope of the Hugoniot curve and flow Mach number

$$j^2 \upsilon_2^2 = u_2^2 = -\frac{p_2 - p_1}{\upsilon_2 - \upsilon_1}\upsilon_2^2 < j_{t,2}^2 \upsilon_2^2 = -\upsilon_2^2 \left(\frac{\partial p}{\partial \upsilon}\right)_{s,2} = a_2^2,$$
$$u_2 < a_2. \tag{4.38}$$

Therefore, the flow upstream of a shock wave is supersonic, while that downstream of a shock wave is subsonic.

4.2.2 *Equations for Calorically Perfect Gas*

If the condition upstream of a shock wave, state 1, is known, the unknown to be solved is that downstream of the shock wave, state 2, which is specified by two thermodynamic properties, such as p_2 and ρ_2, and the flow speed, u_2. For a calorically perfect gas, such a post-shock state is expressed in an explicit form. The equation of state is

$$p = \rho RT. \tag{4.39}$$

The speed of sound, a, and Mach number, M, are given by

$$a = \sqrt{\gamma \frac{p}{\rho}} = \sqrt{\gamma RT}, \tag{4.40}$$

$$\gamma \equiv \frac{C_p}{C_\upsilon}, \tag{4.41}$$

$$M \equiv \frac{u}{a}. \tag{4.42}$$

From (4.19) and (4.20),

$$\frac{p_2}{p_1} = \frac{\rho_1 u_1^2}{p_1}\left(1 - \frac{\rho_1}{\rho_2}\right) + 1.$$

By using (4.40) and (4.42), we obtain

$$\frac{\rho_1 u_1^2}{p_1} = \frac{\gamma u_1^2}{\frac{\gamma p_1}{\rho_1}} = \frac{\gamma u_1^2}{a_1^2} = \gamma M_1^2.$$

M_1 is the Mach number with which a normal shock wave propagates relative to the upstream flow, and it is termed the *shock Mach number*. Hereafter, following our convention, we will write M_s in place of M_1. Because M_s is defined based on the shock-wave propagation velocity relative to the upstream flow, it remains unchanged in any inertial frame of reference.

$$\frac{p_2}{p_1} = -\gamma M_s^2\left(\frac{\rho_1}{\rho_2} - 1\right) + 1 = -\gamma M_s^2\left(\frac{\upsilon_2}{\upsilon_1} - 1\right) + 1. \tag{4.43}$$

Now, we consider variations past a normal shock wave in $\upsilon/\upsilon_1 - p/p_1$ coordinates, as shown in Figs. 4.11 and 4.12. Equation (4.43) gives a Rayleigh line. Its slope is $-\gamma M_s^2$ (Fig. 4.11).

Next, from (4.24) and

$$e = \frac{1}{\gamma - 1}\frac{p}{\rho}, \tag{2.96}$$

$$\frac{p_2}{p_1} = \frac{\frac{\rho_1}{\rho_2} - \frac{\gamma+1}{\gamma-1}}{1 - \frac{\gamma+1}{\gamma-1}\frac{\rho_1}{\rho_2}} = \frac{\frac{\upsilon_2}{\upsilon_1} - \frac{\gamma+1}{\gamma-1}}{1 - \frac{\gamma+1}{\gamma-1}\frac{\upsilon_2}{\upsilon_1}}. \tag{4.44}$$

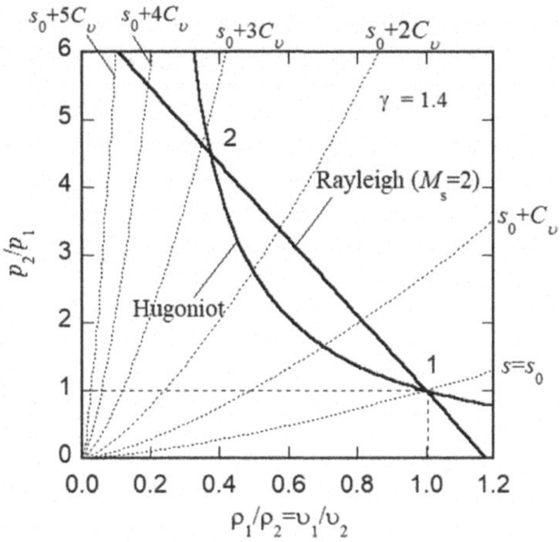

Fig. 4.11 Variation of flow condition in pressure–volume coordinates

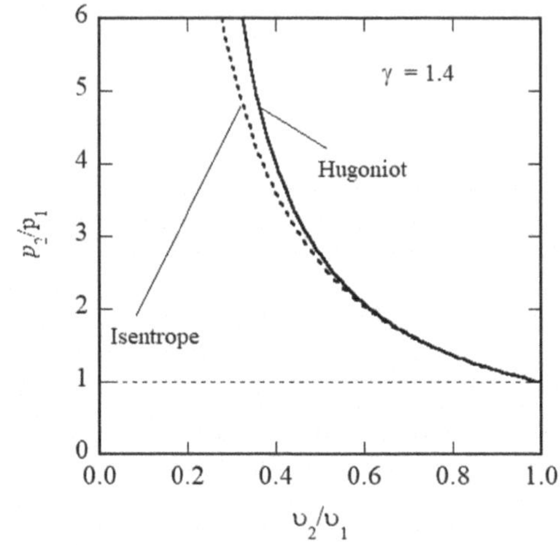

Fig. 4.12 Comparison between shock and isentropic compressions in pressure–volume coordinates

This is an explicit form of the Hugoniot curve, giving the post-shock state without including flow velocity. In general, to specify a gas condition, two independent parameters need to be specified. However, with the condition of shock compression, the post-shock condition is uniquely determined only by (4.44).

Let us compare the shock compression, given by (4.44), with isentropic compression:

$$\frac{p_2}{p_1} = \left(\frac{\rho_2}{\rho_1}\right)^{\gamma} = \left(\frac{\upsilon_1}{\upsilon_2}\right)^{\gamma}. \tag{2,91}$$

As shown in Fig. 4.12, the shock compression $\upsilon_2/\upsilon_1 < 1$ yields a higher pressure than the isentropic compression. As will be shown in Sect. 4.2.2.3, the latter corresponds to quasi-static processes; the compression is gradually performed with the pressure in the gas being uniformly distributed. However, in the shock compression, the local pressure on the contact surface becomes higher than that in the quasi-static processes; consequently, excess work is performed. From (4.33), the pressure increases owing to an increase in entropy.

From (4.44) with $1 < p_2/p_1 < \infty$, the density range is $1 < \frac{\rho_2}{\rho_1} < \frac{\gamma+1}{\gamma-1}$.

For a calorically perfect gas, the post-shock condition (state 2) is obtained as the intersection between the Rayleigh line and Hugoniot curve, as shown in Fig. 4.11. From (4.43),

$$\frac{\rho_1}{\rho_2} = \frac{\upsilon_2}{\upsilon_1} = \frac{1}{\gamma M_s^2}\left(1 - \frac{p_2}{p_1}\right) + 1. \tag{4.45}$$

By substituting the above in (4.44), we obtain

$$\frac{p_2}{p_1} = \frac{\frac{1}{\gamma M_s^2}\left(1 - \frac{p_2}{p_1}\right) + 1 - \frac{\gamma+1}{\gamma-1}}{1 - \frac{\gamma+1}{\gamma-1}\left\{\frac{1}{\gamma M_s^2}\left(1 - \frac{p_2}{p_1}\right) + 1\right\}},$$

$$\left[(\gamma + 1)\frac{p_2}{p_1} - \left\{2\gamma M_s^2 - (\gamma - 1)\right\}\right]\left(\frac{p_2}{p_1} - 1\right) = 0. \tag{4.46}$$

The solution for $\frac{p_2}{p_1} \neq 1$ is

$$\frac{p_2}{p_1} = 1 + \frac{2\gamma}{\gamma + 1}\left(M_s^2 - 1\right). \tag{4.47}$$

The specific heat ratio, γ, is constant for a specified gas species. Equation (4.47) gives the pressure ratio across a normal shock wave as a function of the shock Mach number, M_s, alone. From (4.45) and (4.47),

$$\frac{\rho_2}{\rho_1} = \frac{v_1}{v_2} = \frac{(\gamma + 1)M_s^2}{(\gamma - 1)M_s^2 + 2}. \tag{4.48}$$

By combining the above with (4.19), we obtain

$$u_1 - u_2 = \underline{u}_2 - \underline{u}_1 = \frac{2a_1}{\gamma + 1}\left(M_s - \frac{1}{M_s}\right), \tag{4.49}$$

$$M_s = \frac{u_1}{a_1} = \frac{\underline{U}_s - \underline{u}_1}{a_1}. \tag{4.50}$$

Other thermodynamic properties are obtained using the equation of state:

$$\frac{T_2}{T_1} = \left(\frac{a_2}{a_1}\right)^2 = \frac{p_2/\rho_2}{p_1/\rho_1} = \frac{p_2}{p_1}\left(\frac{\rho_2}{\rho_1}\right)^{-1} = \frac{(2\gamma M_s^2 - \gamma + 1)\{(\gamma - 1)M_s^2 + 2\}}{(\gamma + 1)^2 M_s^2}. \tag{4.51}$$

A shock wave is a nonlinear wave, in which the pressures of two waves cannot be superimposed. With a high shock Mach number, this nonlinearity becomes significant. If we utilize these characteristics in favorable ways, shock waves can become a useful tool to generate high-pressure, high-temperature, and high-speed flows. On the other hand, shock waves can accompany serious hazards such as by explosions.

Figure 4.13 shows the ratio of thermodynamic properties across a normal shock wave as a function of the shock Mach number. $M_s = 1$ corresponds to an infinitesimally weak shock wave, that is, a sound wave in which the time average of the properties remains unchanged. The pressure ratio, p_2/p_1, sharply increases with increasing M_s almost quadratically for a large M_s. On the other hand, the density ratio ρ_2/ρ_1, given by (4.48), saturates with increasing M_s and asymptotically approaches $(\gamma + 1)/(\gamma - 1)$, which is equal to 6 with $\gamma = 1.4$ (gases such as

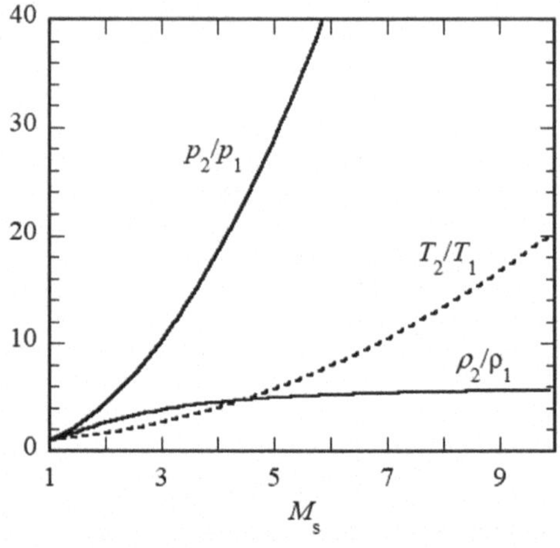

Fig. 4.13 Ratio of thermodynamic properties across a normal shock wave for a calorically perfect gas ($\gamma = 1.4$)

air) and 4 with $\gamma = 5/3$ (monoatomic gases such as helium and argon). This is a result of the heating-dominant characteristic of shock compression, as explained in Sect. 4.2.1.2. Shock compression results in energy input mainly to thermal energy. With compression, the temperature and, subsequently, the pressure increase, limiting the compression work. Since the temperature scales with the pressure-to-density ratio, with a large M_s, it behaves in a similar manner as the pressure.

Next, let us consider entropy variation. By denoting a variation from state 1 to 2 by Δ, from the first law of thermodynamics, we have

$$T\Delta s = \Delta e + p\Delta\left(\frac{1}{\rho}\right). \tag{4.52}$$

By using the equation of state of a calorically perfect gas, we obtain

$$\begin{aligned}\Delta s = s_2 - s_1 &= \frac{1}{T}\Delta\left(\frac{1}{\gamma-1}\frac{p}{\rho}\right) + \frac{p}{T}\Delta\left(\frac{1}{\rho}\right) = \frac{1}{\gamma-1}\frac{1}{\rho T}\Delta p + \frac{1}{\gamma-1}\frac{p}{T}\Delta\left(\frac{1}{\rho}\right) + \frac{p}{T}\Delta\left(\frac{1}{\rho}\right)\\ &= \frac{R}{\gamma-1}\frac{\Delta p}{p} - \frac{\gamma R}{\gamma-1}\frac{\Delta\rho}{\rho} = C_v\left[\ln\left\{\frac{p_2}{p_1}\left(\frac{\rho_2}{\rho_1}\right)^{-\gamma}\right\}\right].\end{aligned} \tag{4.53}$$

By substituting the above in (4.47) and (4.48), we obtain

$$\Delta s = C_v \ln\left[\left\{1 + \frac{2\gamma}{\gamma+1}\left(M_s^2 - 1\right)\right\}\left\{\frac{(\gamma+1)M_s^2}{(\gamma-1)M_s^2+2}\right\}^{-\gamma}\right]. \tag{4.54}$$

As shown in Fig. 4.14, Δs is an increasing function of M_s.

Let us derive a relation between flow velocities. From (4.19) and (4.20),

$$u_1 + \frac{p_1}{\rho_1 u_1} = u_2 + \frac{p_2}{\rho_2 u_2},$$

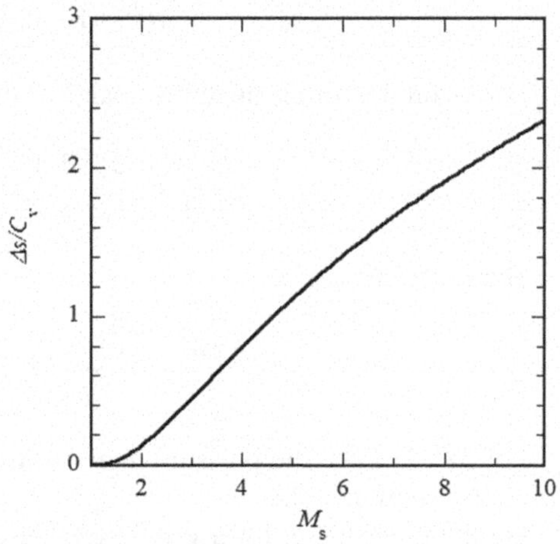

Fig. 4.14 Entropy increment, Δs, as a function of M_s ($\gamma = 1.4$)

$$u_1 + \frac{a_1^2}{\gamma u_1} = u_2 + \frac{a_2^2}{\gamma u_2}. \tag{4.55}$$

From (4.21),

$$\frac{a_1^2}{\gamma - 1} + \frac{1}{2}u_1^2 = \frac{a_2^2}{\gamma - 1} + \frac{1}{2}u_2^2 \equiv \frac{a_*^2}{\gamma - 1} + \frac{1}{2}a_*^2 = \frac{\gamma + 1}{2(\gamma - 1)}a_*^2, \tag{4.56}$$

$$a_1^2 = \frac{\gamma + 1}{2}a_*^2 - \frac{\gamma - 1}{2}u_1^2,$$

$$a_2^2 = \frac{\gamma + 1}{2}a_*^2 - \frac{\gamma - 1}{2}u_2^2,$$

where a_* is the value of the speed of sound when it is equal to the flow speed. Because the total enthalpy is conserved across the shock wave, this value is also kept constant. By inputting the above equations in (4.55), we obtain

$$u_1 u_2 = a_*^2. \tag{4.57}$$

This is called the *Prandtl relation*, which expresses the relation between the flow velocities across the shock wave in a simple form.

4.2.2.1 Relations Under Ultimate Conditions

For very weak shock waves, in which $M_s \cong 1$, the shock-wave relations can be approximated by simpler equations, which are convenient for quickly determining values such as the pressure increment. Under this condition,

$$M_s^2 - 1 \cong 2(M_s - 1). \tag{4.58}$$

For example, for air under the standard condition ($\gamma = 1.4$),

$$\frac{2\gamma}{\gamma + 1} = \frac{2 \times 1.4}{1.4 + 1} = \frac{2.8}{2.4} = \frac{7}{6} \sim 1.$$

Hence, from (4.43),

$$\frac{\Delta p}{p_1} = \frac{p_2 - p_1}{p_1} \cong 2(M_s - 1).$$

For $M_s = 1.05$, the pressure increment is approximately $2 \times (1.05 - 1) = 0.1$ times the ambient value.

On the other hand, for a very strong shock wave, in which $M_S \to \infty$, we have from (4.43) and (4.45),

$$\frac{p_2}{p_1} \approx \frac{2\gamma}{\gamma + 1} M_s^2, \quad \frac{\rho_2}{\rho_1} \approx \frac{\gamma + 1}{\gamma - 1}. \tag{4.59}$$

With increasing M_s, the pressure ratio will increase without limitation, but the density ratio will saturate to a constant value.

4.2.2.2 Mach Number Behind Normal Shock Wave

Let us confirm (4.38), which implies that the flow behind a normal shock wave is subsonic. From (4.47) and (4.48),

$$\begin{aligned} M_2 &= \frac{u_2}{a_2} = \frac{\rho_2 u_2}{\rho_2 a_2} = \frac{\rho_1 u_1}{\rho_2 a_2} = \frac{\rho_1 a_1 u_1}{\rho_2 a_2 a_1} = \left(\frac{\rho_1}{\rho_2}\right)\left(\frac{a_1}{a_2}\right) M_s = \left(\frac{p_2}{p_1}\right)^{-1/2} \left(\frac{\rho_2}{\rho_1}\right)^{-1/2} M_s \\ &= \left\{ \frac{(\gamma - 1) M_s^2 + 2}{2\gamma M_s^2 - \gamma + 1} \right\}^{1/2}. \end{aligned} \tag{4.60}$$

As shown in Fig. 4.15, $M_2 < 1$ for $M_s > 1$. Therefore, the flow behind a normal shock wave is subsonic.

After a normal shock wave propagates in quiescent air, a post-shock flow is induced. From (4.49) and (4.51), the post-shock flow Mach number is obtained such that

$$\underline{M}_2 = \frac{\underline{u}_2}{a_2} = 2(M_s^2 - 1)(2\gamma M_s^2 - \gamma + 1)^{-1/2}\{(\gamma - 1)M_s^2 + 2\}^{-1/2}. \tag{4.61}$$

As shown in Fig. 4.16, $\underline{M}_2$ is an increasing function of M_s. With

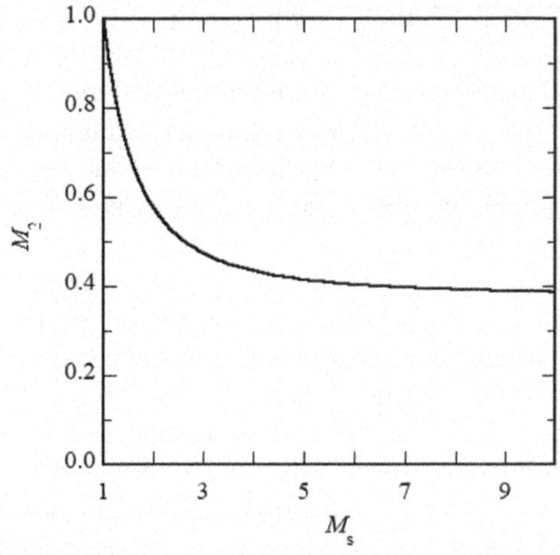

Fig. 4.15 Relation between M_2 and M_s for $\gamma = 1.4$

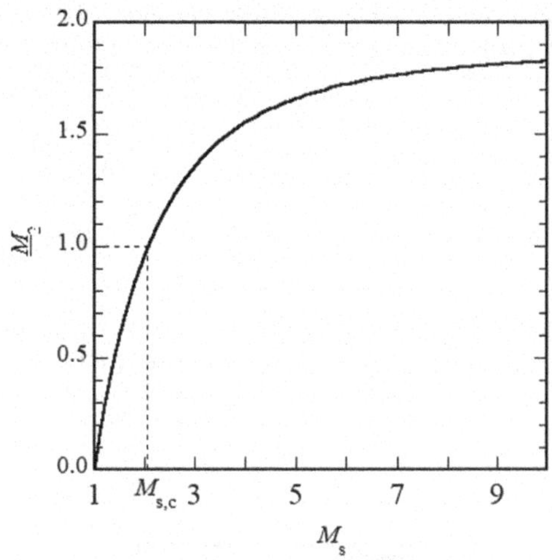

Fig. 4.16 $\underline{M}_2$ versus M_s for $\gamma = 1.4$ and $\underline{M}_1 = 0$

$$M_s = M_{s,c} \equiv \left\{ \frac{7 - \gamma + \sqrt{\gamma^2 + 2\gamma + 17}}{4(2 - \gamma)} \right\}^{\frac{1}{2}}, \tag{4.62}$$

$\underline{M}_2 = 1$. With a high shock Mach number, we can generate a supersonic flow. For $\gamma = 1.4$, this threshold value is $M_{s,c} \cong 2.068$. Supersonic flow can be generated with even higher shock Mach numbers.

4.2.2.3 Shock Compression

Usually, we start learning thermodynamic cycles in an engine as quasi-static processes in which the working gas is compressed and heated slowly enough such that the system is in equilibrium without any nonuniformity. What will happen if we quickly compress the gas? Will any difference appear in the processes and engine performance?

Consider situations in which we compress the gas in the cylinder in Fig. 4.17. In Fig. 4.17a, we compress the gas in the cylinder by slowly pushing the piston. Through the piston motion, pressure waves propagate back and forth in the cylinder, and the pressure is kept almost uniform. Without the addition of external heat, the process is *isentropic*. On the other hand, in Fig. 4.17b, when we rapidly push the piston to compress the gas, a normal shock wave is formed in front of it. Is there any difference in the force required to push the piston? In the quasi-static processes of Fig. 4.17a, the pressure in the cylinder is kept constant, thereby following the isentropic relation of (2.91). In the case of Fig. 4.17b, the piston with a speed of U_p pushes the gas at an initial pressure of p_1, driving a normal shock wave ahead with a post-shock pressure of p_2. From (4.49) with $\underline{u}_1 = 0$ and $\underline{u}_2 = U_\mathrm{p}$,

$$U_\mathrm{p} = \frac{2a_1}{\gamma + 1}\left(M_\mathrm{s} - \frac{1}{M_\mathrm{s}}\right), \tag{4.63}$$

$$M_\mathrm{p} = \frac{U_\mathrm{p}}{a_1}, \tag{4.64}$$

where a_1 and M_p are the speed of sound in the initial condition and the Mach number of the piston, respectively. From (4.63),

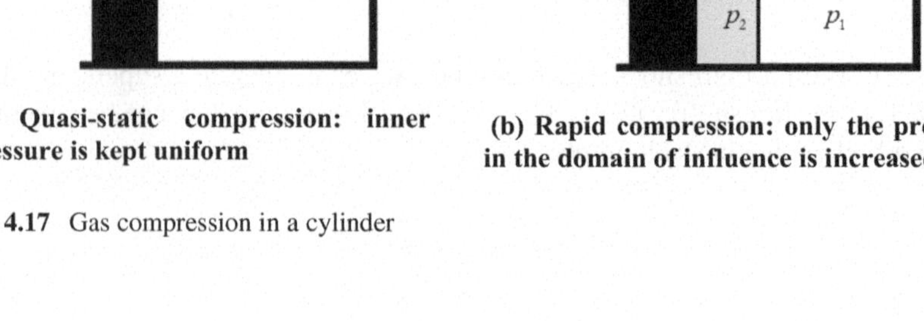

(a) Quasi-static compression: inner pressure is kept uniform

(b) Rapid compression: only the pressure in the domain of influence is increased

Fig. 4.17 Gas compression in a cylinder

$$M_s = \frac{1}{2}\left[\frac{\gamma+1}{2}M_p + \sqrt{\left(\frac{\gamma+1}{2}M_p\right)^2 + 4}\right]. \tag{4.65}$$

By substituting (4.47) with (4.65), we obtain

$$\frac{p_2}{p_1} = 1 + \frac{\gamma(\gamma+1)M_p^2}{4}\left[1 + \sqrt{1 + \left(\frac{4}{(\gamma+1)M_p}\right)^2}\right]. \tag{4.66}$$

The dynamic-to-static pressure ratio behind the shock wave is

$$\frac{\rho_2 U_p^2}{2p_1} = \frac{\gamma M_p^2}{2}\frac{\rho_2}{\rho_1} = \frac{\gamma M_p^2}{2}\frac{(\gamma+1)M_s^2}{(\gamma-1)M_s^2+2}. \tag{4.67}$$

After the shock wave is reflected on the right-end wall, a reflected shock wave propagates to the left, and the pressure increases further. By using (9.48) of Chap. 9, the pressure ratio across the reflected shock wave is

$$\frac{p_5}{p_1} = \frac{p_5}{p_2}\frac{p_2}{p_1} = \frac{\left(\frac{3\gamma-1}{\gamma-1}\right)\frac{p_2}{p_1} - 1}{\frac{p_2}{p_1} + \frac{\gamma+1}{\gamma-1}}\frac{p_2}{p_1}, \tag{4.68}$$

where the subscript "5" refers to the state behind the reflected shock wave. Equation (4.68) is obtained using the boundary condition $\underline{u}_5 = 0$. Variations of the post-shock pressures are shown in Fig. 4.18. At a low $M_p(\leq 0.3)$, the dynamic pressure is negligible. The faster the piston moves, the higher is the post-shock pressure, p_2, becomes. This leads to an increase in the work done by the piston on the gas. Therefore, the gas gains larger energy with rapid compression. With repeated shock-wave

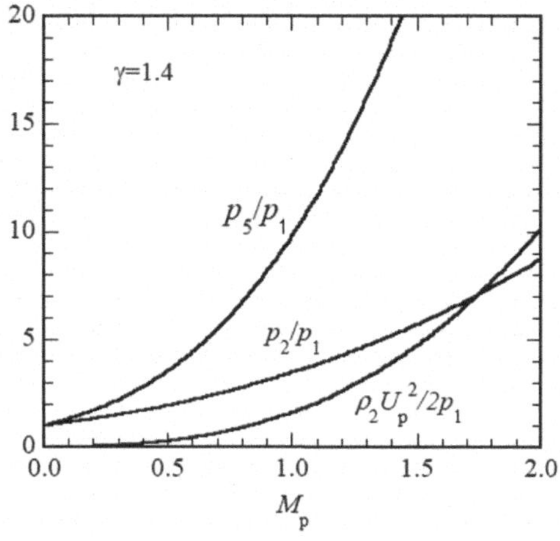

Fig. 4.18 Post-shock pressures as functions of the piston Mach number, M_p

reflections, the static pressure is further increased. With the same compression ratio, that is, the density ratio, the temperature scales with the pressure. Therefore, this result suggests that shock compression is effective also to generate high-temperature gases. In particular, for $M_p \ll 1$, the static pressure is much higher than the dynamics pressure, implying that the static pressure can be substantially increased by the shock compression even with a low speed. In other words, "quick" motion enables high-pressure/high-temperature generation through shock compression.

Let us compare temperature–pressure variation between shock and isentropic compressions. From (4.47) and (4.48),

$$\frac{\frac{T_2}{T_1}}{\frac{p_2}{p_1}} = \frac{(\gamma-1)\frac{p_2}{p_1}+\gamma+1}{(\gamma+1)\frac{p_2}{p_1}+\gamma-1},$$

$$\frac{T_2}{T_1} = \frac{(\gamma-1)\frac{p_2}{p_1}+\gamma+1}{(\gamma+1)\frac{p_2}{p_1}+\gamma-1}\left(\frac{p_2}{p_1}\right). \tag{4.69}$$

The variation in isentropic processes is obtained from (2.91) as

$$\frac{T_2}{T_1} = \left(\frac{p_2}{p_1}\right)^{\frac{\gamma-1}{\gamma}}.$$

As shown in Fig. 4.19, in isentropic compression, the temperature follows the power law of $T_2/T_1 = (p_2/p_1)^{0.29}$ with $\gamma = 1.4$. On the other hand, in the case of shock compression, they have an almost linear dependence, as expressed by (4.69), thereby increasing the temperature much more efficiently.

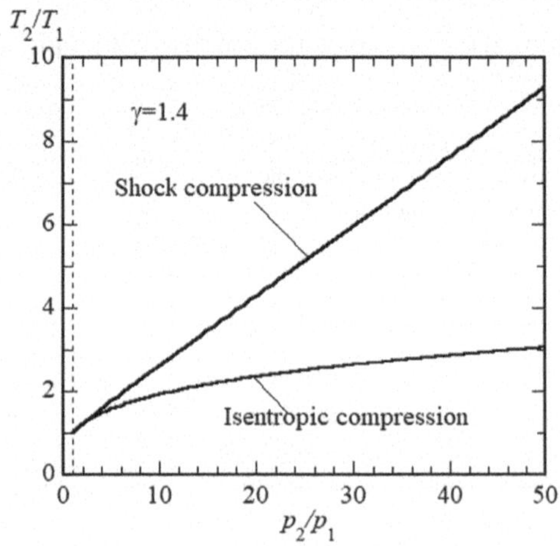

Fig. 4.19 Comparison between shock and isentropic compressions in terms of temperature and pressure ratios

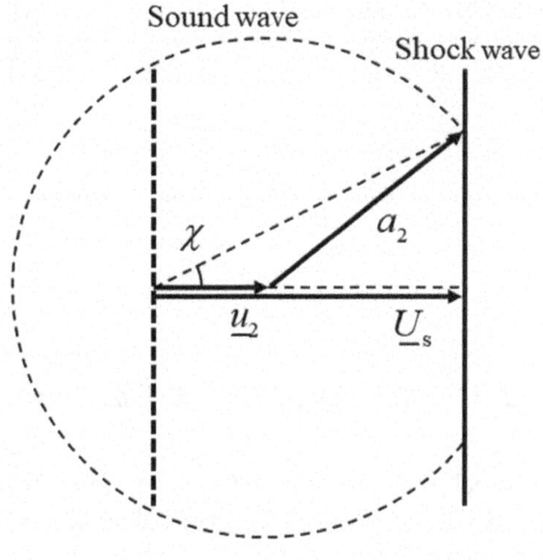

Fig. 4.20 Glancing incidence

4.2.3 Glancing Incidence

When a normal shock wave experiences a local fluctuation, at what velocity will it propagate along the wave front? Figure 4.20 illustrates the propagation of fluctuation behind a normal shock wave. If a fluctuation occurs on the wave front with a velocity of $\underline{U}_s$, its influence will propagate at the speed of sound behind the shock wave, a_2. In this case, the fluctuation propagates along the shock wave front at a *glancing incidence*, χ. From (4.49) and (4.51),

$$\underline{u}_2 = \frac{2a_1}{\gamma+1}\left(M_s - \frac{1}{M_s}\right),$$
$$\frac{a_2}{a_1} = \frac{\sqrt{(2\gamma M_s^2 - \gamma + 1)\{(\gamma-1)M_s^2 + 2\}}}{(\gamma+1)M_s}. \tag{4.70}$$

Therefore,

$$\tan\chi = \frac{\sqrt{a_2^2 - (\underline{U}_s - \underline{u}_2)^2}}{\underline{U}_s}$$
$$= \frac{1}{M_s^2}\sqrt{\frac{\{(\gamma-1)M_s^2 + 2\}(M_s^2 - 1)}{\gamma+1}}. \tag{4.71}$$

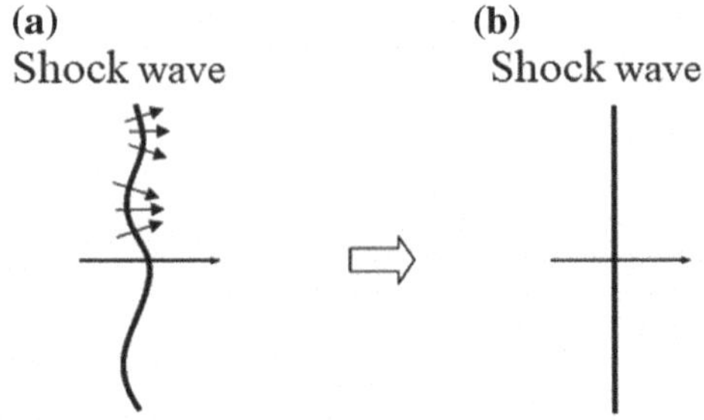

Fig. 4.21 Stability of a shock wave front

4.2.4 *Stability of Shock Wave Front*

A shock wave front is stable; even if the front is disturbed for some reason, the wave front will be restored to its initial shape. A disturbed plane shock wave will restore to a plane as shown in Fig. 4.21 [2]. In order to understand this characteristic, we apply a result from Sect. 8.6 that the shock-wave propagation speed is a decreasing function of the cross-sectional area of a passage. Therefore, if a part of the shock wave front deforms in a convex shape toward the propagation direction, as shown in the uppermost part of Fig. 4.21a, the propagation speed decreases with the expansion of the passage. On the other hand, if the front deforms in a concave shape, the propagation speed increases so that the wave front will catch up with the surroundings. In any case, the wave front will behave to alleviate its deformation.

4.2.5 *Shock-Wave Propagation with Boundary Layer*

When a normal shock wave propagates along a wall in viscous flow, a boundary layer is induced behind. As shown in Fig. 4.22a, behind a shock wave with a propagation velocity of $\underline{U}_s$, a post-shock flow with a velocity of $\underline{u}_2$ is induced. However, in a viscous flow, the flow velocity should vanish on the wall. Thus, the flow-velocity distribution shown in Fig. 4.22a is formed. On observing this flow on a frame fixed to the shock wave, the flow-velocity distribution behind the shock wave takes the

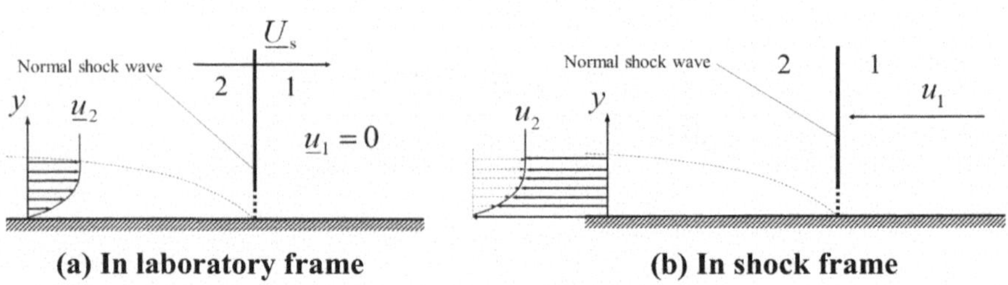

Fig. 4.22 Shock-wave propagation with boundary-layer formation

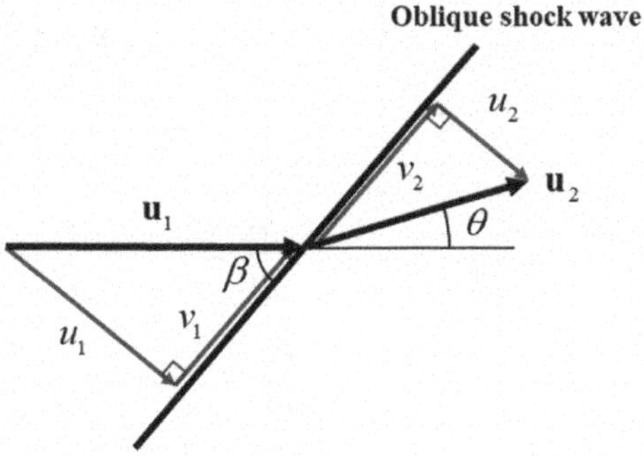

Fig. 4.23 Oblique shock wave and flow velocity vector

form shown in Fig. 4.22b. On the wall, the flow comes in at a speed of $u_1 = \underline{U}_s$ and leaves at the same speed: $u_2(y = 0) = u_1 = \underline{U}_s$. Outside the boundary layer, the flow follows the Rankine–Hugoniot relations. The y distribution of $\underline{u}_2$ and u_2 can be obtained based on the boundary-layer equation. The reader should refer to other articles[4] for further details.

4.3 Oblique Shock Wave

In an oblique shock wave, although only a tangential velocity component to the wave front is added to a normal shock wave, various characteristics are added, as will be seen in the following. Here, we will deal with a calorically perfect gas to quantify the characteristics.

4.3.1 Oblique Shock Relations

Let us consider the oblique shock wave shown in Fig. 4.23. The states upstream and downstream of the shock wave are referred to as "1" and "2," respectively. The flow velocity vector $\mathbf{u}$ is decomposed into a normal component, u, and a tangential component, v, with respect to the wave front. The angle between $\mathbf{u}_1$ and the shock wave is denoted by β, and the *deflection angle* past the shock wave is denoted by θ.

Combining the shock-wave relations given by (4.13)–(4.15) and (4.17), the same equation as (4.19)–(4.21) remain applicable to the normal velocity component, u. Therefore, in order to obtain the relations for the oblique shock wave, the following geometrical relation should be input to the normal shock wave equations given by (4.47), (4.48), and (4.49)[5]:

[4] For example, P. A. Thompson: Compressible-fluid dynamics, McGraw-Hill, 1972, Chap. 10

[5] In the oblique-shock-wave relations, the upstream flow Mach number is designated by M_1.

$$M_1 = \frac{|\mathbf{u_1}|}{a_1}, \tag{4.72}$$

$$M_s = \frac{u_1}{a_1} = M_1 \sin\beta, \tag{4.73}$$

$$\frac{p_2}{p_1} = 1 + \frac{2\gamma}{\gamma+1}\left(M_1^2 \sin^2\beta - 1\right), \tag{4.74}$$

$$\frac{\rho_2}{\rho_1} = \frac{\upsilon_1}{\upsilon_2} = \frac{(\gamma+1)M_1^2 \sin^2\beta}{(\gamma-1)M_1^2 \sin^2\beta + 2}, \tag{4.75}$$

$$u_1 - u_2 = \frac{2a_1}{\gamma+1}\left(M_1 \sin\beta - \frac{1}{M_1 \sin\beta}\right). \tag{4.76}$$

In the tangential direction, no force is exerted. Hence, the tangential velocity does not change across the shock wave, as expressed by (4.15).

$$v_1 = v_2 = |\mathbf{u_1}| \cos\beta. \tag{4.77}$$

From (4.73) and (4.77), and the geometrical relation shown in Fig. 4.23,

$$M_1 a_1 \cos\beta = \frac{u_2}{\tan(\beta - \theta)}. \tag{4.78}$$

By substituting (4.76) with (4.73) and (4.78), we obtain

$$M_1 \sin\beta - M_1 \cos\beta \tan(\beta - \theta) = \frac{2}{\gamma+1}\left(M_1 \sin\beta - \frac{1}{M_1 \sin\beta}\right),$$

$$\tan\theta = \frac{2\cot\beta\left(M_1^2 \sin^2\beta - 1\right)}{M_1^2(\gamma + \cos 2\beta) + 2}. \tag{4.79}$$

Equation (4.79) gives the relation between β and θ with constant γ and M_1, as shown in Fig. 4.24.

As shown in Figs. 4.24 and 4.25, with β increasing from β_M (see the next section), θ first increases and then decreases, whereas the pressure ratio p_2/p_1 monotonically increases and M_2 monotonically decreases, even becoming smaller than unity. For any M_1 value, β has a maximum of 90° with $\theta = 0^\circ$.

4.3.2 *Mach Wave*

As shown in Chap. 1, pressure waves emitted from a fixed place in a supersonic flow forms form a *Mach wave* as their envelope. The Mach wave is the boundary of the domain of influence and is regarded as the weakest shock wave across which the

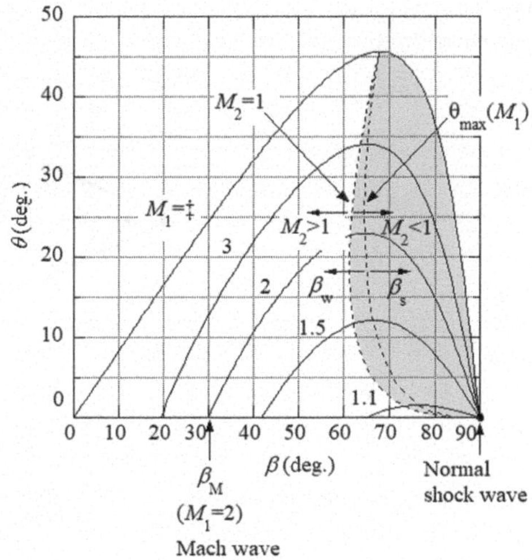

Fig. 4.24 Oblique-shock-wave relations among M_1, β, and θ for $\gamma = 1.4$. The gray domain on the right-hand side of the $M_2 = 1$ curve corresponds to $M_2 < 1$, and the region on the left-hand side corresponds to $M_2 > 1$

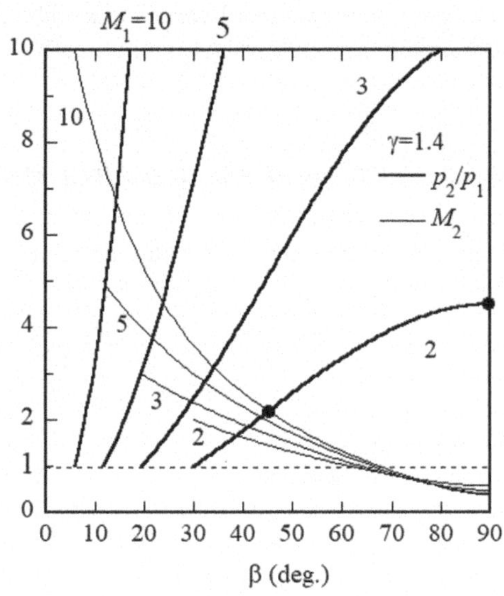

Fig. 4.25 Oblique-shock-wave relations: the variation of pressure ratio and M_2 as functions of β for $\gamma = 1.4$

time averages of flow quantities remain unchanged. Figure 4.26 shows Mach waves that are equivalent to those in Fig. 1.5 but in the form of Fig. 4.23. The *Mach angle*, β_M, is given by

$$\sin\beta_M = \frac{at}{ut} = \frac{a}{u} = \frac{1}{M}, \tag{4.80}$$

$$\beta_M = \sin^{-1}\left(\frac{1}{M}\right), \tag{4.81}$$

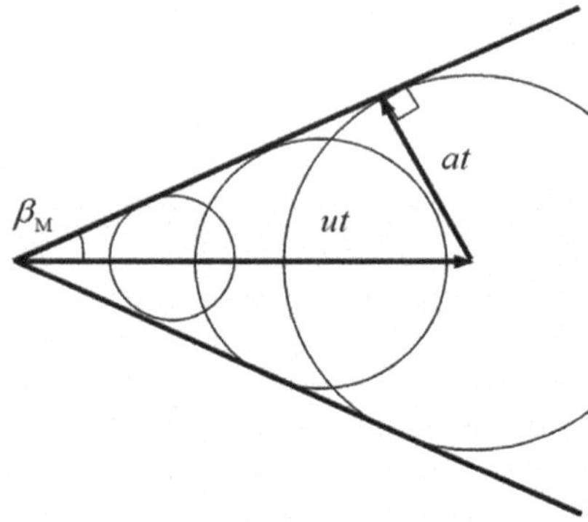

Fig. 4.26 Mach wave in supersonic flow

$$M = \frac{u}{a}. \tag{4.82}$$

β_M is uniquely determined by the Mach number, M. In the far field of a supersonically flying body, disturbances induced by the body become so weak that the angle of the oblique shock wave asymptotically approaches β_M.

4.3.3 Two Solutions and Their Post-shock Mach Numbers

With a constant value of M_1, the maximum value of θ is obtained by applying the condition

$$\frac{d\theta}{d\beta} = 0 \tag{4.83}$$

to (4.79):

$$2\gamma M_1^4 \sin^4\beta + \left\{4 - (\gamma+1)M_1^2\right\} M_1^2 \sin^2\beta - (\gamma+1)M_1^2 - 2 = 0,$$

$$\sin^2\beta = \frac{1}{4\gamma M_1^2}\left[(\gamma+1)M_1^2 - 4 + (\gamma+1)^{\frac{1}{2}}\left\{(\gamma+1)M_1^4 + 8(\gamma-1)M_1^2 + 16\right\}^{\frac{1}{2}}\right],$$

$$\sin\beta = \frac{1}{2\sqrt{\gamma}M_1}\left[(\gamma+1)M_1^2 - 4 + (\gamma+1)^{\frac{1}{2}}\left\{M_1^4(\gamma+1) + 8(\gamma-1)M_1^2 + 16\right\}^{\frac{1}{2}}\right]^{\frac{1}{2}}. \tag{4.84}$$

By substituting (4.79) with (4.84), $\theta = \theta_{max}(M_1)$ is obtained, as shown in Fig. 4.24. For $\theta < \theta_{max}(M_1)$, two solutions exist for a value of θ. We will name the solutions $\beta = \beta_w$ and $\beta = \beta_s$ $\left(\beta_w < \beta_s,\ p_{2,w} < p_{2,s}\right)$ as the *weak* and *strong solution*, respectively.

In the case of a normal shock wave, the post-shock flow is always subsonic. However, because the tangential velocity to a shock wave is conserved, the post-shock flow velocity can be higher than the speed of sound. Let us obtain the speed of sound behind an oblique shock wave.

$$a_2 = \left(\frac{T_2}{T_1}\right)^{\frac{1}{2}} a_1 = \left(\frac{\frac{p_2}{p_1}}{\frac{\rho_2}{\rho_1}}\right)^{\frac{1}{2}} a_1 = \left[\frac{1 + \frac{2\gamma}{\gamma+1}\left(M_1^2 \sin^2\beta - 1\right)}{\frac{(\gamma+1)M_1^2 \sin^2\beta}{(\gamma-1)M_1^2 \sin^2\beta + 2}}\right]^{\frac{1}{2}} a_1$$
$$= \frac{a_1}{(\gamma+1)M_1 \sin\beta}\left[\left\{2\gamma M_1^2 \sin^2\beta - (\gamma-1)\right\}\left\{(\gamma-1)M_1^2 \sin^2\beta + 2\right\}\right]^{\frac{1}{2}}. \tag{4.85}$$

By using the above equation, the post-shock flow Mach number is obtained as

$$M_2 = \frac{|\mathbf{u_2}|}{a_2} = \frac{u_2}{a_2 \sin(\beta-\theta)} = \frac{\rho_1}{\rho_2} \frac{u_1}{a_2 \sin(\beta-\theta)} = \frac{\rho_1}{\rho_2} \frac{a_1}{a_2} \frac{u_1}{a_1 \sin(\beta-\theta)}$$
$$= \frac{\rho_1}{\rho_2} \frac{a_1}{a_2} \frac{M_1 \sin\beta}{\sin(\beta-\theta)} = \frac{1}{\sin(\beta-\theta)}\left[\frac{(\gamma-1)M_1^2 \sin^2\beta + 2}{2\gamma M_1^2 \sin^2\beta - (\gamma-1)}\right]^{\frac{1}{2}}. \tag{4.86}$$

By using (4.79),

$$M_2 = \left[\frac{\left\{(\gamma-1)M_1^2 \sin^2\beta + 2\right\}^2 + (\gamma+1)^2 M_1^4 \sin^2\beta \cos^2\beta}{\left\{2\gamma M_1^2 \sin^2\beta - (\gamma-1)\right\}\left\{(\gamma-1)M_1^2 \sin^2\beta + 2\right\}}\right]^{\frac{1}{2}}. \tag{4.87}$$

In Fig. 4.24, the solution for $M_2 = 1$ is also plotted. In the domain right of this curve, shown in gray, the post-shock flow is subsonic ($M_2 < 1$), while the flow is supersonic in the left domain. In the weak shock domain ($\beta = \beta_w$), the post-shock flow is supersonic ($M_2 > 1$) in most cases, but becomes subsonic under limited conditions. However, in the strong shock domain ($\beta = \beta_s$), the post-shock flow is always subsonic ($M_2 < 1$). As shown in Fig. 4.24, the condition for $\theta = \theta_{max}$ and that for $M_2 = 1$ is very close to each other. In practice, we can neglect the difference and regard a strong shock wave to have subsonic post-shock flow and a weak one to have supersonic post-shock flow.

The solution in real flows depends on the boundary conditions and flow history. As will be shown later, in general, for weak disturbances due to objects such as a slender body, a weak shock solution is obtained. When disturbances are strong, as in the case of stagnated flow on a blunt body, a strong shock solution is obtained.

Column: Newtonian Flow Approximation

Isaac Newton, in his famous article "Principia (1687)," dealt with a force on a flat plate with an oblique angle to a uniform flow. His model is referred to as the *Newtonian flow approximation*.[6] Consider an incoming flow with a speed of u_1 on a flat plate with an inclination angle of θ. In Newton's era, when fluid dynamics was not yet well developed, the concept of shock waves did not exist. Newton assumed that the flow was bent to the direction parallel to the plate; that is, the velocity component normal to the plate was assumed to vanish, and the force, F, exerted normal to the plate was assumed equal to the rate of variation in the momentum component. Let ρ, l, and w be the density of the fluid, plate length, and plate width, respectively.

$$F = \rho u_1 lw \sin\theta \cdot u_1 \sin\theta = \rho u_1^2 lw \sin^2\theta. \quad (4.88)$$

Therefore, the overpressure, that is, the excessive pressure value relative to the upstream one, is

$$\Delta p = \frac{F}{lw} = \rho u_1^2 \sin^2\theta. \quad (4.89)$$

By using the pressure coefficient, C_p, we obtain

$$C_p \equiv \frac{\Delta p}{\frac{1}{2}\rho u_1^2} = 2\sin^2\theta, \quad (4.90)$$

$$\frac{\Delta p}{p_1} = \frac{\gamma u_1^2}{\gamma \frac{p_1}{\rho}} \sin^2\theta = \gamma M_1^2 \sin^2\theta. \quad (4.91)$$

As shown in Fig. 4.28, the Newtonian flow approximation underestimates the pressure increment, particularly at a low Mach number. The difference is less at a high Mach number (Fig. 4.27).

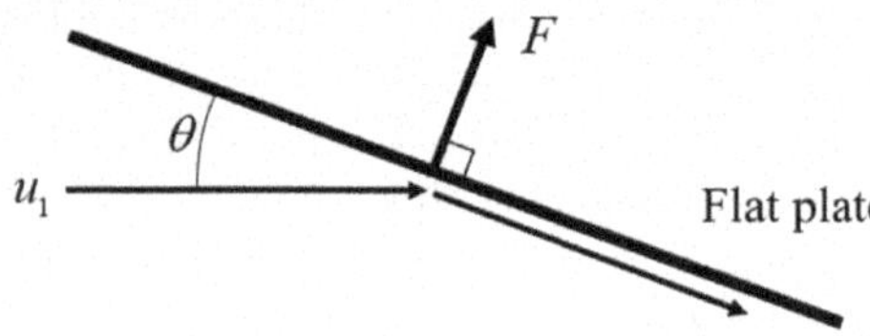

Fig. 4.27 Newtonian flow approximation

[6]Note that this approximation is different from a "Newtonian fluid," in which viscous stress is proportional to the gradient of tangential velocity.

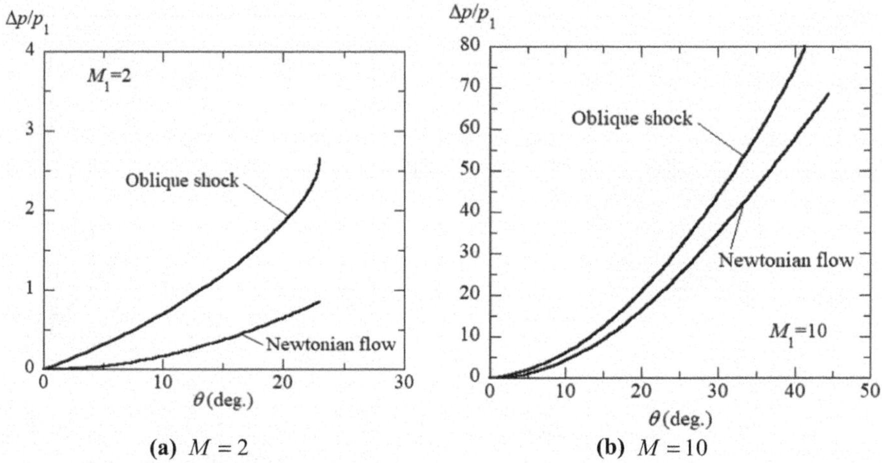

Fig. 4.28 Comparison of pressure increment between the Newtonian flow approximation and an oblique shock wave

4.3.4 Attached and Detached Shock Waves

When the flow behind an oblique shock wave is directed along a wall, a solution of an *attached shock wave* exists for $\theta < \theta_{\max}$. In this case, a weak shock wave, $\beta = \beta_w$, occurs; as shown in Fig. 4.29a, the shock wave attaches to the leading edge of a sharp body, and the flow behind it is supersonic. If the body has an infinite length with a constant deflection angle of θ, the shock wave is maintained. If a strong solution,

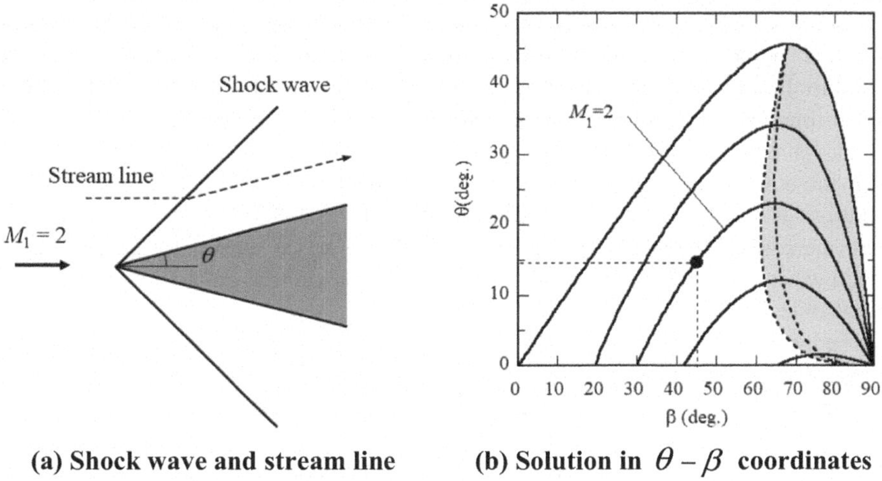

Fig. 4.29 Example of an attached shock wave

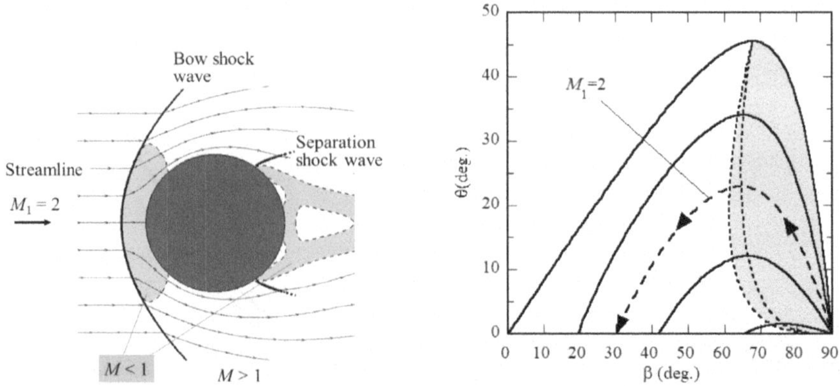

(a) Time-averaged flow field. **(b) Variation of post-shock condition in $\theta - \beta$ coordinates for $M_1 = 2$**

Fig. 4.30 Example of a detached shock wave in front of a circular cylinder for $\gamma = 1.4$, $M_1 = 2$, and $p_2/p_1 = 4.50$

$\beta = \beta_s$, occurs under the flow condition of Fig. 4.29, the flow behind the shock wave would become subsonic, which cannot be maintained. In short, only a weak shock solution occurs under the condition of Fig. 4.29a.

If $\theta > \theta_{max}$, a *detached shock wave* occurs, forming a *shock layer* behind it. Figure 4.30 shows Mach 2 supersonic flow around a circular cylinder. The flow behind the shock wave is bifurcated by a *stagnation line*.[7] The stagnation line intersects normal to the shock wave, across which the normal-shock-wave relations are applied. For a Mach 2 flow ($M_1 = 2$) with $\gamma = 1.4$, the pressure ratio across the oblique shock wave in Fig. 4.29 is equal to 2.17, whereas that for the normal shock wave in Fig. 4.30 is as large as 4.5. The flow speed further decreases along the stagnation and then vanishes at the so-called *stagnation point* on the wall, where the pressure is equal to the pitot pressure given by (2.113): $p_{Pitot}/p_1 = 5.64$ for $M_1 = 2.0$, which is 2.6 times that behind the attached shock wave. This implies that, if a detached shock wave occurs, the drag is strongly enhanced. As shown in Fig. 4.30b, the flow near the stagnation line in the shock layer corresponds to the strong shock wave solution, $\beta = \beta_s$; therefore, it is subsonic. The bow shock wave becomes weaker with increasing distance from the stagnation line, and it eventually approaches a Mach wave asymptotically. In this manner, around a circular cylinder or a sphere in supersonic flows, a *bow shock wave* is formed.

Figure 4.31 shows a Schlieren image of a flow around a scaled model of the "Hayabusa" re-entry capsule. It comprises a hemisphere and a truncated cone with an apex angle of 45°. Strong radiation emission from the shock layer is observed.

The most important parameters for determining the shape of a supersonically flying object are the drag and heat transfer. On one hand, to make the drag small, a

[7]In the book, it is shown as a line, but in reality, it is a plane in two-dimensional flow.

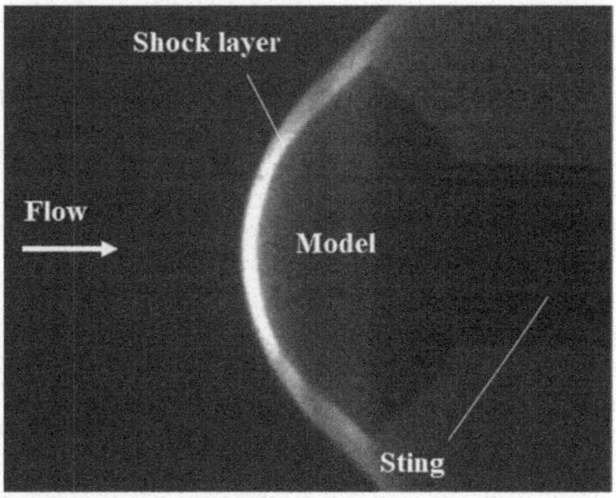

Fig. 4.31 Bow shock wave around a 1/16-scale model of the "Hayabusa" reentry capsule model experimentally observed in an expansion tube. The shadow on the right-hand side is the sting used to support the model. $M_1 = 8.4$

sharp nose is favorable so that the standing shock wave is weak. On the other hand, to decelerate a body such as a re-entry capsule or parachute quickly, a strong shock wave should be generated with a blunt body.

How is the heat transfer to a body scaled? The primary cause for the evaporation of an asteroid entering the atmosphere is the heat transfer from the shock layer, which is technically not due to the friction. Such *aerodynamic heating* is subdivided to *convective* and *radiative* heat transfer. The former corresponds to the heat conduction through a thermal boundary layer over the body, and the latter corresponds to the heat from the radiative emission of the high-temperature gas in the shock layer, which is significant only in hypersonic[8] flows. If we assume that the thickness of the shock layer is proportional to the characteristic dimension of the body, the effective temperature gradient is inversely proportional to the body dimension. Therefore, the heat transfer to a unit area is alleviated by increasing the body size. In other words, to decrease the heat transfer, the body shape should be blunt.

In summary, a sharp nose is favorable to decrease aerodynamic drag, whereas a blunt nose is favorable to decrease aerodynamic heating. Furthermore, a blunt nose in a re-entry body enhances the deceleration of the flight speed. However, if the flow Mach number is so high that the radiative heat transfer is dominant, this scaling is not applicable.

4.4 Interface and Its Stability

As is shown in Sect. 4.2.4, a shock wave front is stable against disturbances. However, in many cases, an interface is unstable without any restoring force [3]. The instability of a contact surface or slip surface can induce mixing, transition to turbulence, or other complicated flow evolutions. If such flow instability occurs in compressible flows, the flow behavior will become complicated through interaction with shock waves.

[8]Flow with a flow Mach number of approximately 5 or higher.

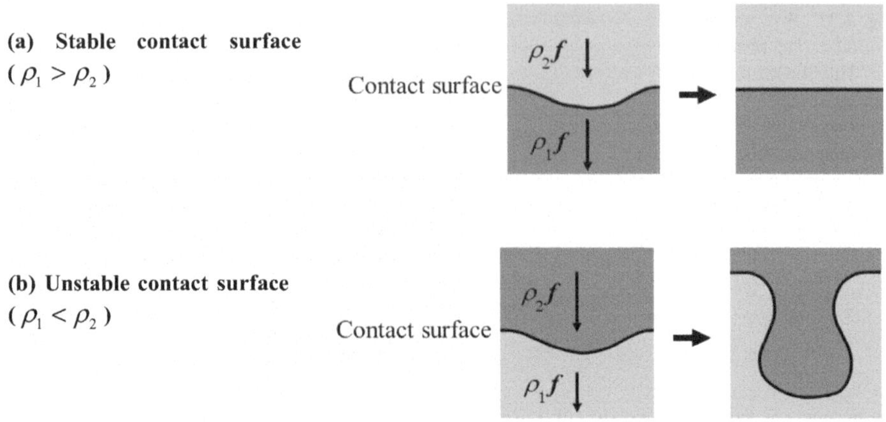

Fig. 4.32 Rayleigh–Taylor instability with a downward body force

4.5 Rayleigh–Taylor Instability

Rayleigh–Taylor (R–T) instability appears when a body exerts force over a contact surface from a heavy to light fluid. For example, under gravity, a contact surface with a light fluid (e.g., oil) on the upper side and a heavy fluid (e.g., water) on the lower side is stable (Fig. 4.32a). However, if the locations of the fluids are swapped, the contact discontinuity becomes unstable (Fig. 4.32b).

4.6 Richtmyer–Meshkov (R–M) Instability

Richtmyer–Meshkov (R–M) instability refers to the instability of a contact surface that experiences a sudden acceleration. The phenomenon occurs when a shock wave is incident on a contact surface. An important difference of R-M instability from R-T instability is that the contact discontinuity is unstable irrespective of the direction of shock-wave incidence.

Figure 4.33 schematically illustrates shock-wave incidences on a curved contact surface, which will hereafter be referred to as the "interface." In the case of incidence from the light fluid to the heavy one, as shown in Fig. 4.33a, the curvature of the interface is further enhanced. On the other hand, in the case of the heavy to light incidence, as shown in Fig. 4.33b, the sign of the curvature of the interface becomes inversed, and its amplitude increases. In both cases, after the large-scale instability grows, smaller scale instabilities will be developed, thereby inducing turbulent mixing.

The mechanisms of the interface deformation in R-M instability is explained by the *baroclinic effect*. By taking the rotation of (3.15), an equation for the generation of vorticity, $\boldsymbol{\omega}$, is obtained as follows:

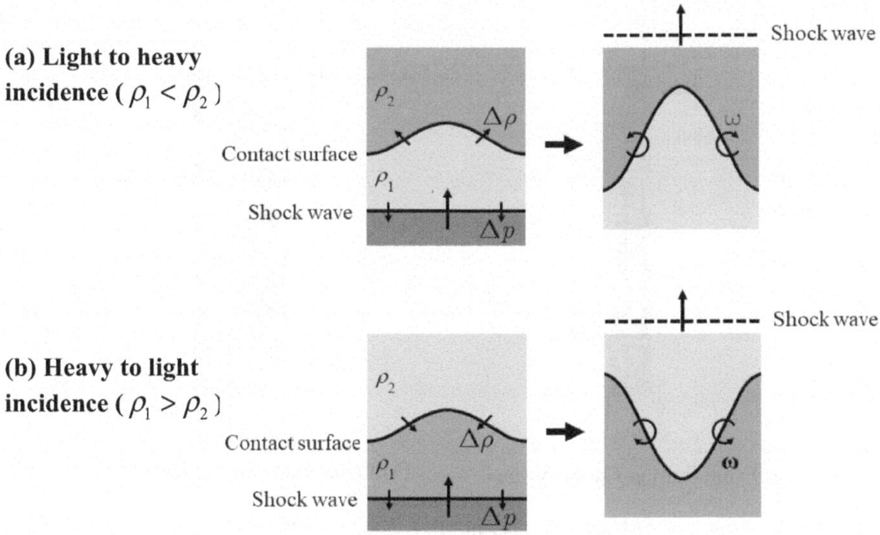

Fig. 4.33 Richtmyer–Meshkov instability. A body force does not exist, but the contact surface experiences the passage of a shock wave

$$\frac{d\boldsymbol{\omega}}{dt} = (\boldsymbol{\omega} \cdot \nabla)\mathbf{u} - \boldsymbol{\omega}(\nabla \cdot \mathbf{u}) + \frac{1}{\rho^2}\nabla\rho \times \nabla p. \tag{4.92}$$

The third term on the right-hand side is called the *baroclinic* term, which implies that a vorticity is produced by the vector product of a density gradient and a pressure gradient. The former corresponds to a density jump across the interface, while the latter corresponds to a pressure jump across the shock wave.[9] In Fig. 4.33, a planar shock wave, in which the pressure increases downward, propagates from the bottom to the top and is incident on a curved interface. In the incidence from a light fluid to a heavy one, as shown in Fig. 4.33a, the curvature of the interface is enhanced in the same phase owing to the induced vorticity. However, in the heavy to light incidence, as shown in Fig. 4.33b, a vorticity in the opposite direction is induced; the interface instability grows with the reversal of sign of the curvature.

In supernova explosion, inertia fusion, or other ultimately high-energy physics, it is critical how long and how high the pressure and temperature are kept in *implosion* processes. In such processes, R-H instability that breaks up the symmetry of the high-pressure/high-temperature domain determines the reaction time and whether explosive heat release proceeds or not.

Column: Discovery of Inversed Phase Deformation (Meshkov's Experiment)
Dr. Evgeny. E. Meshkov from the Union of Soviet Socialist Republics (USSR; currently the Sarov Physics and Technology Institute of the National Research Nuclear University in the Russian Federation) used a shock tube to investigate the behavior

[9]Here, a gradient and a jump or discontinuity have the same role.

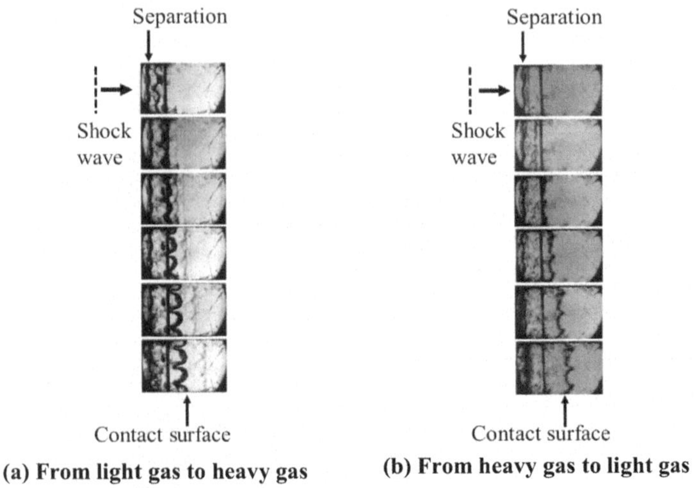

Fig. 4.34 The first series of Meshkov's experiments, conducted in 1966. The shock wave propagates from left to right (courtesy of Dr. E. E. Meshkov)

of a normal shock wave interacting with a curved interface. Figure 4.34 shows the results obtained in his first series of experiments. Two types of gases were separated by a 0.5-μm-thick polymer sheet on which a shock wave was incident from left to right. The separation is sinusoidally curved so that the shock wave is incident at an oblique angle. In the experiment of Fig. 4.34a, the shock wave propagates from a light gas to a heavy gas, and the curvature of the interface increased after the shock wave was transmitted. This result is consistent with the theoretical prediction proposed in Richtmyer's linear theory [4]. Next, he investigated shock-wave incidence from the heavy gas to the light gas and was surprised to observe, for the first time in the world, that the curvature of the interface became reversed, as shown in Fig. 4.34b. Dr. Meshkov remembered that he conducted this series of experiments in December, and he was so surprised by this discovery that he felt it to be a Christmas gift from God.

In his second series of experiments, which were performed with better instrumentation, the behavior of the interface was better visualized, as shown in Fig. 4.35. Later, these phenomena of interface instability and the resulting mixture of fluids appearing solely in compressible fluid dynamics were named *Richtmyer–Meshkov instability*.

Column: Experiments of Mushroom Cloud

Let us consider another example of the behaviors of an interface and a shock wave. Figure 4.36 shows framing Schlieren images of the interaction between a contact surface and a shock wave generated by focusing a collimated, pulsed laser beam reflected from a parabolic mirror set at the bottom of the images [5]. The gas that directly absorbed the laser pulse was rapidly heated and is recognized as the "bright" zone, named "laser-heated plasma." When the laser-heated plasma expands outward,

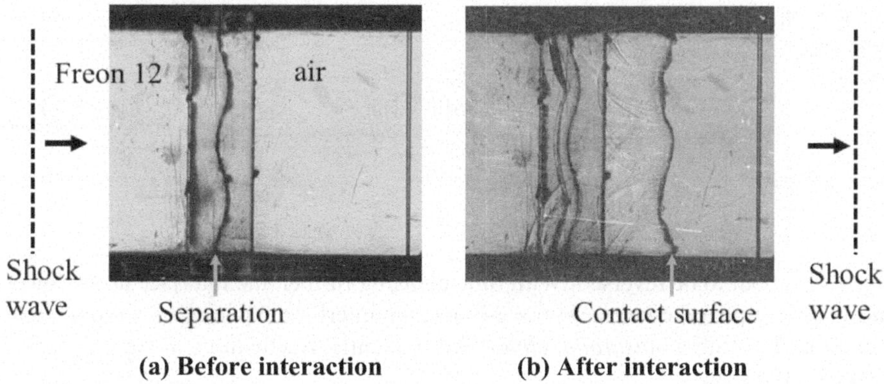

Fig. 4.35 Result of Meshkov's second series of experiments, conducted in 1968. A clear image of the interface phase inversion in heavy to light incidence was obtained

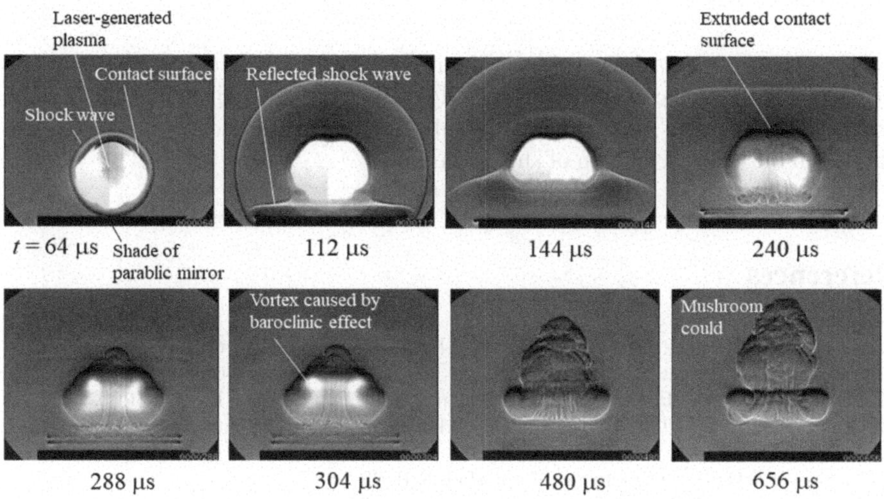

Fig. 4.36 Experimentally visualized behavior of a shock wave and an interface of laser-heated plasma. The time, t, starts at the moment of laser-pulse initiation. The gas species is krypton, initial pressure is 40 kPa, and laser pulse energy is 3.7 J

it produces a spherical shock wave, as shown in the image at $t = 64$ μs, where t represents the time measured from the moment of laser-pulse initiation. The energy output from the irradiation lasts for approximately 3 μs. Even in the first frame, a nearly spherical shock wave driven by the laser-heated plasma is clearly recognized. The shock wave has a smooth surface. The boundary between the bright zone and the surroundings corresponds to an interface. At $t = 112$ μs, the shock wave reflected from the parabolic mirror at the bottom of each frame starts the interaction with the laser-heated plasma. At $t = 144$ μs, the curvature of the lower half of the laser-heated

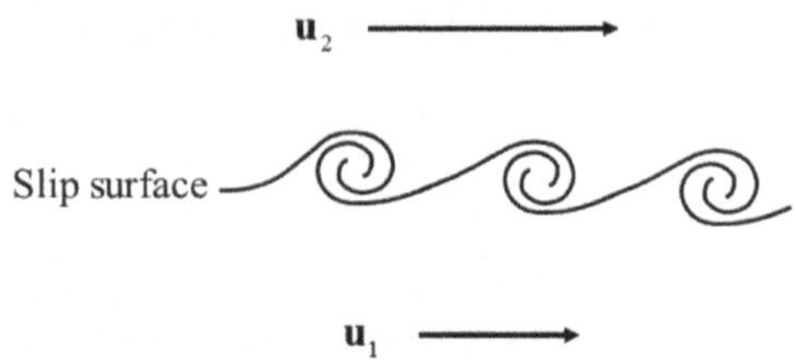

Fig. 4.37 Kelvin–Helmholtz instability

plasma is about to be reversed. With time elapsing further, the interface at the lower half became upward convex. At $t = 240$ μs, the interface experiences a topological transition, forming a *mushroom cloud*. Subsequently, fluctuations of smaller scales dominate the flow.

4.6.1 Kelvin–Helmholtz (K–H) Instability

Kelvin–Helmholtz instability refers to the instability at a slip surface. The interface instability grows in the manner shown in Fig. 4.37. This instability appears in a wake from a supersonically flying object, fuel injection in an aerospace engine, and a slip surface in the Mach reflection of shock waves.

References

1. Zel'dovich YB, Raizer, YP (1967) Physics of shock waves and high-temperature hydrodynamic phenomena. In: Hayes WD, Probstein RF (eds). Academic Press, pp 468–477
2. Landau LD, Lifshitz EM (1987) Fluid mechanics, 2nd edn. Pergamon Press, Oxford, U. K.
3. Chandrasekhar S (1981) Hydrodynamic and hydromagnetic stability. Dover Publications, Chap. 10
4. Richtmyer, RD (1960) Comm Pure Appl Math XIII, 297
5. Sasoh A, Ohtani T, Mori K (2006) Phys Rev Lett 97:205004

Chapter 5
Quasi-One-Dimensional Flows

Why do we make a small mouth when blowing out candle flames or playing with a mouth whistle? The same question can be applied to a shower, spray, or rocket nozzle, in which high-speed flow is generated by varying the cross-sectional area of a flow passage. When heated, is a flow accelerated or decelerated? Is a flow really accelerated when a force is exerted in the same direction as the flow? In this chapter, we will study *quasi-one-dimensional flow* with a variable cross-sectional area, exchange of heat, and an external force, as well as its important applications (Fig. 5.1).

5.1 Control Volume and Basic Equations

5.1.1 Control Volume and Associated Equations

Let us consider a one-dimensional, steady flow with a variable cross-sectional area along the x-axis. Let us write the conservation equations for a control volume with a thickness of Δx, which is shown as a gray zone in Fig. 5.2. Here, a viscous force is also included by assuming that the boundary layer is infinitesimally thin. The flow velocity, u, and thermodynamic properties are assumed to be uniform over the cross section at x. On the flow element, a pressure acts on walls and boundaries, and a friction force acts on walls[1]; a heat of $\mathrm{d}Q$ and a body force of f are input per unit mass. Note here that the pressure and friction force act only on boundary surfaces including walls, whereas the heat and body force act directly on the entire volume of the gas element. At the entrance of the control volume at $x = x$, the cross-sectional area is equal to A; at the exit at $x = x + \mathrm{d}x$, the cross-sectional area is equal to $A + \mathrm{d}A$. The same applies to other parameters.

---

[1] Although this book mostly deals with inviscid flows, friction force is taken into consideration in this chapter because it is often important in internal flows.

A. Sasoh, *Compressible Fluid Dynamics and Shock Waves*,
https://doi.org/10.1007/978-981-15-0504-1_5

Fig. 5.1 Blowing out candle flames

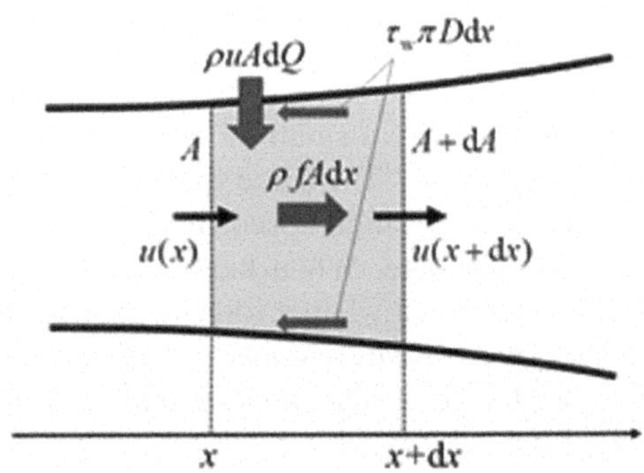

Fig. 5.2 Control volume

Let us apply the conservation equations in Sect. 3.1 to an ideal, calorically perfect gas.

(1) Conservation of mass: From (3.2), by neglecting the time derivative,

$$\int_{CS} \rho(\mathbf{u} \cdot \mathbf{n})dA' = 0. \tag{5.1}$$

Here, for distinction from the cross-sectional area, let a surface element in the control surface enclosing the control volume (CS) be designated by dA'. Since the mass flux through the wall is zero, (5.1) yields

$$\mathrm{d}(\rho u A) = 0. \tag{5.2}$$

On integrating the above, we obtain

$$\rho u A = \text{const.} \equiv \dot{m}, \tag{5.3}$$

where $\dot{m}$ is the mass flow rate.

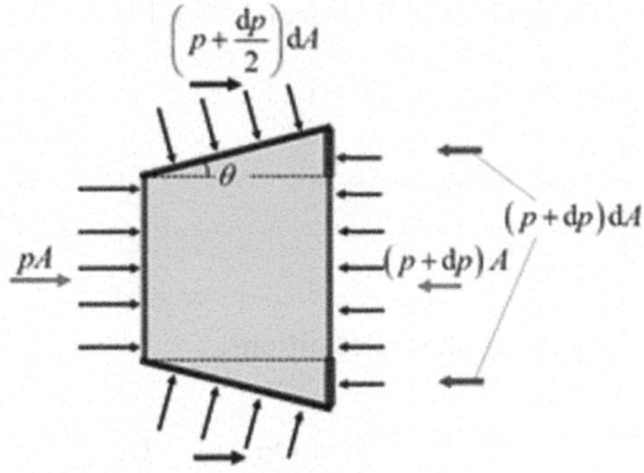

Fig. 5.3 Pressure balance over the control volume

(2) Conservation of momentum: By applying (3.6) to the control volume of Fig. 5.2 and neglecting the time derivative, we obtain

$$\int_{CS} \rho \mathbf{u}(\mathbf{u} \cdot \mathbf{n}) \mathrm{d}A' = \int_{CS} \boldsymbol{\sigma} \cdot \mathbf{n} \mathrm{d}A' + \int_{CV} \rho \mathbf{f} \mathrm{d}V. \tag{5.4}$$

Here, the first and second terms on the right-hand side correspond to the stresses, that is pressure and friction force on the control surface, and the body force exerted in the control volume. The x-component of (5.4) is

$$\mathrm{d}\rho u^2 A = \rho u A \mathrm{d}u = \int_{CS} [\boldsymbol{\sigma} \cdot \mathbf{n}]_x \mathrm{d}A' + \int_{CV} \rho f \mathrm{d}V, \tag{5.5}$$

where f is the x-component of $\mathbf{f}$. Figure 5.3 illustrates the pressure balance over the control volume. Note here that the force exerted by the pressure on the sidewall has an x-component, and the sidewall has a diverging angle of θ. The reaction force on the sidewall has a positive sign with a diverging cross section, that is, for a positive value of θ; the reaction force has a negative sign with a negative θ. For an infinitesimally small dx, the force exerted on the control volume as a result of the pressure balance is

$$pA - (p + \mathrm{d}p)(A + \mathrm{d}A) + \left(p + \frac{\mathrm{d}p}{2}\right)\frac{\mathrm{d}A}{\sin\theta}\sin\theta \cong$$
$$-p\mathrm{d}A - A\mathrm{d}p + p\mathrm{d}A = -A\mathrm{d}p. \tag{5.6}$$

Equation (5.6) implies that the direction of the resultant force over the control volume depends solely on the sign of the pressure gradient; the flow is accelerated if the pressure decreases with x, whereas it is decelerated if the pressure increases.

The hydraulic diameter, D, and friction coefficient, c_f, are defined as

$$\frac{\pi D^2}{4} = A, \tag{5.7}$$

$$c_\mathrm{f} \equiv \frac{\tau_w}{\frac{1}{2}\rho u^2}. \tag{5.8}$$

By using these definitions,

$$\tau_w \pi D \mathrm{d}x = \pi D c_\mathrm{f} \frac{1}{2}\rho u^2 \mathrm{d}x = c_\mathrm{f} \frac{1}{2}\rho u^2 \frac{4A}{D}\mathrm{d}x, \tag{5.9}$$

$$\int_\mathrm{CS} [\boldsymbol{\sigma} \cdot \mathbf{n}]_x \mathrm{d}A' \cong -A\mathrm{d}p - c_\mathrm{f} \frac{1}{2}\rho u^2 \frac{4A}{D}\mathrm{d}x. \tag{5.10}$$

The body force is approximately given by

$$\int_\mathrm{CV} \rho f \mathrm{d}V \cong \rho f A \mathrm{d}x. \tag{5.11}$$

Combining (5.5)–(5.11), the following first-order equation is obtained:

$$\rho u \mathrm{d}u = -\mathrm{d}p - c_\mathrm{f} \frac{1}{2}\rho u^2 \frac{4}{D}\mathrm{d}x + \rho f \mathrm{d}x. \tag{5.12}$$

It should be noted here that, except for the viscous term, which is the second term on the right-hand side of (5.12), the terms do not contain A or D.

(3) Conservation of energy: We apply (3.20) to steady-state flow without heat conduction. The friction force on the wall does not contribute to any energy exchange, because the flow velocity vanishes on the wall.

$$\int_\mathrm{CS} \rho \left(e_\mathrm{t} + \frac{p}{\rho} \right) \mathbf{u} \cdot \mathbf{n} \mathrm{d}A' = \int_\mathrm{CV} \rho \dot{Q} \mathrm{d}V + \int_\mathrm{CV} \rho \mathbf{f} \cdot \mathbf{u} \mathrm{d}V. \tag{5.13}$$

As the velocity component normal to the wall is equal to zero,

$$\mathrm{d}\left\{ \rho u A \left(e_\mathrm{t} + \frac{p}{\rho} \right) \right\} = \dot{m} \mathrm{d}\left(e_\mathrm{t} + \frac{p}{\rho} \right) = \rho (f u + \dot{Q}) A \mathrm{d}x, \tag{5.14}$$

$$\dot{Q} \equiv u \mathrm{d}Q/\mathrm{d}x. \tag{5.15}$$

Since

$$e_{\mathrm{t}} + \frac{p}{\rho} = e + \frac{1}{2}u^2 + \frac{p}{\rho} = h + \frac{1}{2}u^2,$$

$$\mathrm{d}\left(h + \frac{u^2}{2}\right) = \mathrm{d}Q + f\mathrm{d}x. \tag{5.16}$$

(4) Equation of State

$$p = \rho RT, h = \frac{\gamma}{\gamma - 1}\frac{p}{\rho}, \text{ etc.} \tag{5.17}$$

(5) Speed of Sound

$$a^2 = \gamma RT = \gamma \frac{p}{\rho} = (\gamma - 1)h. \tag{5.18}$$

(6) Mach number

$$M \equiv \frac{u}{a}. \tag{5.19}$$

5.1.2 Equations in Derivative Form

Let us transform the equations obtained so far into their respective derivative forms. From (5.3),

$$\frac{\mathrm{d}u}{u} + \frac{\mathrm{d}\rho}{\rho} + \frac{\mathrm{d}A}{A} = 0. \tag{5.20}$$

From (5.12),

$$\gamma M^2 \frac{\mathrm{d}u}{u} + \frac{\mathrm{d}p}{p} + \frac{\gamma M^2}{2}\frac{4c_{\mathrm{f}}}{D}\mathrm{d}x - \frac{\rho f \mathrm{d}x}{p} = 0. \tag{5.21}$$

From (5.16),

$$\mathrm{d}\left(\frac{\gamma}{\gamma - 1}\frac{p}{\rho} + \frac{u^2}{2}\right) = \mathrm{d}Q + f\mathrm{d}x,$$

$$(\gamma-1)M^2\frac{\mathrm{d}u}{u}-\frac{\mathrm{d}\rho}{\rho}+\frac{\mathrm{d}p}{p}-\frac{\mathrm{d}Q}{h}-\frac{\gamma-1}{\gamma}\frac{\rho f\mathrm{d}x}{p}=0. \tag{5.22}$$

From (5.17)–(5.19),

$$\frac{\mathrm{d}T}{T}+\frac{\mathrm{d}\rho}{\rho}-\frac{\mathrm{d}p}{p}=0, \tag{5.23}$$

$$\frac{\mathrm{d}a^2}{a^2}+\frac{\mathrm{d}\rho}{\rho}-\frac{\mathrm{d}p}{p}=0, \tag{5.24}$$

$$\frac{\mathrm{d}M^2}{M^2}-\frac{2\mathrm{d}u}{u}+\frac{\mathrm{d}a^2}{a^2}=0. \tag{5.25}$$

Equations (5.20)–(5.25) are expressed in a matrix form as follows:

$$\begin{pmatrix} \cdot & 1 & \cdot & \cdot & 1 & \cdot & 1 & \cdot & \cdot & \cdot \\ \cdot & \gamma M^2 & \cdot & \cdot & \cdot & 1 & \cdot & \cdot & -1 & \gamma M^2/2 \\ \cdot & (\gamma-1)M^2 & \cdot & \cdot & -1 & 1 & \cdot & -1 & -(\gamma-1)/\gamma & \cdot \\ \cdot & \cdot & \cdot & 1 & 1 & -1 & \cdot & \cdot & \cdot & \cdot \\ \cdot & \cdot & 1 & \cdot & 1 & -1 & \cdot & \cdot & \cdot & \cdot \\ 1 & -2 & 1 & \cdot & \cdot & \cdot & \cdot & \cdot & \cdot & \cdot \end{pmatrix} \begin{pmatrix} \mathrm{d}M^2/M^2 \\ \mathrm{d}u/u \\ \mathrm{d}a^2/a^2 \\ \mathrm{d}T/T \\ \mathrm{d}\rho/\rho \\ \mathrm{d}p/p \\ \mathrm{d}A/A \\ \mathrm{d}Q/h \\ \rho f\mathrm{d}x/p \\ \frac{4c_\mathrm{f}}{D}\mathrm{d}x \end{pmatrix} = \mathbf{O}$$

By transforming this equation, we obtain

$$\begin{pmatrix} 1 & \cdot & \cdot & \cdot & \cdot & \cdot & -\left\{(\gamma-1)M^2+2\right\} & \gamma M^2+1 & -(\gamma+1)/\gamma & \left\{(\gamma-1)M^2+2\right\}\gamma M^2/2 \\ \cdot & 1 & \cdot & \cdot & \cdot & \cdot & -1 & 1 & -1/\gamma & \gamma M^2/2 \\ \cdot & \cdot & 1 & \cdot & \cdot & \cdot & (\gamma-1)M^2 & -\gamma M^2+1 & (\gamma-1)/\gamma & -\left\{(\gamma-1)M^2\right\}\gamma M^2/2 \\ \cdot & \cdot & \cdot & 1 & \cdot & \cdot & (\gamma-1)M^2 & -\gamma M^2+1 & (\gamma-1)/\gamma & -\left\{(\gamma-1)M^2\right\}\gamma M^2/2 \\ \cdot & \cdot & \cdot & \cdot & 1 & \cdot & M^2 & -1 & 1/\gamma & -\gamma M^2/2 \\ \cdot & \cdot & \cdot & \cdot & \cdot & 1 & \gamma M^2 & -\gamma M^2 & 1 & -\left\{(\gamma-1)M^2+1\right\}\gamma M^2/2 \end{pmatrix}$$

$$\begin{pmatrix} \mathrm{d}M^2/M^2 \\ \mathrm{d}u/u \\ \mathrm{d}a^2/a^2 \\ \mathrm{d}T/T \\ \mathrm{d}\rho/\rho \\ \mathrm{d}p/p \\ \mathrm{d}A/A(M^2-1) \\ \mathrm{d}Q/h(M^2-1) \\ \rho f\mathrm{d}x/p(M^2-1) \\ 4c_\mathrm{f}\mathrm{d}x/D(M^2-1) \end{pmatrix} = \mathbf{O}$$

$$\begin{pmatrix} dM^2/M^2 \\ du/u \\ da^2/a^2 \\ dT/T \\ d\rho/\rho \\ dp/p \end{pmatrix} = \begin{pmatrix} -(\gamma-1)M^2-2 & \gamma M^2+1 & -(\gamma+1)/\gamma & (\gamma-1)M^2+2 \\ -1 & 1 & -1/\gamma & 1 \\ (\gamma-1)M^2 & -\gamma M^2+1 & (\gamma-1)/\gamma & -(\gamma-1)M^2 \\ (\gamma-1)M^2 & -\gamma M^2+1 & (\gamma-1)/\gamma & -(\gamma-1)M^2 \\ M^2 & -1 & 1/\gamma & -1 \\ \gamma M^2 & -\gamma M^2 & 1 & -(\gamma-1)M^2-1 \end{pmatrix} \begin{pmatrix} \frac{1}{1-M^2}\frac{dA}{A} \\ \frac{1}{1-M^2}\frac{dQ}{h} \\ \frac{1}{1-M^2}\frac{\rho f dx}{p} \\ \frac{\gamma M^2/2}{1-M^2}\frac{4c_f dx}{D} \end{pmatrix} \tag{5.26}$$

5.2 Flow Characteristics

5.2.1 *Influence Coefficients*

From (5.26), the variation in a fluid-dynamics quantity, X, is given by the following linear function:

$$\frac{dX}{X} = C_1\frac{dA}{A} + C_2\frac{dQ}{h} + C_3\frac{\rho f dx}{p} + C_4\frac{4c_f dx}{D}, \tag{5.27}$$

where C_i $(i = 1, 2, 3, 4)$ are *influence coefficients*.

Note here that (5.27) holds only if appropriate boundary conditions are satisfied. For example, a candle flame is not blown out by merely protruding one's mouth; to blow out the flame, high-pressure airflow needs to be supplied from the lungs.

The signs of all influence coefficients are reversed with $M = 1$. Let us summarize the effects of these actions.

5.2.2 *Effects of Variation in Cross-Sectional Area*

From Table 5.1, a variation in the cross-sectional area has the following effects:

(1) Subsonic flow ($M < 1$)

(1-1) The flow is accelerated through a converging duct:

$$dA < 0 \rightarrow dM > 0,\ du > 0.$$

Table 5.1 Influence coefficients

X	C_1	C_2	C_3	C_4
M^2	$-\frac{(\gamma-1)M^2+2}{1-M^2}$	$\frac{1+\gamma M^2}{1-M^2}$	$-\frac{\gamma+1}{\gamma}\frac{1}{1-M^2}$	$\frac{(\gamma-1)M^2+2}{1-M^2}\cdot\frac{\gamma M^2}{2}$
u	$-\frac{1}{1-M^2}$	$\frac{1}{1-M^2}$	$-\frac{1}{\gamma}\frac{1}{1-M^2}$	$\frac{1}{1-M^2}\cdot\frac{\gamma M^2}{2}$
a^2,T	$\frac{(\gamma-1)M^2}{1-M^2}$	$\frac{1-\gamma M^2}{1-M^2}$	$\frac{\gamma-1}{\gamma}\frac{1}{1-M^2}$	$-\frac{(\gamma-1)M^2}{1-M^2}\cdot\frac{\gamma M^2}{2}$
ρ	$\frac{M^2}{1-M^2}$	$-\frac{1}{1-M^2}$	$\frac{1}{\gamma}\frac{1}{1-M^2}$	$-\frac{1}{1-M^2}\cdot\frac{\gamma M^2}{2}$
p	$\frac{\gamma M^2}{1-M^2}$	$-\frac{\gamma M^2}{1-M^2}$	$\frac{1}{1-M^2}$	$-\frac{(\gamma-1)M^2+1}{1-M^2}\cdot\frac{\gamma M^2}{2}$

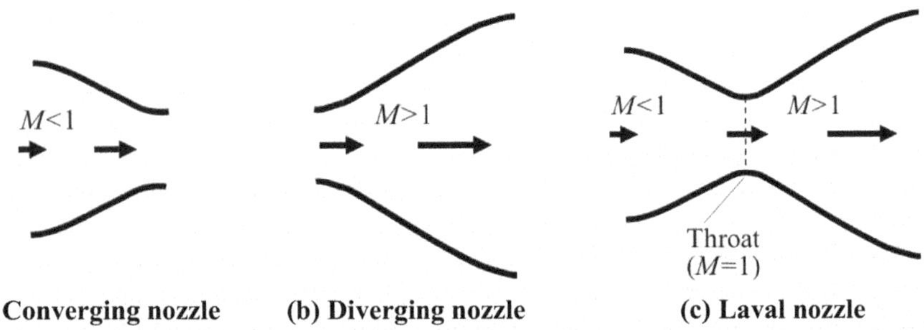

Fig. 5.4 Different types of nozzles

(1-2) The flow is decelerated through a diverging duct:

$$dA > 0 \rightarrow dM < 0,\ du < 0.$$

(2) Supersonic flow ($M > 1$)

(2-1) The flow is decelerated through a converging duct:

$$dA < 0 \rightarrow dM < 0,\ du < 0.$$

(2-2) The flow is accelerated through a diverging duct:

$$dA > 0 \rightarrow dM > 0,\ du > 0.$$

We unconsciously perform (1-1) when we blow out a candle flame. The same principle applies to a shower, spray, etc. A *nozzle* is a device that accelerates flows with a variable cross-sectional area. A nozzle for subsonic flow has a converging shape (Fig. 5.4a), and that for supersonic flow has a diverging shape (Fig. 5.4b). In order to accelerate a flow from subsonic to supersonic, a converging–diverging nozzle called the *Laval nozzle* (Fig. 5.4c) is used. In a Laval nozzle, the connection part, which has the minimum cross-sectional area, is termed as a *throat*, where the flow becomes sonic ($M = 1$).

A device that decelerates flow is termed a *diffuser*.[2] Variations in the cross-sectional area of a diffuser are opposite to those of a nozzle.

A subsonic flow is accelerated through a converging duct. In this case, the flow experiences a force in the opposite direction from the duct wall. Why is the flow accelerated even with the force in the opposite direction? In (5.12) with $c_f = 0$ and $f = 0$, the direction of flow acceleration is determined solely by the sign of the pressure gradient. In subsonic flow, the pressure gradient has a negative sign with $\mathrm{d}A < 0$, which accelerates the flow.

5.2.3 *Effects of Heating/Cooling*

(1) Subsonic flow ($M < 1$)
 (1-1) The flow is accelerated with heating: $\mathrm{d}Q > 0 \rightarrow \mathrm{d}M > 0,\ \mathrm{d}u > 0.$
 (1-2) The flow is decelerated with cooling: $\mathrm{d}Q < 0 \rightarrow \mathrm{d}M < 0,\ \mathrm{d}u < 0.$
(2) Supersonic flow ($M > 1$)
 (2-1) The flow is decelerated with heating: $\mathrm{d}Q > 0 \rightarrow \mathrm{d}M < 0,\ \mathrm{d}u < 0.$
 (2-2) The flow is accelerated with cooling: $\mathrm{d}Q < 0 \rightarrow \mathrm{d}M > 0,\ \mathrm{d}u > 0.$

In Table 5.1, the variation in temperature, T, changes its sign at two Mach numbers. In almost all regimes, T increases with heating ($\mathrm{d}Q > 0$). However, in the regime of $1/\sqrt{\gamma} < M < 1$, T decreases with heating! We do not experience this interesting variation in daily life. However, this behavior occasionally appears in the numerical simulation of transonic flows.

5.2.4 *Effects of Friction*

Friction force is always exerted in the direction opposite to the flow direction. From Table 5.1,

(1) Subsonic flow ($M < 1$)
 The flow is accelerated with friction:

$$c_f > 0 \rightarrow \mathrm{d}M > 0,\ \mathrm{d}u > 0.$$

(2) Supersonic flow ($M > 1$)
 The flow is decelerated with friction:

$$c_f > 0 \rightarrow \mathrm{d}M < 0,\ \mathrm{d}u < 0.$$

---

[2]For further details, refer to Sect. 11.2.

Friction force affects the flow in the same manner as a flow through a converging duct. However, unlike a body force, it is not involved in an external energy input.

5.2.5 *Effects of Body Force*

(1) Subsonic flow ($M < 1$)

(1-1) The flow is decelerated with a body force in the same direction:

$$f > 0 \rightarrow \mathrm{d}M < 0,\ \mathrm{d}u < 0.$$

(1-2) The flow is accelerated with a body force in the opposite direction:

$$f < 0 \rightarrow \mathrm{d}M > 0,\ \mathrm{d}u > 0.$$

(2) Supersonic flow ($M > 1$)

(2-1) The flow is accelerated with a body force in the same direction:

$$f > 0 \rightarrow \mathrm{d}M > 0,\ \mathrm{d}u > 0.$$

(2-2) The flow is decelerated with a body force in the opposite direction:

$$f < 0 \rightarrow \mathrm{d}M < 0,\ \mathrm{d}u < 0.$$

(1-1) is explained in the same manner as the variation in cross-sectional area. By substituting (5.12), $\mathrm{d}A = 0$, $c_\mathrm{f} = 0$, and $\mathrm{d}Q = 0$ with the influence coefficients, we obtain

$$\rho u \mathrm{d}u = -\mathrm{d}p + \rho f \mathrm{d}x = -\mathrm{d}p + \left(1 - M^2\right)\mathrm{d}p = -M^2 \mathrm{d}p. \qquad (5.28)$$

Therefore, whether the flow is accelerated depends only on the sign of the pressure gradient. In subsonic flows, the flow is accelerated with $f < 0$ if the pressure difference past the duct is sufficiently large.

5.2.6 *Choking Condition*

The condition under which a quasi-one-dimensional flow becomes sonic is termed as the *choking condition*. From Table 5.1,

$$\frac{dM^2}{M^2} = \frac{-\left\{2+(\gamma-1)M^2\right\}\frac{dA}{A} + \left(1+\gamma M^2\right)\frac{dQ}{h} - \frac{\gamma+1}{\gamma}\frac{\rho f dx}{p} + \frac{\gamma M^2\left\{(\gamma-1)M^2+2\right\}}{2}\frac{4c_f dx}{D}}{1-M^2} \tag{5.29}$$

For the right-hand side of (5.29) to have a finite value with $M = 1$, the numerator needs to vanish. That is,

$$-\frac{dA}{A} + \frac{dQ}{h} - \frac{1}{\gamma}\frac{\rho f dx}{p} + \frac{\gamma}{2}\frac{4c_f dx}{D} = 0 \quad \text{for} \quad M = 1. \tag{5.30}$$

The above is the equation of the choking condition, which is a linear function of the variations of cross-sectional area, heating, body force, and friction. With only the variation in cross-sectional area,

$$dA = 0 \quad \text{for} \quad M = 1. \tag{5.31}$$

The above corresponds to the fact that the flow becomes sonic at the throat in a Laval nozzle, as shown in Fig. 5.4c.

Choking with heating alone is termed *thermal choking*.

$$dQ = 0 \quad \text{for} \quad M = 1. \tag{5.32}$$

For subsonic flow to be thermally chocked, the heating needs to be terminated at the sonic point.

For solving a flow with choking, the choking condition gives an additional condition.

5.3 Duct Flow with Friction

In many fluid devices, fluid passes through narrow passages, and such a flow is accompanied by a pressure loss due to friction force. This section focuses on the flow in a duct with a constant cross-sectional area and friction force, which is termed *Fanno flow*.

From (5.27),

$$\frac{dM^2}{M^2} = \frac{(\gamma-1)M^2+2}{1-M^2} \cdot \frac{\gamma M^2}{2}\frac{4c_f dx}{D}, \tag{5.33}$$

$$\frac{4\gamma c_f dx}{D} = -\frac{\gamma+1}{2} d\ln M^2 + \frac{\gamma+1}{2} d\ln\left(M^2 + \frac{2}{\gamma-1}\right) - d\left(\frac{1}{M^2}\right). \tag{5.34}$$

Here, D is the effective inner diameter of the duct, usually termed the *hydraulic diameter*. By integrating (5.34) for a constant value of c_f with respect to the location

x from the inlet ($x = 0$), we obtain

$$\frac{4\gamma c_{\mathrm{f}}}{D}x = \frac{1}{M_1^2} - \frac{1}{M^2} + \frac{\gamma+1}{2}\ln\frac{M^2 + \frac{2}{\gamma-1}}{M_1^2 + \frac{2}{\gamma-1}} \cdot \frac{M_1^2}{M^2}. \tag{5.35}$$

As summarized in Table 5.1, with a friction force, the Mach number increases in subsonic flow and decreases in supersonic flow. In either case, it approaches the sonic speed. By applying (5.35) up to a sonic point x^*, we obtain

$$\frac{4\gamma c_{\mathrm{f}}}{D}x^* = \frac{1 - M_1^2}{M_1^2} + \frac{\gamma+1}{2}\ln\frac{(\gamma+1)M_1^2}{(\gamma-1)M_1^2 + 2}. \tag{5.36}$$

From the choking condition expressed by (5.29),

$$\mathrm{d}x = 0 \quad \text{for} \quad M = 1. \tag{5.37}$$

This implies that the flow can be sonic only at the exit.
From (5.27),

$$\frac{\mathrm{d}p}{p} = -\frac{(\gamma-1)M^2 + 1}{1 - M^2} \cdot \frac{\gamma M^2}{2}\frac{4c_{\mathrm{f}}\mathrm{d}x}{D}. \tag{5.38}$$

By substituting (5.38) with (5.33), we obtain

$$\frac{\mathrm{d}p}{p} = -\frac{(\gamma-1)M^2 + 1}{(\gamma-1)M^2 + 2}\frac{dM^2}{M^2}. \tag{5.39}$$

By integrating the above to the sonic point, we obtain

$$\frac{p}{p_*} = \frac{1}{M}\sqrt{\frac{\gamma+1}{(\gamma-1)M^2 + 2}}, \tag{5.40}$$

where p_* is the pressure at the sonic point. Equation (5.40) holds even with a normal shock wave in the duct.

Figure 5.5 shows the flow variation of a Fanno flow with an inlet Mach number of 3. The abscissa is the normalized distance from the inlet. Without a normal shock wave, the supersonic flow monotonically decelerated owing to the friction force. If the dimensionless duct length is less than 0.73, the flow exits at a supersonic speed. If the length is equal to 0.73, the flow becomes sonic at the inlet. If the duct is even longer, a normal shock wave is generated so that the post-shock, subsonic flow becomes sonic at the exit. The location of the normal shock wave is such that the choking condition is satisfied. In Fig. 5.5, flows with a normal shock wave at $M =$ 1.5, 2.0, 2.5, and 3.0 are shown. With a length of 1.81, the normal shock wave occurs at the inlet. With an even longer duct, the flow upstream of the inlet is affected.

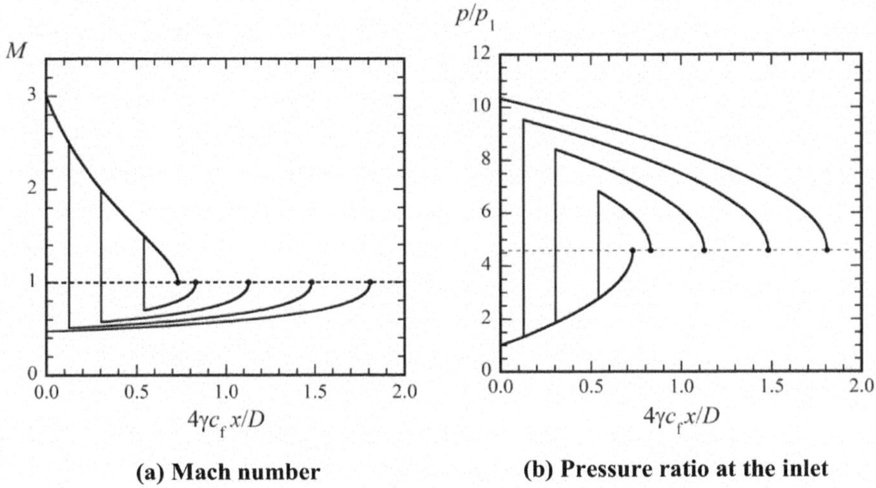

Fig. 5.5 Variations of the Mach number and pressure in a Fanno flow for $\gamma = 1.4$, $c_{\mathrm{f}} = 0.005$, and $M_1 = 3.0$

As shown in Fig. 5.5b, in the supersonic flow on the upstream side of a shock wave, the pressure increases with the friction force. Further, the pressure jumps across the shock wave. Subsequently, the flow becomes subsonic, and the pressure decreases in the downstream. Irrespective of where a normal shock wave appears, the pressure at the exit remains constant.

Chapter 6
Systems with Source Terms

A system accompanying a shock wave that exchanges energy and/or momentum with the external environment can enhance the strength of the shock wave or generate a thrust. In this chapter, we will study such a system based on the generalized Rankin–Hugoniot relations that contain source terms.

6.1 Generalized Rankine–Hugoniot Relations

The conservation equations in Chap. 4 can be applied to a control volume with a finite thickness, if fluxes through the lateral surface are absent. Let us consider the control volume in Fig. 6.1a. The flow is steady and uniform at its inlet and at the exit. Because the control volume is in a black box, the details of the inside state are unknown. We only know that the box exchanges a net mass of W', momentum I', and energy Q' with the external environment. For example, the distribution of pressure and friction force on the walls contribute to the net momentum exchange.

We use subscripts 1 and 2 for quantities at the inlet and exit of the control volume, respectively. The variation of the cross-sectional area in the control volume does not need to be determined. The flow passage may change, such as in Fig. 6.1b or a center body may exist in the control volume, like in Fig. 6.1c.

For steady-state flow, the following conservation relations hold:

$$\textbf{Mass conservation}: \quad \rho_2 u_2 = \rho_1 u_1 + W' \tag{6.1}$$

$$\textbf{Momentum conservation}: \quad p_2 + \rho_2 u_2^2 = p_1 + \rho_1 u_1^2 + I' \tag{6.2}$$

$$\textbf{Energy conservation}: \quad h_2 + \frac{1}{2}u_2^2 = h_1 + \frac{1}{2}u_1^2 + Q' \tag{6.3}$$

A. Sasoh, *Compressible Fluid Dynamics and Shock Waves*,
https://doi.org/10.1007/978-981-15-0504-1_6

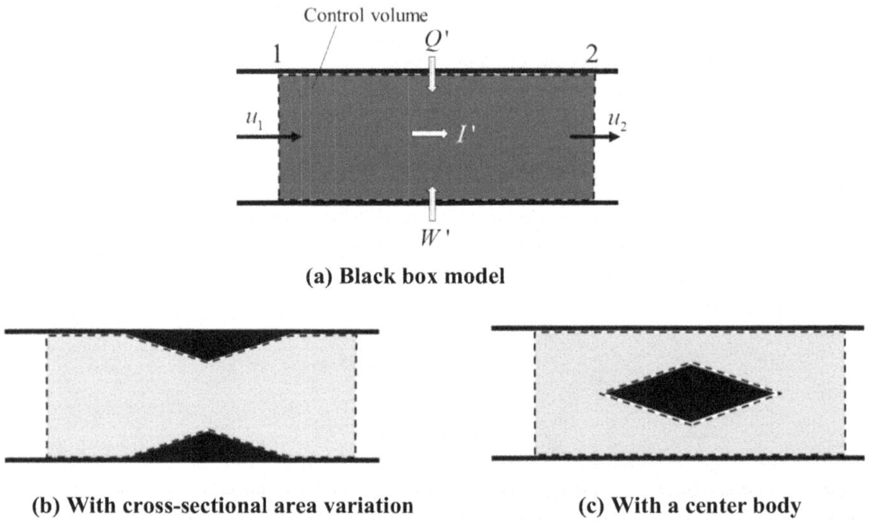

Fig. 6.1 Control volume with mass, momentum, and energy exchange with the external environment. The gray regime is enclosed with dashed lines. Walls and internal objects are not included

Here W', I', and Q' denote the source terms. Note here that parts of I' and Q' may be contributed by the mass W' introduced into the system. Equation (6.2) is derived for inviscid flows. However, it is applicable even to viscous flows. The flow behavior is determined uniquely by the net value of the momentum exchange; the respective contributions of the pressure and viscous stress do not affect the result of the present analyses. Since at the wall, the flow does not have a velocity component in the direction of the pressure or friction force, these forces do not contribute to the energy source term. Equations (6.1)–(6.3), in which the source terms are added to the Rankine–Hugoniot relations of (4.19)–(4.21), are *generalized Rankine–Hugoniot relations*. We deal with detonation and deflagration occurring with the energy source term, Q' in Sect. 6.2. Moreover, by combining with the momentum source term, I', we model the ram accelerator operation in Sect. 6.3.

When heat is generated by chemical reactions, the fluid is a mixture comprised of multiple chemical species, as described in Sect. 2.8. In this chapter, an enthalpy, h, is defined as the sum of a static enthalpy, h_s, and a *standard enthalpy of formation*, h_f, with a reference temperature of T_{ref}.[1] For a gas with N chemical components and a mass fraction of $Y_i (i = 1 \cdots N)$,

$$h = \sum_{i=1}^{N} Y_i h_i, \quad \sum_{i=1}^{N} Y_i = 1 \tag{6.4}$$

[1]Chemical reactions are explicitly dealt with only in this chapter. In the other chapters, the static enthalpy is designated by h.

$$h_i = h_{f,i} + h_{s,i}(T) \tag{6.5}$$

$$h_{s,i} \equiv \int_{T_{ref}}^{T} C_{p,i} \mathrm{d}T \tag{6.6}$$

From Sect. 2.8,

$$\overline{C}_p = \sum_{i=1}^{N} Y_i C_{p,i} \tag{2.122}$$

$$\gamma = \frac{\overline{C}_p}{\overline{C}_p - \bar{R}} = \frac{\overline{C}_p}{\overline{C}_v} \tag{2.125}$$

Here $\overline{C}$ denotes a mass-averaged quantity. From (6.3) to (6.6),

$$h_{s,2} + \frac{1}{2}u_2^2 = h_{s,1} + \frac{1}{2}u_1^2 + Q'' \tag{6.7}$$

$$h_{s,j} \equiv \sum_{i=1}^{N} Y_{i,j} \int_{T_{ref}}^{T_j} C_{p,i} \mathrm{d}T \tag{6.8}$$

$$Q'' \equiv \sum_{i=1}^{N} (Y_{i,1} - Y_{i,2}) h_{f,i} + Q' \tag{6.9}$$

The first term on the right-hand side of (6.9) corresponds to a heat due to chemical reactions, the second term corresponds to the other heating.

Here, we assume that $W' = 0$. From (6.1) and (6.2),

$$\frac{p_2 - p_1 - I'}{\frac{1}{\rho_1} - \frac{1}{\rho_2}} = j^2 \tag{6.10}$$

$$j \equiv \rho_1 u_1 = \rho_2 u_2 \tag{6.11}$$

Transforming (6.10),

$$\frac{p_2}{p_1} = \frac{u_1^2}{\frac{p_1}{\rho_1}} \left(1 - \frac{\rho_1}{\rho_2}\right) + 1 + I \tag{6.12}$$

$$I \equiv \frac{I'}{p_1} \tag{6.13}$$

Equation (6.12), which is applicable irrespective to the equation of state of the gas, is the equation for the *generalized Rayleigh line*. Since the energy conservation

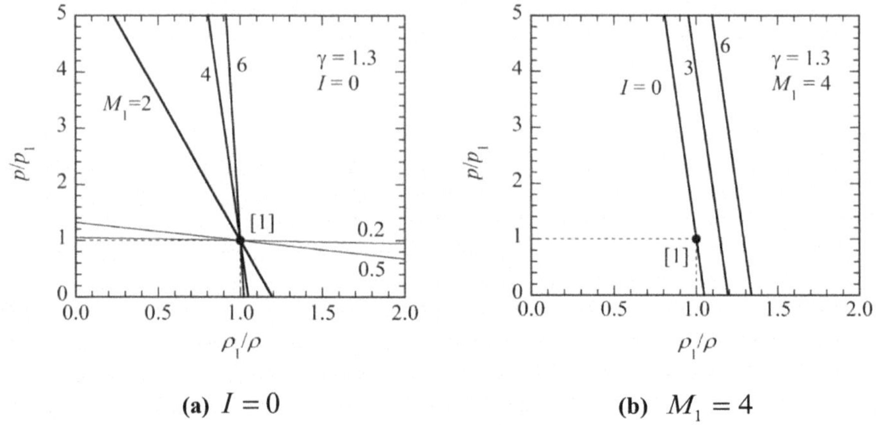

Fig. 6.2 Generalized Rayleigh lines

Eq. (6.3) is not used, it does not contain Q. Defining a Mach number based on frozen speed of sound,

$$a = \sqrt{\gamma \frac{p}{\rho}} = \sqrt{\gamma \bar{R} T} \tag{6.14}$$

$$M = \frac{u}{a} \tag{6.15}$$

$$\frac{p_2}{p_1} = -\gamma M_1^2 \left(\frac{\rho_1}{\rho_2} - 1 \right) + 1 + I \tag{6.16}$$

This corresponds to a line that passes at $(1, 1 + I)$ and has a slope of $-\gamma M_1^2$. For $I = 0$ (Fig. 6.2a), it is the same as in (4.43), and a solution with $M_1 < 1$ does not exist. However, as shown later, with $Q > 0$, solutions with $M_1 < 1$ also exist. The corresponding lines are plotted in the figure. The line is shifted by I if $I \neq 0$.

From (6.1) to (6.3),

$$h_{s,2} - h_{s,1} - \frac{1}{2}(p_2 - p_1 - I')\left(\frac{1}{\rho_1} + \frac{1}{\rho_2} \right) = Q''. \tag{6.17}$$

Otherwise, using $h_s = e + \frac{p}{\rho}$,

$$e_2 - e_1 + \frac{1}{2}(p_2 + p_1)\left(\frac{1}{\rho_2} - \frac{1}{\rho_1} \right) + \frac{1}{2}\left(\frac{1}{\rho_1} + \frac{1}{\rho_2} \right) I' = Q''. \tag{6.18}$$

Fig. 6.3 Generalized Hugoniot curves

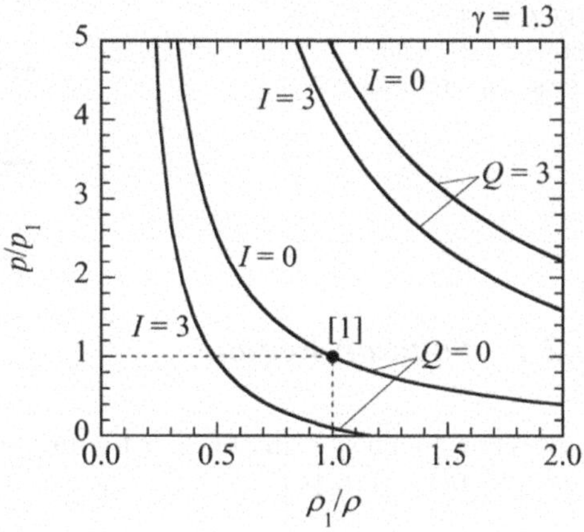

Equations (6.17) and (6.18) hold irrespective of the equation of state of the gas, corresponding to the *generalized Hugoniot curve*.[2] h_s and e are given as the functions of p and $v = 1/\rho$. Once the form of the functions and the value of Q'' are given, the curve can be plotted. The intersections between the Rayleigh line and the Hugoniot curve correspond to the states at the inlet 1 and exit 2. Later we show which corresponds to which.

For a calorically perfect gas, the Hugoniot curve is given as an explicit formula.

$$\frac{p_2}{p_1} = \frac{1 - \frac{\gamma-1}{\gamma+1}(1+I)\frac{\rho_1}{\rho_2} + \frac{2\gamma}{\gamma+1}Q - \frac{\gamma-1}{\gamma+1}I}{\frac{\rho_1}{\rho_2} - \frac{\gamma-1}{\gamma+1}} \tag{6.19}$$

$$Q \equiv \frac{Q''}{\bar{C}_p T_1} = \frac{Q''}{h_{s,1}} = \frac{Q''}{\frac{\gamma}{\gamma-1}\frac{p_1}{\rho_1}} \tag{6.20}$$

For $I = Q = 0$, (6.19) coincides with (4.44). Equation (6.19) is shifted positively by Q and negatively by I, as shown in Fig. 6.3.

6.2 Detonation/Deflagration

Let us assume a system with only an energy source term, $W' = I = 0$ and $Q \neq 0$. The solution of the Rankine–Hugoniot relations is obtained by combining (6.12) and (6.17). However, it is difficult to understand the characteristics of the solution based on such general forms. Here, we analyze the basic behavior of detonation and deflagration based on a thermally perfect gas with constant specific heat.

[2]Hereafter, for simplicity, we will omit the notation "generalized."

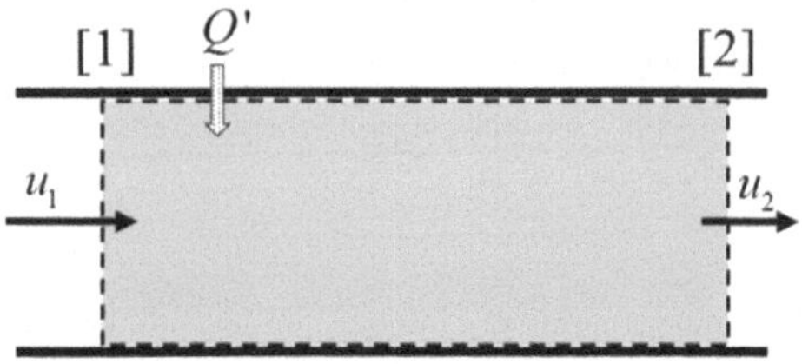

Fig. 6.4 Flow with heating through a constant cross-sectional area

6.2.1 Solution Regime

Let us consider a system of Fig. 6.1 with $I = 0$ and $Q > 0$, that is shown in Fig. 6.4. From (6.16) and (6.19),

$$\text{The Rayleigh line}: \ \frac{p_2}{p_1} = -\gamma M_1^2\left(\frac{\rho_1}{\rho_2} - 1\right) + 1 \tag{6.21}$$

is not affected by Q.

$$\text{Meanwhile, the Hugoniot curve}: \ \frac{p_2}{p_1} = \frac{1 - \frac{\gamma-1}{\gamma+1}\frac{\rho_1}{\rho_2} + \frac{2\gamma}{\gamma+1}Q}{\frac{\rho_1}{\rho_2} - \frac{\gamma-1}{\gamma+1}} \tag{6.22}$$

is shifted up with $Q > 0$.

The solutions are subdivided, as shown in Table 6.1 and Fig. 6.5. The solution with supersonic flow at the inlet ($M_1 > 1$, Regimes I–III) corresponds to *detonation*, where the shock wave is enhanced by heating. The solution with subsonic flow at the inlet corresponds to *deflagration*, where the flow expands and is accelerated with the heating. Because a Rayleigh line does not have a positive gradient, no solution exists in the gray portion of the Huroniot curve in Fig. 6.5. As will be shown later, in the regime of III and IV, a physically meaningful solution does not exist.

Table 6.1 Types of detonation and deflagration

		M_1	M_2	Solution
I	Overdriven/Strong detonation	$M_1 > 1$	$M_2 < 1$	Backup driver is necessary
II	C–J detonation	$M_1 > 1$	$M_2 = 1$	Possible
III	(Weak detonation)	$(M_1 > 1)$	$(M_2 > 1)$	Impossible
IV	(Strong deflagration)	$(M_1 < 1)$	$(M_2 > 1)$	Impossible
V	C–J deflagration	$M_1 < 1$	$M_2 = 1$	Possible
VI	Weak deflagration	$M_1 < 1$	$M_2 < 1$	Possible

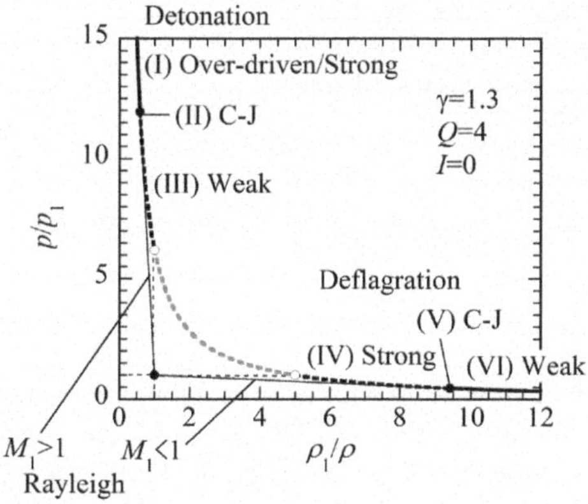

Fig. 6.5 Solution regimes of detonation and deflagration

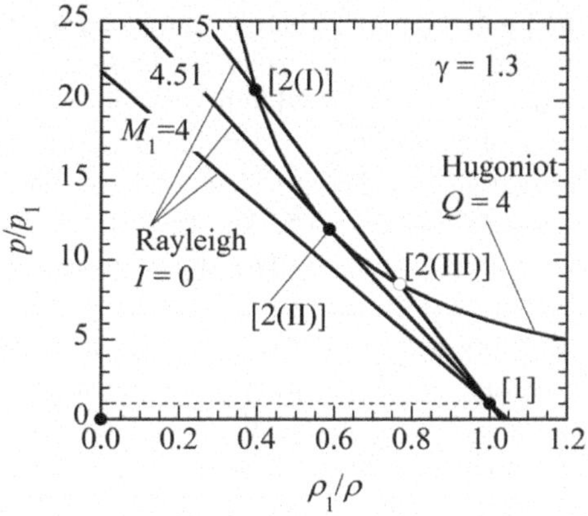

Fig. 6.6 Examples of detonation solution, [1] inlet, [2] exit, [2(I)] overdriven detonation, [2(II)] C–J detonation, [2(III)] physically meaningless solution, refer to Table 6.1

6.2.2 *Detonation*

The solution of the Rankine–Hugoniot relation provides the states only at the inlet and the outlet of the control volume. However, the flow inside of the control volume also satisfies the fluid dynamics conservation relations. In a detonation, a normal shock wave stands at the inlet of the control volume. Behind the shock wave, the flow becomes subsonic and accelerates with the heating. Figure 6.6 shows examples of the detonation solution. The point [1] at (1, 1) corresponds to the state at the inlet. The state at the exit corresponds to another intersection. In this figure, three Rayleigh lines against a Hugoniot curve for $Q = 4$ associated with the respective values of M_1 are plotted. With $M_1 = 4$, no solution exists.

The Rayleigh line with $M_1 = 5$ has two intersections, Points 2(I) and 2(III). Point 2(I) corresponds to the condition where the flow becomes subsonic at the exit. Such a solution is illustrated in Fig. 6.7, where the shock wave propagated in the quiescent mixture to the right. The speed of the shock wave propagation, U_S, equals to the absolute value of u_1 in Fig. 6.4, but in the opposite direction. Since the flow at

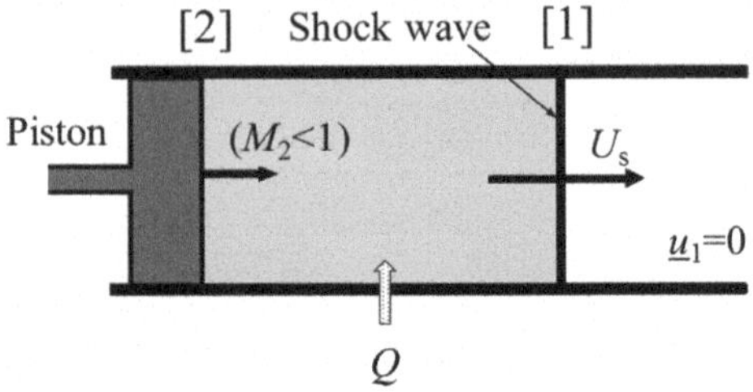

Fig. 6.7 Overdriven detonation propagating in quiescent gas in a laboratory flame

the exit is subsonic ($M_2 < 1$), the flow inside the control volume is affected by the condition at the exit. The shock Mach number, M_1, equals to U_s/a_1. The shock wave is driven by a piston or equivalent, thereby propagating at a Mach number higher than that of the C–J detonation (which will be explained later). The higher the driving pressure or the higher the piston speed, the higher the shock Mach number becomes. Such a detonation, which is enhanced by the motion of a piston or its equivalent, is referred to as an *overdriven detonation.*

In Fig. 6.6, the Rayleigh line with $M_1 \simeq 4.51$ becomes tangent to the Hugoniot curve at Point 2(II), (0.588, 11.9), corresponding to the *Chapman–Jouguet (C-J) detonation.* As shown in Sect. 4.2.1.4, the flow becomes sonic at the exit, $M_2 = 1$, regime (II) in Fig. 6.5. The flow downstream of the exit does not affect the flow in the control system. The shock wave can propagate without any driver. The C–J detonation is the unique self-sustained solution. Almost all one-dimensional detonations appearing in laboratories, artificial products, and nature are C–J detonations. In reality, the shock wave front has a complicated three-dimensional shape that is caused by unsteady fluid dynamics associated with chemical reactions. Nevertheless, its propagation Mach number is very close to the solution of C–J detonation, M_{CJ}. For a calorically perfect gas, this has a closed form.[3]

$$M_{\text{CJ, detonation}} = \left[1 + (\gamma + 1)Q + \sqrt{\{1 + (\gamma + 1)Q\}^2 - 1} \right]^{\frac{1}{2}} \tag{6.23}$$

At Point 2(III) in Fig. 6.6, comparing the slopes of the Rayleigh line and the Hugoniot curve, the flow at the exit should become supersonic. However, as described in Chap. 5, the Mach number of the subsonic flow behind the shock wave cannot exceed unity just by heating. Therefore, such a solution does not have a physical sense. For this reason, in regime III in Fig. 6.5, no solution exists, depicted by the Hugoniot curve with dashed lines.

Experimentally observed detonation phenomena are further complicated by the influence of finite-rate chemical reactions, multidimensionality, etc. The detonation propagation speeds, experimentally measured and calculated in chemical equilibrium under typical conditions, are shown in Table 6.2 [1]. In spite of complicated phenomena in practice, the C–J model provides a fairly accurate estimation. The stoichiometric ratio of the hydrogen–oxygen mixture is equal to 2, corresponding to

[3]See Sect. 6.3.2.

Table 6.2 Comparison between the experimentally measured detonation speed, U_1, and the calculated value for the chemical equilibrium, U_{CJ}, at $p_1 = 100$ kPa, $T_1 = 298$ K. $p_{2,CJ}$ and $T_{2,CJ}$ are depict calculated values

Mixture		U_1 (km/s)	U_{CJ} (km/s)	$p_{2,CJ}$ (MPa)	$T_{2,CJ}$ (K)
A	$2H_2 + O_2$	2.82	2.81	1.80	3580
B	$2H_2 + O_2 + 5N_2$	1.82	1.85	1.44	2690
C	$8H_2 + O_2$	3.53	3.75	1.42	2650

mixture A. Mixtures B and C contain diluting molecules, thereby decreasing the specific heat release, and subsequently the pressure and temperature. On the one hand, the detonation speed is also decreased in mixture B, while on the other hand that of mixture C is increased, because the diluted gas is hydrogen itself, thus increasing the speed of sound.

6.2.3 *Deflagration*

As described in Chap. 5, one-dimensional. subsonic flow expands and is accelerated with heating. At the frame fixed to the inlet of the control volume, where the heating starts (state 1), the Mach number increases and the pressure decreases toward the downstream. If we label the state at the exit where the heating terminates, we can apply Rankine–Hugoniot relations to this system even though they are not accompanied by a shock wave. This solution is referred to as *deflagration*. The exit Mach number, M_2, depends on the amount of input heat, Q. With a relatively small Q, M_2 remains smaller than unity, thereby indicating weak deflagration, i.e., the regime VI in Fig. 6.5 and Table 6.1. Increasing Q past a certain value yields a sonic condition of $M_2 = 1$, corresponding to *Chapman–Jouguet (C–J) deflagration,* i.e., the regime V in Fig. 6.5 and Table 6.1. For a given value of Q, the value of $M_1 (< 1)$ is uniquely determined. In C–J deflagration, the flow becomes sonic at the exit, not being affected by the further downstream. As shown in Table 6.1 and Fig. 6.5, the regime in which the deflagration solution has a physical meaning is limited to the regime of $M_2 \leq 1$, which are only the regimes V and VI. Since the supersonic flow cannot be generated only by heating, a solution with $M_2 > 1$ (the regime IV) does not exist. The Mach number of C–J deflagration is given in a closed form for a calorically perfect gas.

$$M_{\mathrm{CJ,\,deflagration}} = \left[1 + (\gamma + 1)Q - \sqrt{\{1 + (\gamma + 1)Q\}^2 - 1}\right]^{\frac{1}{2}} \tag{6.24}$$

6.2.4 Entropy Variation

Let us investigate how entropy varies behind a detonation and deflagration. Here, we apply the same approach as in Chap. 4 for $Q = 0$. Substituting (6.18) with $\upsilon = 1/\rho$ and $I' = 0$,

$$e_2 - e_1 = \frac{1}{2}(p_2 + p_1)\left(\frac{1}{\rho_1} - \frac{1}{\rho_2}\right) + Q' = \frac{1}{2}(p_2 + p_1)(\upsilon_1 - \upsilon_2) + Q'. \tag{6.25}$$

Differentiating this expression,

$$\mathrm{d}e_2 = \frac{1}{2}(\upsilon_1 - \upsilon_2)\mathrm{d}p_2 - \frac{1}{2}(p_2 + p_1)\mathrm{d}\upsilon_2 + \mathrm{d}Q'. \tag{6.26}$$

From the first law of thermodynamics,

$$T_2\mathrm{d}s_2 = \mathrm{d}e_2 + p_2\mathrm{d}\upsilon_2. \tag{6.27}$$

Combining (6.26) and (6.27),

$$T_2\frac{\mathrm{d}s_2}{\mathrm{d}\upsilon_2} = \frac{1}{2}(\upsilon_1 - \upsilon_2)\frac{\mathrm{d}p_2}{\mathrm{d}\upsilon_2} + \frac{1}{2}(p_2 - p_1) + \frac{\mathrm{d}Q'}{\mathrm{d}\upsilon_2}. \tag{6.28}$$

Since on a Hugoniot curve (HG) $\mathrm{d}Q' = 0$,

$$T_2\left[\frac{\mathrm{d}s_2}{\mathrm{d}\upsilon_2}\right]_{\mathrm{HG}} = \frac{1}{2}(\upsilon_1 - \upsilon_2)\left[\frac{\mathrm{d}p_2}{\mathrm{d}\upsilon_2}\right]_{\mathrm{HG}} + \frac{1}{2}(p_2 - p_1). \tag{6.29}$$

Equation (6.29) gives the variation in entropy on the Hugoniot curve. Further differentiating this expression,

$$\begin{aligned} T_2\left[\frac{\mathrm{d}^2 s_2}{\mathrm{d}\upsilon_2^2}\right]_{\mathrm{HG}} &= \frac{1}{2}\left(-\frac{1}{p_2}\right)(\upsilon_1 - \upsilon_2)\left[\frac{\mathrm{d}p_2}{\mathrm{d}\upsilon_2}\right]_{\mathrm{HG}}^2 + \left\{\frac{1}{2}\frac{p_1}{p_2} - \frac{1}{2}\left(\frac{\upsilon_1}{\upsilon_2}\right)\right\}\left[\frac{\mathrm{d}p_2}{\mathrm{d}\upsilon_2}\right]_{\mathrm{HG}} \\ &+ \frac{1}{2}(\upsilon_1 - \upsilon_2)\left[\frac{\mathrm{d}^2 p_2}{\mathrm{d}\upsilon_2^2}\right]_{\mathrm{HG}} + \frac{1}{2}\left(-\frac{1}{\upsilon_2}\right)(p_2 - p_1). \end{aligned} \tag{6.30}$$

Because at the C–J point, (6.10) and

$$u_2 = a_2 = \sqrt{-\upsilon_2^2\frac{\mathrm{d}p_2}{\mathrm{d}\upsilon_2}} \tag{6.31}$$

are applied,

$$\frac{p_2 - p_1}{v_1 - v_2} = \dot{m}^2 = \frac{1}{v_2^2}\left(-v_2^2\left[\frac{dp_2}{dv_2}\right]_{CJ}\right) = -\left[\frac{dp_2}{dv_2}\right]_{CJ}. \tag{6.32}$$

Substituting (6.32) into (6.29) and (6.30),

$$T_2\left[\frac{ds_2}{dv_2}\right]_{CJ} = 0, \tag{6.33}$$

$$T_2\left[\frac{d^2 s_2}{dv_2^2}\right]_{CJ} = \frac{1}{2}(v_1 - v_2)\left[\frac{d^2 p_2}{dv_2^2}\right]_{CJ}. \tag{6.34}$$

From (6.33), the entropy has a maximum at the C–J point. For ordinal gases,

$$\frac{d^2 p_2}{dv_2^2} > 0. \tag{6.35}$$

Therefore, in the C–J detonation where $v_1 - v_2 > 0$, $\left[\frac{d^2 s_2}{dv_2^2}\right]_{CJ} > 0$.

In C–J deflagration, where $v_1 - v_2 < 0$, $\left[\frac{d^2 s_2}{dv_2^2}\right]_{CJ} < 0$.

It follows from these results that along a Hugoniot curve, the entropy has a minimum at the C–J detonation, and a maximum at the C–J deflagration.

6.2.5 *Energy Variation*

Let us determine how the internal and kinetic energy of a flow change by heating, shock compression, or expansion. Substituting (6.18) with $v = 1/\rho$ and $I' = 0$,

$$e_2 - e_1 = \frac{1}{2}(p_2 + p_1)(v_1 - v_2) + Q'. \tag{6.36}$$

The kinetic energy induced behind the detonation from a quiescent state is obtained from (4.26).

$$\frac{1}{2}\left(\underline{u}_2 - \underline{u}_1\right)^2 = \frac{1}{2}(u_1 - u_2)^2 = \frac{1}{2}(p_2 - p_1)(v_1 - v_2). \tag{6.37}$$

Therefore, the variation of the total energy, e_t, becomes

$$e_{t,2} - e_{t,1} = e_2 - e_1 + \frac{1}{2}\left(\underline{u}_2 - \underline{u}_1\right)^2 = p_2(v_1 - v_2) + Q'. \tag{6.38}$$

Figure 6.8 shows the energy variations induced by a detonation on pressure-specific volume coordinates. The area C equals to the second term on the right-hand side of (6.36). Equations (6.36)–(6.38) correspond to the respective areas.

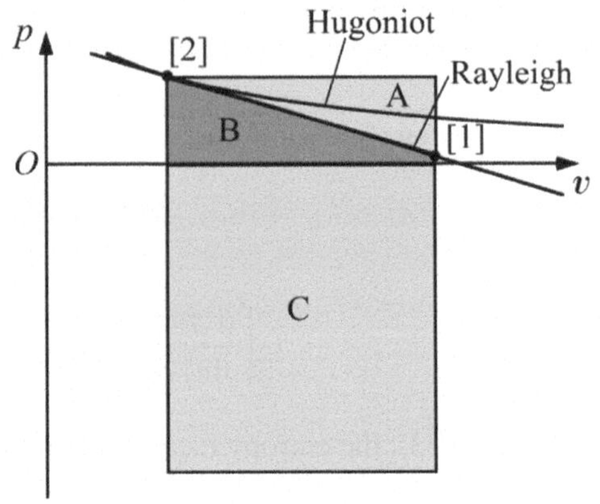

Fig. 6.8 Energy increments by detonation

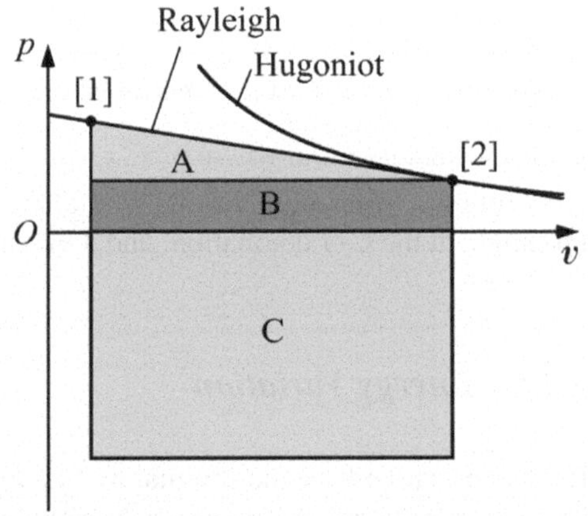

Fig. 6.9 Energy increments by deflagration

Increment in the internal energy (6.36) = Area B + C
Increment in the kinetic energy (6.37) = Area A
Increment in the total energy (6.38) = Area A + B + C

e is further increased than the input heat (Area C) by compression (Area B). The kinetic energy is generated as a consequence of the acceleration of the gas resulting from the pressure imbalance (Area A). In the detonation, the increment in e dominates over the increment in the kinetic energy due to heating.

At the same, the energy variation behind a deflagration is analyzed in Fig. 6.9.

Increment in the internal energy (6.36) = Area – (A + B) + C
Increment in the kinetic energy (6.37) = Area A
Increment in the total energy (6.38) = Area – B + C

In deflagration, the gas expands ($v_2 > v_1$), and the pressure decreases ($p_2 < p_1$). Therefore, the increment in e equals the difference from the input heat (Area C), and the work done to the external environment (Area A + B). The increment in the kinetic energy is depicted by the Area A. However, the direction of the induced flow is opposite to that of the detonation.

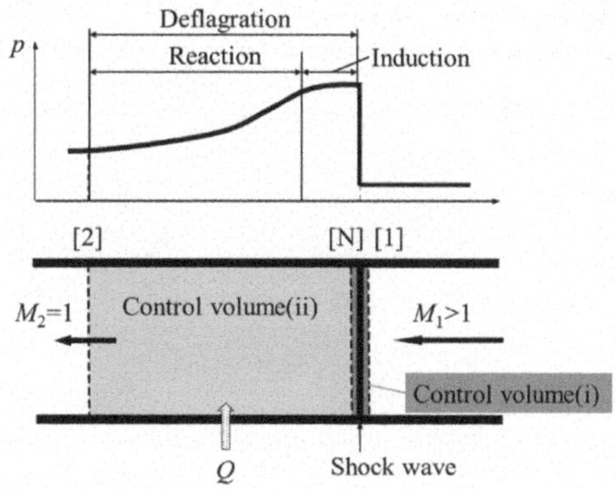

Fig. 6.10 Schematic of the ZDN model

6.2.6 *ZND Model*

Let us consider internal flow in the detonation. We have analyzed the system only by considering the flow states at the inlet and the exit of the control volume. Zel'dovich, von Neumann, and Döring developed a model in which the flow regime is subdivided into two sub-regimes (*ZND model*), see Fig. 6.10. The sub-regime i entails the shock wave, and is not accompanied by heating ($Q = 0$). The sub-regime ii has a thickness with heating and covers the subsonic flow between the shock wave and the exit. If the heat is input by exothermic chemical reactions, the thickness equals the convection distance for completing the reactions. Immediately behind the shock wave, there is an *induction zone* where the pressure and temperature are kept almost constant, which is followed by a *reaction zone* where the pressure and temperature vary. At the inlet of the sub-regime ii, the flow is subsonic; and at the exit, the flow becomes sonic. Therefore, the flow in the latter is equivalent to C–J deflagration. In summary, we can treat a C–J detonation in two ways:

(1) As a single system, this is C–J detonation.
(2) We can subdivide it to a shock wave and a subsequent C–J deflagration.

Let us trace the processes on $p - \upsilon$ coordinates, see Fig. 6.11. Note here that the processes of the pressure and the specific volume variation do not exactly follow these paths, but experience the states of [1], ([N]), and [2]. In the process (1), the state changes directly from [1] on the Hugoniot curve with $Q = 0$ to the tangent point [2] to the Hugoniot curve with $Q = 4$. In the process (2), the process once changes from [1] to [N] on the Hugoniot curve with $Q = 0$, then transitions to the tangent point [2] of the subsonic Rayleigh line to the Hugoniot curve with $Q = 4$. The pressure at [N] is higher than that at [2], is referred to as the *von Neumann spike*.

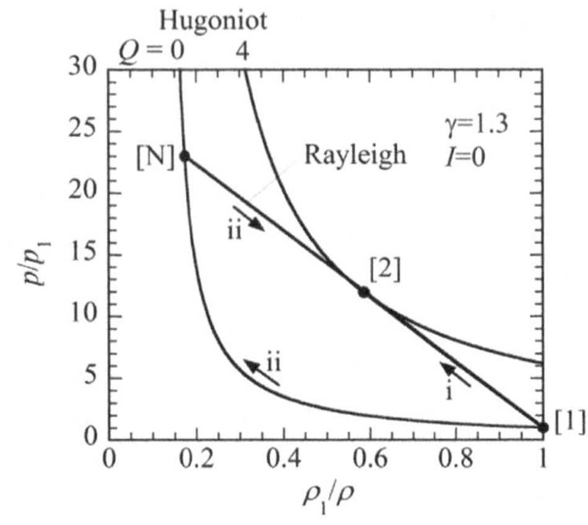

Fig. 6.11 ZDN model on $p - \upsilon$ coordinates

6.2.7 *Cellular Structure*

Detonation involves in the coupling between shock waves and chemical reactions. Usually chemical reaction rates are sensitive to the pressure and the temperature of the mixture, which in turn is strongly affected by the strength of a shock wave. Often a chemical reaction rate, k_r, is assumed to follow the Arrhenius equation,

$$k_r = A \exp\left(-\frac{E_a}{kT}\right), \tag{6.39}$$

where the activation energy, E_a, differs depending on elemental chemical reaction. From (6.39), k_r is an increasing function of T and is sensitive to it. Usually, a shock wave appearing in detonation has a shock Mach number on the order of five or even higher. Since the post-shock temperature is almost proportional to the square of the shock Mach number, this chemical reaction system coupled by the shock wave is considerably sensitive to the variation of the local behavior of the shock wave. If the shock wave becomes locally stronger, the post-shock temperature and then the chemical reaction rate increase, thereby further enhancing the shock wave. Because of such unstable mechanisms, the shock wave in detonation hardly keeps its planar wave front, but has a complicated wave structure as shown in the left Schlieren images of Fig. 6.12a and b. In the regime where the shock wave is locally stronger, behind M in Fig. 6.12b, the exothermic chemical reaction progresses faster, and thereby the shock wave is further enhanced. In contrast, in the regime where the shock wave is relatively weak, behind I in Fig. 6.12b, the exothermic chemical reaction takes place at a distant location from the shock wave. The generated pressure nonuniformity in the transversal direction induces the transversal shock waves. Coupling these shock waves, the concave and convex wave fronts transversely move in the direction of the detonation.

(a) -1 Regular pattern, $2H_2 + O_2 + 12Ar$

(a) -2 Irregular pattern, $2H_2 + N_2O + 1.33N_2$

(a) Schlieren images (left) and soot traces (right) of detonation propagation, initial pressure, 20 kPa; high of test section, 150 mm, courtesy of Prof. J. Austin, California Institute of Technology.

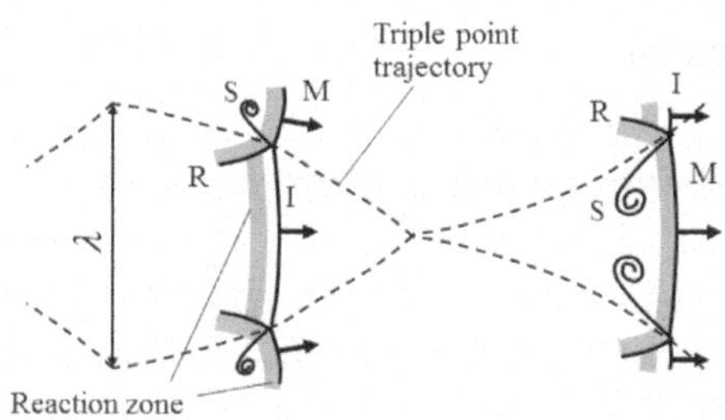

(b) Schematic illustration of shock wave propagation with chemical reaction to form cellular structure; I, incident shock wave; R, reflected shock wave; M, Mach stem; S, slip line.

Fig. 6.12 Cellular structure of detonation waves

The right pictures in Fig. 6.12a are photographs of the tube wall observed after detonation propagation. Before the experiment, the wall was covered by soot. After the detonation propagation, the trajectories of a triple point are recorded as the cellular structure. This structure is caused by the above-mentioned transversal wave propagation. The shock wave reflections as shown in Fig. 6.12b are repeated [2]. Along the leading front, the fast portion and slow portion are not smoothly connected, but form a *Mach reflection*.[4] The slow portion is the *incident shock wave* (I), the fast portion is the *Mach stem* (M), and the transversal wave corresponds to the *reflected shock wave* (R). The intersection of these waves is the *triple point*. Because I is weaker

[4]Refer to Chap. 7.

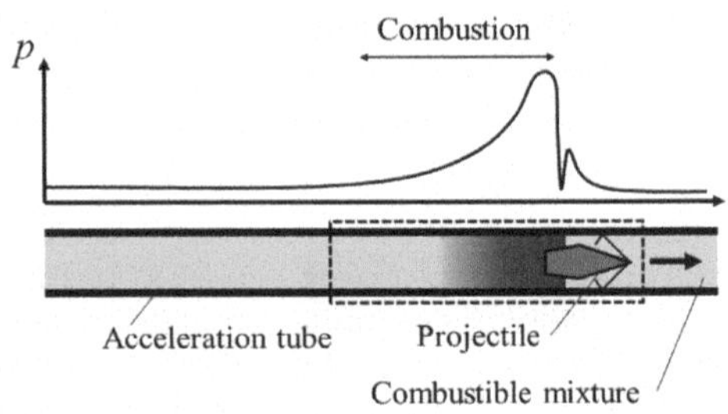

Fig. 6.13 Schematic illustration of ram accelerator, the region enclosed by the dashed lines is the control volume

than M, it has a lower propagation speed, and the chemical reaction behind I takes place behind M. Along the propagation path of R, the location of the triple point also moves in the transverse direction. From the triple point, the *slip line* (S) is generated,[5] then a vorticity for the flow to approach the Mach stem is induced. The trajectory of such triple points is recorded as the soot pattern. The maximum width of the cellular structure is referred to as the *cell size*. In experiments, the cellular structure pattern depends on the type of mixture and the initial pressure. The pattern in Fig. 6.12a-1 has a regular, while that in (a)-2 has an irregular pattern.

6.3 Ram Accelerator

6.3.1 Operation Principle and Characteristics

In Rankine–Hugoniot relations, a thrust is generated if a source term is introduced not only in the energy conservation equation but also in the momentum conservation equation. Employing such relations, we can model a *ram accelerator* [3]. In the ram accelerator, the acceleration tube is filled with combustible mixture. As seen in Fig. 6.13, the projectile is injected into the acceleration tube at a supersonic speed. Around the projectile, a shock wave is formed and repeats reflection between the tube wall and the projectile, thereby compressing and heating the combustible mixture. If appropriate conditions are satisfied, the mixture is ignited on the aftbody of the projectile. If the combustion is sustained there, the local pressure is enhanced, yielding a thrust.

This device has a great advantage over the conventional power gun of obtaining a large thrust over a larger tube length. In a conventional gun (Fig. 6.14), the projectile is accelerated by a high-pressure driver gas at the bottom of the acceleration tube that is produced by combusted propellant power (sometimes in liquid) or pre-charged, high-pressure gas. However, the farther the projectile travels from the bottom, the lower the thrust becomes due to the expansion of the driver gas. Nevertheless, in a ram accelerator, a high pressure immediately behind the projectile and hence a large

[5]We traditionally refer to a "line" instead of "plane" in the three-dimensional phenomena.

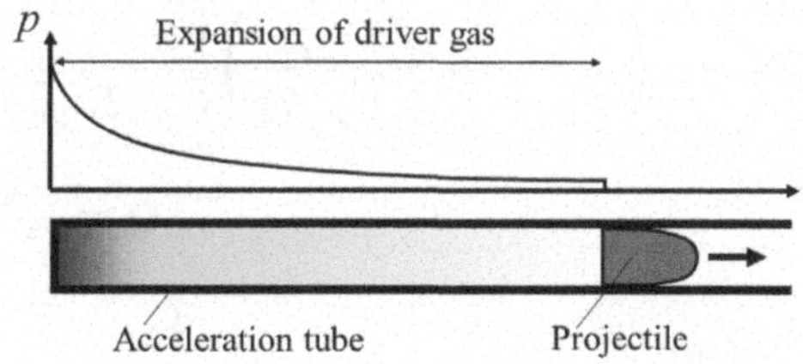

Fig. 6.14 Schematic illustration of a conventional gun

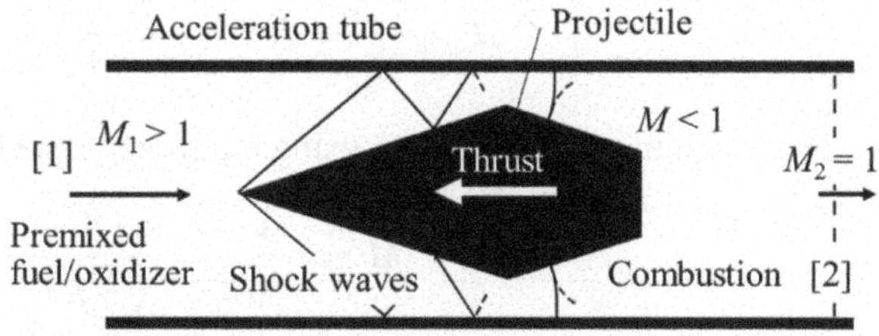

Fig. 6.15 Schematic illustration of thermally choked flow in the control volume (see Fig. 6.13). On the frame fixed to the projectile, the flow direction is reversed

thrust can be kept for a much longer travel distance, thereby obtaining a much larger impulse that is transferred to the momentum of the projectile.

Figure 6.15 illustrates the flow around the projectile on the frame fixed to it. The flow direction is reversed, such that the right direction has a positive sign. The combustible mixture flows at a supersonic speed, M_1, in the space between the projectile and the acceleration tube wall. Around the projectile, oblique shock waves are generated, repeat reflection on the tube wall and the projectile, thereby the pressure and the temperature are increased. In appropriate conditions, the mixture is ignited, and combusted such that the pressure around the aftbody of the projectile is enhanced. A thrust is produced as the surface integration of the pressure on the projectile in the left direction. The combustion is terminated in the downstream at the location 2. As will be shown later, in the thermally choked operation, the Mach number at the exit, M_2, equals unity.

6.3.2 *Derivation of Thrust*

Let us obtain the thrust by applying Rankine–Hugoniot relations to the system of Fig. 6.15. Note here that the general form obtained in this section is applicable not only to the thermally choked operation, $M_2 = 1$, but to other values of M_2. The following equations will be applied to multicomponent gas. A mass-averaged quantity is depicted by $\overline{\{\,\}}$. Dividing (6.7) by $\overline{C}_{p,1} T_1$ using

$$\overline{C}_p = \sum_{i=1}^{N} Y_{i,j} C_{p,i} \tag{6.40}$$

$$\left(\frac{h_{s,2}}{\overline{C}_{p,2}T_2}+\frac{1}{2}\frac{u_2^2}{\overline{C}_{p,2}T_2}\right)\frac{\overline{C}_{p,2}T_2}{\overline{C}_{p,1}T_1}=\frac{h_{s,1}}{\overline{C}_{p,1}T_1}+\frac{1}{2}\frac{u_1^2}{\overline{C}_{p,1}T_1}+Q \tag{6.41}$$

$$Q\equiv\frac{Q''}{\overline{C}_{p,1}T_1} \tag{6.42}$$

Here,

$$\overline{C}_pT=\frac{\bar{\gamma}}{\bar{\gamma}-1}\bar{R}T=\frac{\bar{\gamma}}{\bar{\gamma}-1}\frac{p}{\rho}=\frac{a^2}{\bar{\gamma}-1} \tag{6.43}$$

The Mach number is defined using the frozen speed of sound, a.

$$M=\frac{u}{\sqrt{\bar{\gamma}\frac{p}{\rho}}}=\frac{u}{\sqrt{\bar{\gamma}\bar{R}T}}. \tag{6.44}$$

Substituting (6.41) with (6.43),

$$\left(\frac{h_{s,2}}{\bar{C}_{p,2}T_2}+\frac{\bar{\gamma}_2-1}{2}M_2^2\right)\frac{\frac{\bar{\gamma}_2}{\bar{\gamma}_2-1}}{\frac{\bar{\gamma}_1}{\bar{\gamma}_1-1}}\frac{p_2}{p_1}\frac{\rho_1}{\rho_2}=\frac{h_{s,1}}{\bar{C}_{p,1}T_1}+\frac{\bar{\gamma}_1-1}{2}M_1^2+Q,$$

$$\left(\frac{p_2}{p_1}\right)^2=\frac{\frac{\bar{\gamma}_1}{\bar{\gamma}_1-1}}{\frac{\bar{\gamma}_2}{\bar{\gamma}_2-1}}\frac{p_2}{p_1}\frac{\rho_2}{\rho_1}\frac{\frac{h_{s,1}}{\bar{C}_{p,1}T_1}+\frac{\bar{\gamma}_1-1}{2}M_1^2+Q}{\frac{h_{s,2}}{\bar{C}_{p,2}T_2}+\frac{\bar{\gamma}_2-1}{2}M_2^2}.$$

Since

$$\frac{p_2}{p_1}\frac{\rho_2}{\rho_1}=\frac{\frac{p_2}{\rho_2}}{\frac{p_1}{\rho_1}}\frac{\rho_2^2}{\rho_1^2}=\frac{\bar{\gamma}_2\frac{p_2}{\rho_2}}{\bar{\gamma}_1\frac{p_1}{\rho_1}}\frac{\bar{\gamma}_1u_1^2}{\bar{\gamma}_2u_2^2}=\frac{\bar{\gamma}_1M_1^2}{\bar{\gamma}_2M_2^2},$$

$$\therefore\ \frac{p_2}{p_1}=\frac{\bar{\gamma}_1M_1}{\bar{\gamma}_2M_2}\sqrt{\frac{\bar{\gamma}_2-1}{\bar{\gamma}_1-1}\frac{\frac{h_{s,1}}{\bar{C}_{p,1}T_1}+\frac{\bar{\gamma}_1-1}{2}M_1^2+Q}{\frac{h_{s,2}}{\bar{C}_{p,2}T_2}+\frac{\bar{\gamma}_2-1}{2}M_2^2}}. \tag{6.45}$$

Next, from (6.12) and (6.13),

$$\begin{aligned}\frac{p_2}{p_1}&=\frac{u_1^2}{\frac{p_1}{\rho_1}}-\frac{\rho_1^2u_1^2}{\rho_2}\frac{1}{p_1}+1+I=\frac{u_1^2}{\frac{p_1}{\rho_1}}-\frac{\rho_2^2u_2^2}{\rho_2}\frac{1}{p_1}+1+I\\&=\bar{\gamma}_1M_1^2-\bar{\gamma}_2M_2^2\frac{p_2}{p_1}+1+I,\\I&=\frac{p_2}{p_1}(1+\bar{\gamma}_2M_2^2)-(1+\bar{\gamma}_1M_1^2).\end{aligned} \tag{6.46}$$

$$I\equiv\frac{I'}{p_1}. \tag{6.47}$$

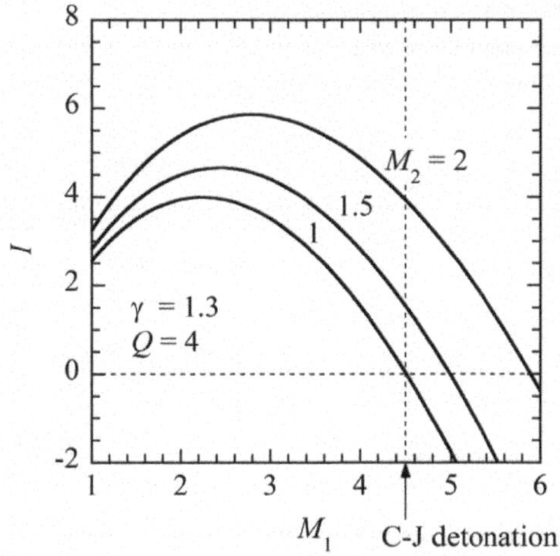

Fig. 6.16 I versus M_1 for calorically perfect gas

From (6.45) and (6.46),

$$I = (1 + \bar{\gamma}_2 M_2^2)\frac{\bar{\gamma}_1 M_1}{\bar{\gamma}_2 M_2}\sqrt{\frac{\bar{\gamma}_2 - 1}{\bar{\gamma}_1 - 1}\,\frac{\frac{h_{s,1}}{\bar{C}_{p,1}T_1} + \frac{\bar{\gamma}_1 - 1}{2}M_1^2 + Q}{\frac{h_{s,2}}{\bar{C}_{p,2}T_2} + \frac{\bar{\gamma}_2 - 1}{2}M_2^2}} - (1 + \bar{\gamma}_1 M_1^2). \qquad (6.48)$$

Equation (6.48) provides the thrust of a ram accelerator, which is applicable even to a variable $\bar{\gamma}$. For a constant value of $\bar{\gamma}$,

$$I = (1 + \gamma\, M_2^2)\frac{M_1}{M_2}\left(\frac{1 + \frac{\gamma - 1}{2}M_1^2 + Q}{1 + \frac{\gamma - 1}{2}M_2^2}\right)^{\frac{1}{2}} - (1 + \gamma\, M_1^2). \qquad (6.49)$$

The dimensionless thrust, I, is given as a function of M_1, M_2, and Q, see Fig. 6.16. Since the value of Q is determined by the initial mixture conditions, I becomes a function of M_1 for a given M_2. As will be described in the next section, a self-sustainable solution is possible only with $M_2 \geq 1$.

6.3.3 Thermally Choked Operation

A *thermally choked* operation corresponds to a ram accelerator operation mode with $M_2 = 1$. As in detonation, this operation mode has special meanings. Figure 6.17 shows an example of the pressure–volume relation in a thermally chock operation for a calorically perfect gas with $\gamma = 1.3$. A Rayleigh line (6.16) is shifted up by I. A Hugoniot curve (6.19) is shifted up with Q and down with I. With a given combination of M_1 and Q, a unique value of I (in Fig. 6.17, $I \simeq 3.514$) exists,

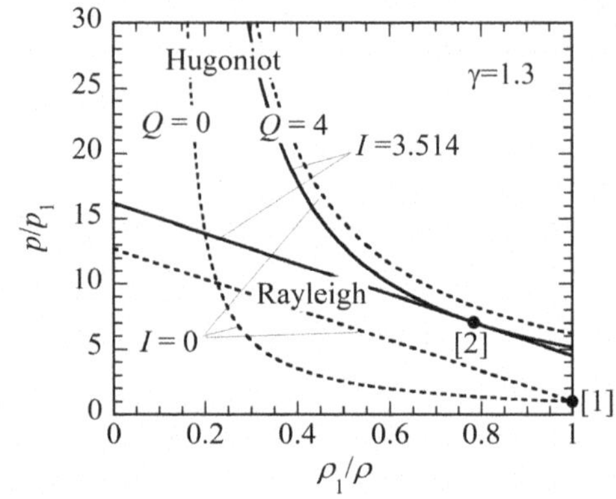

Fig. 6.17 Pressure–volume relation related to a thermally chocked operation

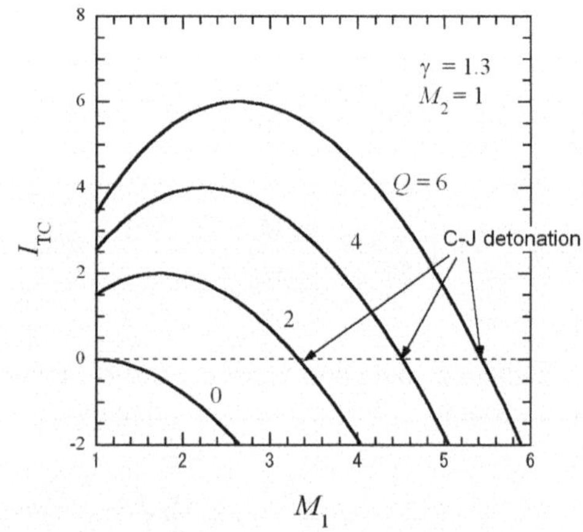

Fig. 6.18 Dimensionless thrust in thermally choked mode, I_{TC}, versus M_1

with which a Rayleigh line becomes a tangent to the Hugoniot curve (at Point [2] in Fig. 6.17). As described in Sect. 4.2.1.4, this tangent condition is equivalent to the sonic condition at exit [2].

In a calorically perfect gas, the following quadratic equation is obtained by combining (6.16) and (6.19).

$$M_1^2\left(\frac{\rho_1}{\rho_2}\right)^2 - \frac{2}{\gamma+1}\left(I + 1 + \gamma M_1^2\right)\left(\frac{\rho_1}{\rho_2}\right) + \frac{(\gamma-1)M_1^2 + 2(Q+1)}{\gamma+1} = 0. \tag{6.50}$$

The solution in the thermally choked mode is obtained by setting its discriminant to 0, as shown in Fig. 6.18.

$$I_{TC} = M_1\sqrt{(\gamma+1)\left\{(\gamma-1)M_1^2 + 2(Q+1)\right\}} - \left(1 + \gamma M_1^2\right). \tag{6.51}$$

I_{TC} does not explicitly depend on the size, shape of the projectile, or details of the flow field around the projectile. Conveniently, it is determined only by the projectile Mach number, the specific heat ratio, and the amount of heat release. In this mode,

$$\left(\frac{\rho_1}{\rho_2}\right)_{TC} = \frac{1 + \gamma M_1^2 + I}{(\gamma + 1)M_1^2} = \sqrt{\frac{(\gamma - 1)M_1^2 + 2(Q + 1)}{(\gamma + 1)M_1^2}}, \tag{6.52}$$

$$\left(\frac{p_2}{p_1}\right)_{TC} = \frac{1 + \gamma M_1^2 + I}{\gamma + 1} = M_1\sqrt{\frac{(\gamma - 1)M_1^2 + 2(Q + 1)}{\gamma + 1}}. \tag{6.53}$$

When the effect of the acceleration of the projectile vanishes, the thermally choked mode becomes a unique, self-sustainable solution as is in C–J detonation. As will be shown in the next section, this solution agrees well with the experimental performance.

For a value of Q, a unique value of M_1 (> 1) exists with $I_{TC} = 0$. In this condition, no momentum is exchanged between the projectile and the gas. This condition is equivalent to the one where a projectile does not exist, and subsequently to the Chapman–Jouguet detonation. The corresponding Mach number, M_{CJ}, is obtained by setting the left-hand side of (6.51) to zero.

I_{TC} has a positive value in the regime $M_{min} < M_1 < M_{CJ}$. The value of M_{min} is determined from the operational condition of the diffuser that is formed between the forebody of the projectile and the acceleration tube wall.[6] In theory, operation with $M_2 > 1$ is possible. In that case, a positive thrust can be obtained even with $M_1 > M_{CJ}$. However, no clear experimental evidence of this super-detonative mode has been obtained.

6.3.4 Experiments on the Ram Accelerator

The ram accelerator was invented by a research group at the University of Washington, Seattle, WA, U.S.A., led by Prof. A. Herzberg in 1986. Intensive investigations were conducted by his group, including the groups of A. P. Bruckner and C. Knowlen, and by other followers in the U.S.A., France, Germany, China, Brazil, Korea, and Japan [4].

Figure 6.19 shows the device developed at the University of Washington with its almost optimal performance. The ram acceleration tube has an inner diameter of 38 mm (1.5 inch), and a total length of 16 m (Fig. 6.19a). Figure 6.19b illustrates the typical projectile, which has a conical nose and a center body with 4 to 5 fins for stabilizing its in-tube attitude. In order to save the mass, it is made of aluminum alloy A7075-T6, and its parts are hollowed and threaded. Figure 6.19c is the representative projectile acceleration profile, where a speed close to the world record of 2.7 km/s

[6] See Sect. 11.2 on the supersonic diffuser.

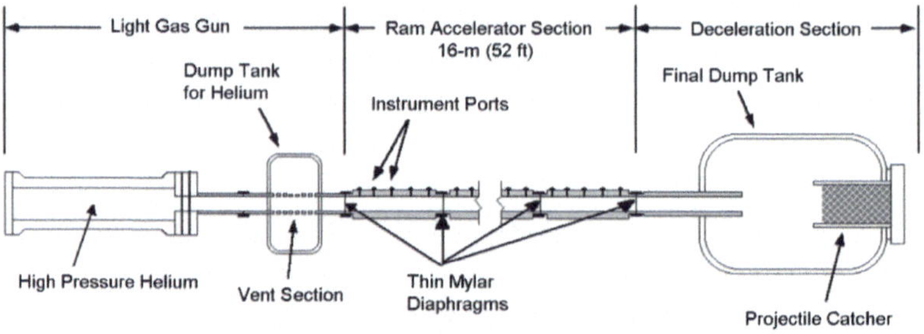

(a) Schematic of the system

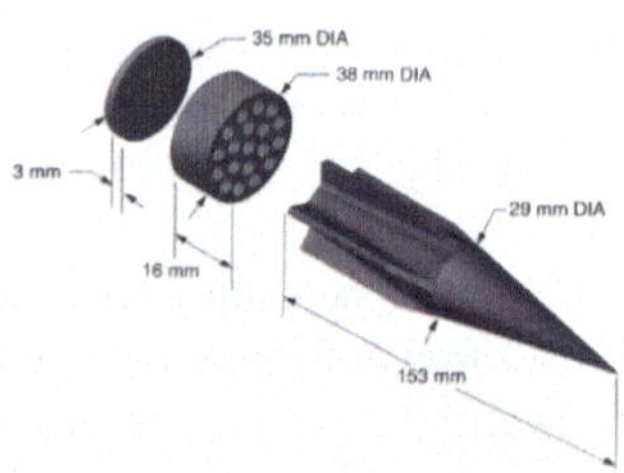

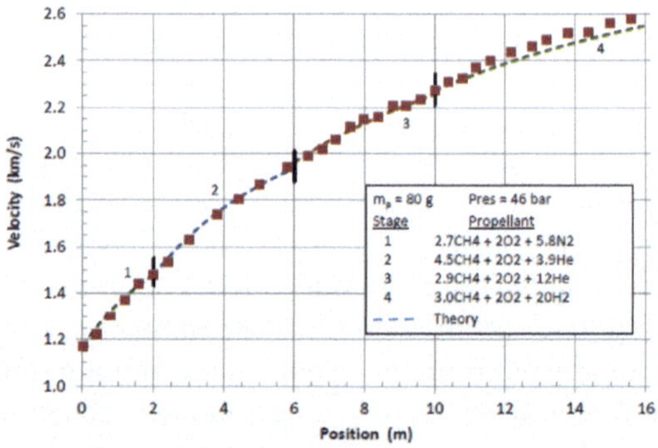

(b) Projectile **(c) Projectile speed profile in a four-stage operation**

Fig. 6.19 Ram accelerator (RAMAC38) at the University of Washington and its performance

was attained [5]. Above this speed level, the projectile itself starts combusting due to the ambient high-pressure oxygen in the shock-heated mixture.

Figure 6.20 shows the ram accelerator (RAMAC25) at Tohoku University. Through the open base of the projectile (Fig. 6.20a), the pressures between the inside and the outside of the projectile wall are almost balanced, such that the light projectile achieved an average acceleration of 4.4×10^4 g (g, gravitational acceleration) through the 6-m-long acceleration tube [6].

6.4 Thrust by Exhaust Jet

Let us add a source term in the mass conservation Rankine–Hugoniot relations to obtain a thrust, F, generated by an exhaust jet. Here, even the cross-sectional area varies. First, let us consider a jet exhausted from a rocket shown in Fig. 6.21. When the rocket flies in quiescent air at a speed u_1 to the left, on the frame fixed to the rocket, a uniform flow enters the control volume through the inlet 1 with a speed u_1 to

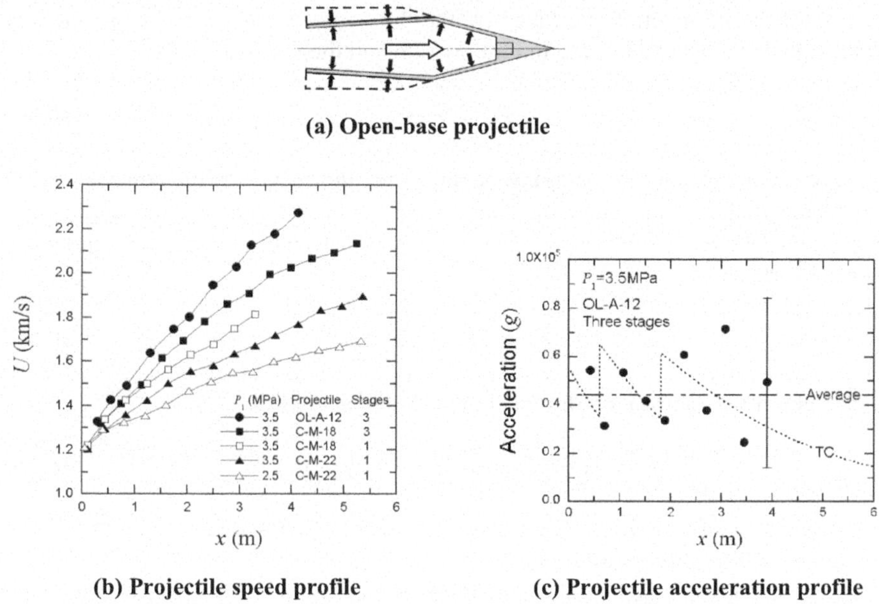

Fig. 6.20 Ram accelerator (RAMAC25) at Tohoku University and its performance

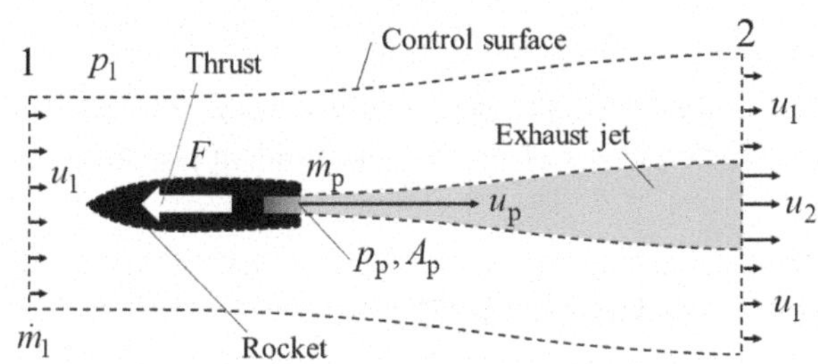

Fig. 6.21 Flow around a rocket on a frame fixed to the rocket. The broken lines correspond to control surfaces. The boundary between the exhaust jet and the surrounding area is a slip surface

the right. The upper and lower boundaries are along the streamline. The downstream exit of the control volume is labeled "2." The static pressure on the boundaries, depicted by dashed lines, equals p_1. The flow in the exhaust jet at the exit boundary 2 is higher than the rocket by u_2. Let the mass flow rate, the exhaust speed, static pressure, and cross-sectional area from or at the rocket exit be designated by $\dot{m}_p$, u_p, p_p, and A_p, respectively. The boundary between the exhaust jet, the gray regime in Fig. 6.21, and the surrounding air is a slip surface with a static pressure of p_1. With the expansion of the exhaust jet gas, its flow cross-sectional area increases. The mass flow rate of the incoming flow is $\dot{m}_1$.

The mass conservation in Fig. 6.21 is satisfied by the fact that the total mass flow rate through the exit 2 equals the sum of $\dot{m}_p$ in the exhaust jet from the rocket and $\dot{m}_1$. The momentum conservation relations are expressed in the following two ways. First, let us consider the control volume enclosing only the rocket. The thrust occurring because of the momentum produced by the flow velocity, F_{m}, i.e., the *momentum thrust*, equals the product of the mass flow rate and the exhaust speed.

$$F_{\mathrm{m}} = \dot{m}_p u_{\mathrm{p}} \tag{6.54}$$

Next, let us consider the pressure balance over the rocket. In the control surface around the rocket, a pressure imbalance exists only at the projection area of the engine exit. The *pressure thrust* is obtained as the surface integration of the pressure over the surfaces.

$$F_{\mathrm{p}} = (p_{\mathrm{p}} - p_1)A_{\mathrm{p}}. \tag{6.55}$$

The net thrust exerting on the rocket equals the sum of these.

$$F = F_{\mathrm{m}} + F_{\mathrm{p}} = \dot{m}_{\mathrm{p}} u_{\mathrm{p}} + (p_{\mathrm{p}} - p_1)A_{\mathrm{p}}. \tag{6.56}$$

By launching a rocket on the ground, the thrust becomes smaller than that in the vacuum by a frontal pressure of p_1.[7]

Then, we consider the overall momentum conservation. The thrust exerted on the rocket is balanced with the momentum balance during a unit time through the control surfaces 1 and 2.

$$F = \dot{m}_1 u_1 + \dot{m}_{\mathrm{p}} u_2 - \dot{m}_1 u_1 = \dot{m}_{\mathrm{p}} u_2 \tag{6.57}$$

Employing (6.57), the thrust is measured by determining the exhaust speed of u_2 over the control surface with the ambient pressure.

6.5 Air-Breathing Engine

A rocket engine obtains thrust from the momentum of the jet exhausted from the propellant preloaded before the launch. If oxygen from the ambient air is used, it is not loaded on board. An engine utilizing ambient air is referred to as an *air-breathing engine*. In automobiles and airplanes, such air breathing is common. Nevertheless, if an engine breathes the ambient without any reciprocating or rotating mechanics, and powers it using chemical, electrical, beam (laser or microwave), or nuclear power, then a device different from the conventional ones can be constructed. Here, let us consider such an air-breathing engine.

[7] In rocket engineering, a loss in a vehicle velocity increment for this reason is referred to as *pressure loss*.

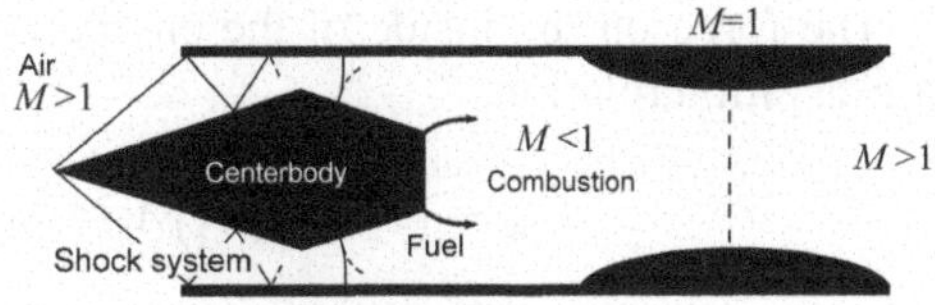

Fig. 6.22 Ramjet

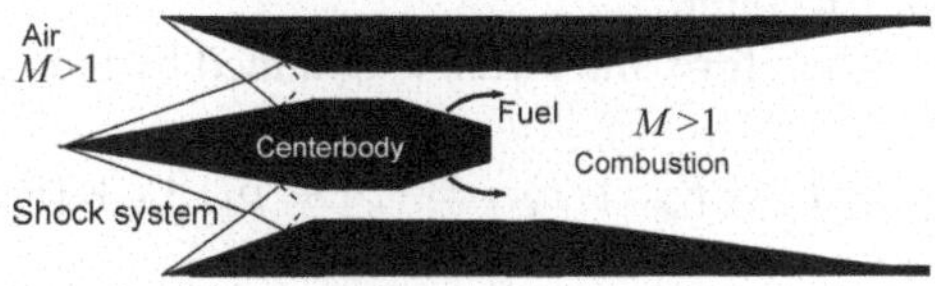

Fig. 6.23 SCRAM jet

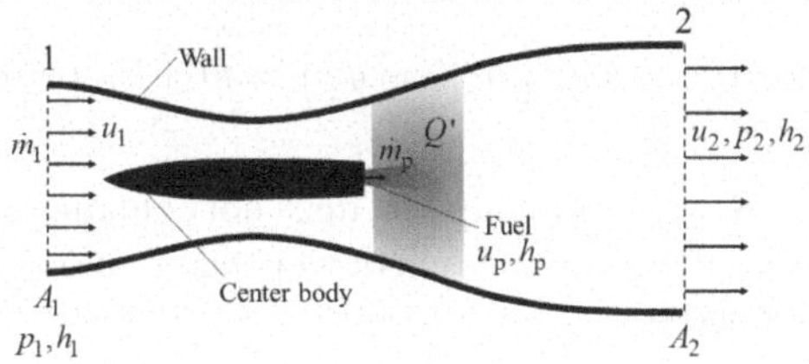

Fig. 6.24 Air-breathing engine model. The pressure in the outside equals p_1

A *ramjet* is a kind of air-breathing engine, capable of generating a thrust even at supersonic speeds without any rotating part. As shown in Fig. 6.22, the supersonic flow coming in through the inlet is compressed and decelerated in the diffuser.[8] After fuel injection and heat addition due to burning, the flow expands and is accelerated in the nozzle. In particular, ramjet, where the flow is kept at supersonic speed even in the combustion zone, is termed SCRAMjet (Supersonic Combustion Ramjet, Fig. 6.23).

Let us obtain the thrust of the air-breathing engine shown in Fig. 6.24. At inlet 1, the flow is uniform and supersonic. In the engine, a fuel is injected with a mass flow rate of $\dot{m}_\mathrm{p}$, a specific enthalpy of h_p, and a pressure that is equal to the surrounding. The flow speed and the pressure at exit 2 is determined from the nozzle design.

The mass conservation equation is

$$\rho_2 u_2 A_2 = \dot{m}_1 + \dot{m}_\mathrm{p}. \tag{6.58}$$

Here, no body force exists. The thrust equals the resultant force acting on the inside and the outside walls of the engine. At the same time, a force of equal magnitude but in the opposite direction exerts on the fluid passing through the engine. The force on the outside of the engine is

$$F_\mathrm{ext} = (A_1 - A_2)p_1. \tag{6.59}$$

[8]The compression due to the dynamic pressure of the flow itself is referred to as *ram compression*.

The force on the inside of the engine is obtained by the following momentum conservation relation.

$$F_{\mathrm{int}} = \left(p_2 + \rho_2 u_2^2\right) A_2 + \dot{m}_{\mathrm{p}} u_{\mathrm{p}} - \left(p_1 + \rho_1 u_1^2\right) A_1. \tag{6.60}$$

Note here that the effect of pressure distribution over the center body is not taken into account.

Therefore, the thrust is obtained by the net force.

$$F = F_{\mathrm{int}} + F_{ext} = (p_2 - p_1) A_2 + \left(\dot{m}_1 + \dot{m}_{\mathrm{p}}\right) u_2 + \dot{m}_{\mathrm{p}} u_{\mathrm{p}} - \dot{m}_1 u_1. \tag{6.61}$$

The energy conservation is given by

$$\left(\dot{m}_1 + \dot{m}_{\mathrm{p}}\right)\left(h_2 + \frac{1}{2} u_2^2\right) = \dot{m}_1 \left(h_1 + \frac{1}{2} u_1^2 + Q'\right) + \dot{m}_{\mathrm{p}} \left(h_{\mathrm{p}} + \frac{1}{2} u_{\mathrm{p}}^2\right). \tag{6.62}$$

Here, the enthalpy, h, does not contain a standard enthalpy of formation. The heat from the chemical reactions is taken into account in Q'. The mass flow ratio, α_{p}, is defined by

$$\alpha_{\mathrm{p}} \equiv \frac{\dot{m}_{\mathrm{p}}}{\dot{m}_1}. \tag{6.63}$$

In many cases of air-breathing engines, the fuel injected in the engine reacts with the oxygen breathed from the intake; the gases are not necessarily calorically nor thermally perfect. Yet, in the following, for simplicity, we will deal with a calorically perfect gas.

$$h = \frac{\gamma}{\gamma - 1} \frac{p}{\rho} = \frac{a^2}{\gamma - 1}. \tag{6.64}$$

$$M^2 = \frac{u^2}{a^2} = \frac{u^2}{\gamma \frac{p}{\rho}}. \tag{6.65}$$

Transforming (6.58) and (6.61)–(6.65),

$$\widetilde{F} \equiv \frac{F}{p_1 A_1} = \left(1 + \gamma M_2^2\right) \frac{M_1}{M_2} \sqrt{(1 + \alpha_{\mathrm{p}}) \frac{1 + \frac{\gamma - 1}{2} M_1^2 + Q + \alpha_{\mathrm{p}} \chi_{\mathrm{p}} \left(1 + \frac{\gamma - 1}{2} M_{\mathrm{p}}^2\right)}{1 + \frac{\gamma - 1}{2} M_2^2}}$$
$$- \left(\frac{A_2}{A_1} + \gamma M_1^2\right) + \widetilde{F}_{\mathrm{p}}, \tag{6.66}$$

$$Q \equiv \frac{Q'}{h_1}, \tag{6.67}$$

$$\chi_p \equiv \frac{h_p}{h_1}, \tag{6.68}$$

$$\widetilde{F}_p = \alpha_p \gamma M_1^2 \frac{u_p}{u_1}. \tag{6.69}$$

In (6.66), with the conditions of the injected propellant, α_p, χ_p, and $\widetilde{F}_p$ being given, the dimensionless thrust, $\widetilde{F}$, becomes a function of M_1, M_2, and Q. M_1 equals the flight Mach number; M_2 is determined from the nozzle design; Q is given from the mixture composition or the specific energy input to the gas. With $\alpha_p = 0$, $\widetilde{F}_p = 0$, and $A_2/A_1 = 1$, (6.66) becomes equivalent to (6.49).

References

1. Kuo KK (2005) Principles of Combustion, 2nd edn, p 379
2. Shepherd JE (2009) Proc Combust Inst 32:83–98
3. Herzberg A, Bruckner AP, Bogdanoff D (1988) AIAA J 26:195–203
4. Knowlen C et al (2004) J Prop Power 20:801–810. Bruckner AP et al (1991) J. Prop. Power 7:828–836. Takayama K, Sasoh A (eds) (1998) Ram accelerators. In: Proceedings of 3rd International workshop on ram accelerators. Springer, Heidelberg
5. Bruckner AP (1998) The ram accelerator: overview and state of the art," In: Takayama K, Sasoh A (eds) Ram accelerators, Proceedings of 3rd international workshop on ram accelerators. Springer, Heidelberg, pp 3–23
6. Hamate Y, Sasoh A, Takayama K (2003) J Prop Power 19:190–195

Chapter 7
Two-Dimensional Flows

Real flows are usually three dimensional. However, many important flow characteristics appear even in two-dimensional flows. In this chapter, we deal with steady state, two dimensional, supersonic flows. In such flows, compression/shock and expansion waves appear. A *Mach wave*, which is an envelope of sound waves, depicts the weakest wave. If a flow is bent toward compression, *compression waves* are generated, which then make a transition to a shock wave after propagating over a certain distance. If a supersonic flow is bent by an object, either an *attached* or a *detached shock wave* appears in front of it. The shape of the shock wave, which for the most part is an oblique shock wave, is determined according to the boundary condition set by the object. If the flow is bent toward expansion, *expansion waves* are generated. In particular, when the flow turns a corner, an *expansion fan* is formed.

7.1 Compression/Expansion Waves

When a flow is compressed moderately such that shock waves do not appear, or experiences expansion, such processes are isentropic. In such a flow, the relation between a flow Mach number and a deflection angle is obtained by the Prandtl–Meyer function. Let us derive and study it.

We apply the conservation equations in Sect. 3.1 to steady, two-dimensional flows. Here, we assume that the processes are isentropic. Body force and heat input are neglected. From (3.4),

$$\begin{aligned} &\nabla \cdot \rho \mathbf{u} = 0 \\ &\rho \frac{\partial u}{\partial x} + u \frac{\partial \rho}{\partial x} + \rho \frac{\partial v}{\partial y} + v \frac{\partial \rho}{\partial y} = 0 \end{aligned} \tag{7.1}$$

The momentum conservation Eq. (3.14) yields

A. Sasoh, *Compressible Fluid Dynamics and Shock Waves*,
https://doi.org/10.1007/978-981-15-0504-1_7

$$\rho(\mathbf{u}\cdot\nabla)\mathbf{u} = -\nabla p$$
$$\rho\begin{pmatrix} u\dfrac{\partial u}{\partial x} + v\dfrac{\partial u}{\partial y} \\ u\dfrac{\partial v}{\partial x} + v\dfrac{\partial v}{\partial y} \end{pmatrix} = -\begin{pmatrix} \dfrac{\partial p}{\partial x} \\ \dfrac{\partial p}{\partial y} \end{pmatrix}. \tag{7.2}$$

By substituting the energy conservation (3.27) with

$$\delta Q = \delta Q_{\text{reversible}} = T\text{d}s \tag{7.3}$$

$$T\text{d}s = \text{d}e + p\text{d}\left(\frac{1}{\rho}\right) = \text{d}h - \frac{\text{d}p}{\rho} = 0 \tag{7.4}$$

Using the result in Chap. 8, in the isentropic flow the speed of sound, a, is given by

$$\frac{\text{d}p}{\text{d}\rho} = a^2. \tag{7.5}$$

Using (7.5), (7.1) is transformed to

$$\rho a^2\frac{\partial u}{\partial x} + u\frac{\partial p}{\partial x} + \rho a^2\frac{\partial v}{\partial y} + v\frac{\partial p}{\partial y} = 0. \tag{7.6}$$

Equations (7.2) and (7.6) are combined into the following form:

$$\left(\bar{\mathbf{A}}\frac{\partial}{\partial x} + \bar{\mathbf{B}}\frac{\partial}{\partial y}\right)\mathbf{V} = \mathbf{O} \tag{7.7}$$

$$\bar{\mathbf{A}} = \begin{pmatrix} \rho a^2 & 0 & u \\ \rho u & 0 & 1 \\ 0 & \rho u & 0 \end{pmatrix} \tag{7.8}$$

$$\bar{\mathbf{B}} = \begin{pmatrix} 0 & \rho a^2 & v \\ \rho v & 0 & 0 \\ 0 & \rho v & 1 \end{pmatrix} \tag{7.9}$$

$$\mathbf{V} = \begin{pmatrix} u \\ v \\ p \end{pmatrix}. \tag{7.10}$$

Multiplying $\bar{\mathbf{A}}^{-1}$ to (7.7),

$$\left(\frac{\partial}{\partial x} + \hat{\mathbf{A}}\frac{\partial}{\partial y}\right)\mathbf{V} = \mathbf{O} \tag{7.11}$$

$$\hat{\mathbf{A}} = \bar{\mathbf{A}}^{-1}\bar{\mathbf{B}} = \begin{pmatrix} -\frac{1}{\rho(u^2-a^2)} & \frac{u}{\rho(u^2-a^2)} & 0 \\ 0 & 0 & \frac{1}{\rho u} \\ \frac{u}{u^2-a^2} & -\frac{a^2}{u^2-a^2} & 0 \end{pmatrix} \begin{pmatrix} 0 & \rho a^2 & v \\ \rho v & 0 & 0 \\ 0 & \rho v & 1 \end{pmatrix}$$
$$= \frac{1}{u^2-a^2} \begin{pmatrix} uv & -a^2 & -\frac{v}{\rho} \\ 0 & \frac{v}{u}(u^2-a^2) & \frac{1}{\rho u}(u^2-a^2) \\ -\rho v a^2 & \rho u a^2 & uv \end{pmatrix} \tag{7.12}$$

Substituting (7.11) with a wave-form solution

$$\mathbf{V} = \mathbf{V_0} e^{i(\omega t - \mathbf{k}\cdot\mathbf{r})} \tag{7.13}$$

$$\left(k_x \mathbf{I} + k_y \hat{\mathbf{A}}\right)\mathbf{V} = \mathbf{O}, \quad \mathbf{I}, \text{ identity matrix} \tag{7.14}$$

with

$$\lambda = -k_x/k_y \tag{7.15}$$

$$\left(\hat{\mathbf{A}} - \lambda\mathbf{I}\right)\mathbf{V} = \mathbf{O}. \tag{7.16}$$

To find a nontrivial solution to (7.16), its determinant should be set to 0.

$$\left|\hat{\mathbf{A}} - \lambda\mathbf{I}\right| = 0 \tag{7.17}$$

$$\begin{vmatrix} uv - (u^2-a^2)\lambda & -a^2 & -\frac{v}{\rho} \\ 0 & \left(\frac{v}{u} - \lambda\right)(u^2-a^2) & \frac{1}{\rho u}(u^2-a^2) \\ -\rho v a^2 & \rho u a^2 & uv - (u^2-a^2)\lambda \end{vmatrix} = 0 \tag{7.18}$$

$$\left(\lambda - \frac{v}{u}\right)\left[(u^2-a^2)\lambda^2 - 2uv\lambda + v^2 - a^2\right] = 0 \tag{7.19}$$

$$\lambda = \frac{v}{u}, \quad \frac{uv \pm \alpha a^2}{u^2-a^2} \tag{7.20}$$

$$\alpha = \sqrt{\frac{u^2+v^2}{a^2} - 1} = \sqrt{M^2-1}. \tag{7.21}$$

Therefore, three eigenvalues are obtained as follows:

$$\lambda_1 = \frac{v}{u} \tag{7.22}$$

$$\lambda_2 = \frac{\alpha v + u}{\alpha u - v} \tag{7.23}$$

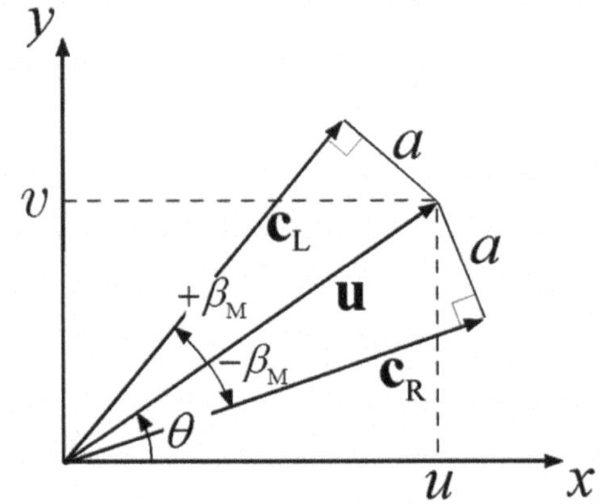

Fig. 7.1 Geometrical wave relations

$$\lambda_3 = \frac{\alpha v - u}{\alpha u + v}. \tag{7.24}$$

Here, we apply the relation between the deflection angle, θ, and the Mach angle, β_M, which is shown in Fig. 7.1, and (7.20).

$$\tan\theta = \frac{v}{u} \tag{7.25}$$

$$\alpha = \sqrt{\left(\frac{1}{\sin\beta_M}\right)^2 - 1} = \frac{1}{\tan\beta_M} \tag{7.26}$$

$$\lambda_1 = \tan\theta \tag{7.27}$$

$$\lambda_2 = \tan(\theta + \beta_M) \tag{7.28}$$

$$\lambda_3 = \tan(\theta - \beta_M). \tag{7.29}$$

As shown in Fig. 7.1, λ_1 is the tangent of the flow velocity vector, $\mathbf{u}$, to the x-axis; λ_2 and λ_3 are the tangents to Mach waves $\mathbf{c}_L$ and $\mathbf{c}_R$, respectively. In other words, in two dimensional, steady state, compressible flows, the characteristic velocity, $\mathbf{u}$, $\mathbf{c}_L$, and $\mathbf{c}_R$ exist.

With

$$\mathbf{\Lambda} = \begin{pmatrix} \lambda_1 & 0 & 0 \\ 0 & \lambda_2 & 0 \\ 0 & 0 & \lambda_3 \end{pmatrix}. \tag{7.30}$$

Let us obtain $\mathbf{L}^{-1}$ for

$$\mathbf{L}^{-1}\hat{\mathbf{A}} = \mathbf{\Lambda}\mathbf{L}^{-1}. \tag{7.31}$$

$$\mathbf{L}^{-1} = \begin{pmatrix} \rho u & \rho v & 1 \\ -\frac{\rho v}{\alpha} & \frac{\rho u}{\alpha} & 1 \\ \frac{\rho v}{\alpha} & -\frac{\rho u}{\alpha} & 1 \end{pmatrix} \tag{7.32}$$

Its inverse matrix is

$$\mathbf{L} = \begin{pmatrix} \frac{u}{\rho(u^2+v^2)} & -\frac{u+\alpha v}{2\rho(u^2+v^2)} & \frac{\alpha v-u}{2\rho(u^2+v^2)} \\ \frac{v}{\rho(u^2+v^2)} & \frac{\alpha u-v}{2\rho(u^2+v^2)} & -\frac{v+\alpha u}{2\rho(u^2+v^2)} \\ 0 & \frac{1}{2} & \frac{1}{2} \end{pmatrix}. \tag{7.33}$$

From (7.11) and (7.31),

$$\mathbf{L}^{-1}\frac{\partial \mathbf{V}}{\partial x} + \mathbf{\Lambda}\mathbf{L}^{-1}\frac{\partial \mathbf{V}}{\partial y} = \mathbf{O} \tag{7.34}$$

Defining

$$\left(\frac{\mathrm{d}y}{\mathrm{d}x}\right)_i = \lambda_i \tag{7.35}$$

Equation (7.34) has a form of

$$\frac{\partial}{\partial x} + \left(\frac{\mathrm{d}y}{\mathrm{d}x}\right)_i \frac{\partial}{\partial y} = 0, \quad i = 1, 2, 3$$

which gives an invariant along a characteristic line with a slope of λ_i.

$$\mathbf{L}^{-1}d\mathbf{V} = \mathbf{O} \text{ along } \frac{dy}{dx} = \lambda_i, \quad i = 1, 2, 3. \tag{7.36}$$

Writing the respective components,

$$\rho u \mathrm{d}u + \rho v \mathrm{d}v + \mathrm{d}p = 0 \text{ along } \frac{\mathrm{d}y}{\mathrm{d}x} = \lambda_1 = \tan\theta. \tag{7.37}$$

Using (7.4),

$$\rho u \mathrm{d}u + \rho v \mathrm{d}v + \mathrm{d}p = \rho u \mathrm{d}u + \rho v \mathrm{d}v + \rho \mathrm{d}h = \rho \mathrm{d}\left(\frac{1}{2}|\mathbf{u}|^2 + h\right) = 0$$

$$\mathrm{d}h_\mathrm{t} = \mathrm{d}\left(h + \frac{1}{2}|\mathbf{u}|^2\right) = 0. \tag{7.38}$$

This is *Bernoulli's equation* in compressible flows,[1] which implies that along the streamline total enthalpy is conserved.

From (7.36) with $i = 2$,

$$-\frac{\rho v}{\alpha}\mathrm{d}u + \frac{\rho u}{\alpha}\mathrm{d}v + \mathrm{d}p = 0 \text{ along } \frac{\mathrm{d}y}{\mathrm{d}x} = \lambda_2 = \tan(\theta + \beta_\mathrm{M}). \tag{7.39}$$

Here the flow is uniform in the upstream, such that (7.38) is applied there. From (7.26), (7.38), and (7.39),

$$(u + v\tan\beta_\mathrm{M})\mathrm{d}u + (-u\tan\beta_\mathrm{M} + v)\mathrm{d}v = 0. \tag{7.40}$$

Using relations shown in Fig. 7.1,

$$(\cos\theta + \sin\theta\tan\beta_\mathrm{M})\mathrm{d}(|\mathbf{u}|\cos\theta) + (-\cos\theta\tan\beta_\mathrm{M} + \sin\theta)\mathrm{d}(|\mathbf{u}|\sin\theta) = 0$$

$$\frac{1}{\tan\beta_\mathrm{M}}\frac{\mathrm{d}|\mathbf{u}|}{|\mathbf{u}|} - \mathrm{d}\theta = 0 \tag{7.41}$$

Defining

$$\mathrm{d}\nu \equiv \frac{1}{\tan\beta_\mathrm{M}}\frac{\mathrm{d}|\mathbf{u}|}{|\mathbf{u}|} \tag{7.42}$$

$$\mathrm{d}(\nu - \theta) = 0 \tag{7.43}$$

ν is the *Prandtl–Meyer function*. From the definition of the Mach number,

$$M = \frac{|\mathbf{u}|}{a} = \frac{1}{\sin\beta_\mathrm{M}}.$$

Applying (7.38) to calorically perfect gas,

$$\frac{2a\mathrm{d}a}{\gamma - 1} + |\mathbf{u}|\mathrm{d}|\mathbf{u}| = 0$$

$$\frac{2}{(\gamma-1)M^2}\frac{\mathrm{d}a}{a} + \frac{\mathrm{d}|\mathbf{u}|}{|\mathbf{u}|} = 0. \tag{7.44}$$

Using these equations, the right hand side of (7.42) is expressed as a function of M.

$$\frac{1}{\tan\beta_\mathrm{M}} = \frac{\cos\beta_\mathrm{M}}{\sin\beta_\mathrm{M}} = \frac{\sqrt{1-\sin^2\beta_\mathrm{M}}}{\sin\beta_\mathrm{M}} = \sqrt{M^2 - 1}$$

$$\frac{\mathrm{d}a}{a} + \frac{\mathrm{d}M}{M} = \frac{\mathrm{d}|\mathbf{u}|}{|\mathbf{u}|} \tag{7.45}$$

[1] For incompressible flows, $p + \frac{1}{2}\rho|\mathbf{u}|^2 = \text{const.}$

$$\frac{2}{(\gamma-1)M^2}\left(\frac{\mathrm{d}|\mathbf{u}|}{|\mathbf{u}|}-\frac{\mathrm{d}M}{M}\right)+\frac{\mathrm{d}|\mathbf{u}|}{|\mathbf{u}|}=0 \tag{7.46}$$

$$\frac{\mathrm{d}|\mathbf{u}|}{|\mathbf{u}|}=\frac{1}{1+\frac{\gamma-1}{2}M^2}\frac{\mathrm{d}M}{M} \tag{7.47}$$

$$\mathrm{d}\nu=\frac{\sqrt{M^2-1}}{1+\frac{\gamma-1}{2}M^2}\frac{\mathrm{d}M}{M}. \tag{7.48}$$

Usually the integral constant is set such that $\nu(M=1)=0$. ν is obtained in the explicit form.

$$\begin{aligned}\nu(M)&=\int_1^M\frac{\sqrt{M^2-1}}{1+\frac{\gamma-1}{2}M^2}\frac{\mathrm{d}M}{M}\\&=\sqrt{\frac{\gamma+1}{\gamma-1}}\tan^{-1}\sqrt{\frac{\gamma-1}{\gamma+1}(M^2-1)}-\tan^{-1}\sqrt{M^2-1}\end{aligned}. \tag{7.49}$$

From (7.36) with $i=3$,

$$\begin{aligned}&\frac{\rho v}{\alpha}\mathrm{d}u-\frac{\rho u}{\alpha}\mathrm{d}v+\mathrm{d}p=0\ \text{along}\ \frac{\mathrm{d}y}{\mathrm{d}x}=\lambda_3=\tan(\theta-\beta_\mathrm{M})\\&(u-v\tan\beta_\mathrm{M})\mathrm{d}u+(u\tan\beta_\mathrm{M}+v)\mathrm{d}v=0\end{aligned} \tag{7.50}$$

$$\begin{aligned}&(\cos\theta-\sin\theta\tan\beta_\mathrm{M})\mathrm{d}(|\mathbf{u}|\cos\theta)+(\cos\theta\tan\beta_\mathrm{M}+\sin\theta)\mathrm{d}(|\mathbf{u}|\sin\theta)=0\\&\mathrm{d}(\nu+\theta)=0\end{aligned}. \tag{7.51}$$

To summarize the above,

$$\text{Along}\ \frac{\mathrm{d}\mathbf{x}}{\mathrm{d}t}=\mathbf{u},\quad\frac{\mathrm{d}y}{\mathrm{d}x}=\tan\theta,\quad\mathrm{d}s=0,\quad\mathrm{d}h_\mathrm{t}=0 \tag{7.52}$$

$$\text{Along}\ \frac{\mathrm{d}\mathbf{x}}{\mathrm{d}t}=\mathbf{c}_\mathrm{L},\quad\frac{\mathrm{d}y}{\mathrm{d}x}=\tan(\theta+\beta_\mathrm{M}),\quad\mathrm{d}R\equiv\mathrm{d}(\nu-\theta)=0 \tag{7.53}$$

$$\text{Along}\ \frac{\mathrm{d}\mathbf{x}}{\mathrm{d}t}=\mathbf{c}_\mathrm{R},\quad\frac{\mathrm{d}y}{\mathrm{d}x}=\tan(\theta-\beta_\mathrm{M}),\quad\mathrm{d}Q\equiv\mathrm{d}(\nu+\theta)=0. \tag{7.54}$$

In (7.52)–(7.54), s, R, Q are *Riemann invariants.*

A steady, supersonic flow is decelerated if it is bent at the compression side, and accelerated at the expansion side. Along a slowing bending wall, the flow is kept isentropic, irrespective of whether it is compressed or expanded. However, at a certain distance from the wall, if compression waves are taken over by proceeding compression waves from behind, the transition to a shock wave occurs. In that case, the isentropic assumption is violated, and thus (7.53) and (7.54) cannot be applied.

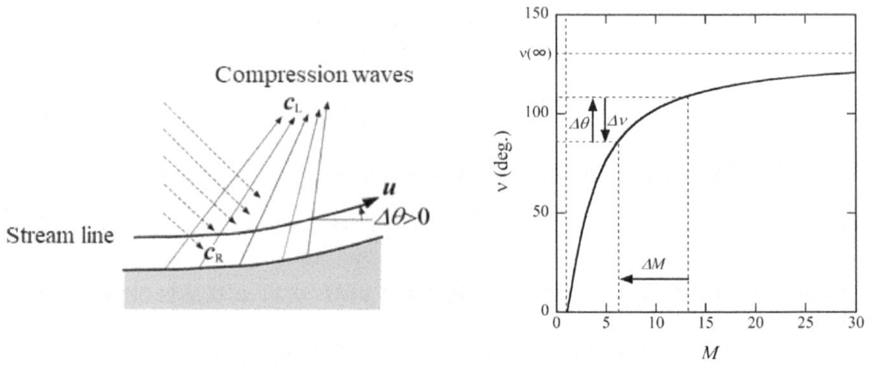

(a) Flow and waves (b) Variation in Prandtl-Meyer function

Fig. 7.2 Supersonic flow with moderate compression, $\Delta\theta > 0$

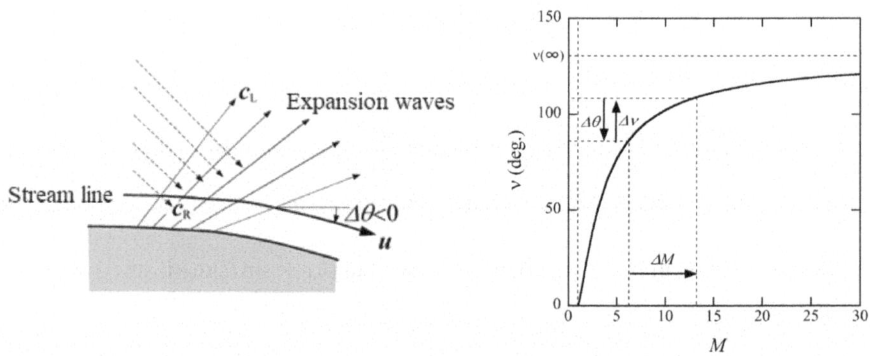

(a) Flow and waves (b) Variation in Prandtl-Meyer function

Fig. 7.3 Supersonic flow with moderate expansion, $\Delta\theta < 0$

Using the Prandtl–Meyer function, (7.49), (7.53), and (7.54), a variation in flow Mach number ΔM is obtained from a deflection angle, $\Delta\theta$. Figure 7.2 shows supersonic flow that is moderately compressed on a concave surface. In the flow field, the invariant $Q = \nu + \theta$, which is kept constant along $\mathbf{c}_R$, equals to that in the upstream.

$$\Delta\nu = -\Delta\theta. \tag{7.55}$$

With $\Delta\theta$ being given, $\Delta\nu$ then ΔM is obtained from (7.49), as shown in Fig. 7.2b. In this case, the flow is decelerated, meaning that $\Delta M < 0$.

When the flow is bent toward expansion, as shown in Fig. 7.3a, the flow is accelerated, that is $\Delta\nu > 0$ and $\Delta M > 0$.

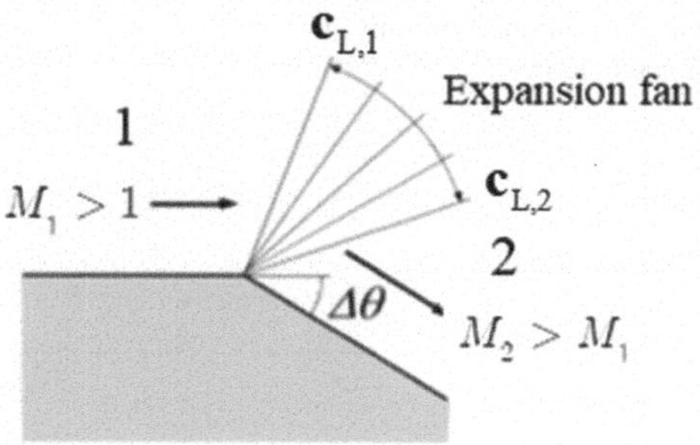

Fig. 7.4 Prandtl–Meyer expansion across an expansion corner

7.2 Prandtl–Meyer Expansion

If a supersonic flow is incident to a compression corner, a shock wave, either attached or detached, is formed. Figure 7.4 illustrates the behavior of a supersonic flow in the case where it is incident to an expansion corner. After a supersonic flow at state 1 turns the corner by an angle of $\Delta\theta$, its condition changes to state 2. Resultantly, the flow is accelerated due to expansion, $M_1 < M_2$. This expansion occurs only in the *expansion fan*, which is sandwiched by $\mathbf{c}_{L,1}$ and $\mathbf{c}_{L,2}$ waves originating in the corner. Such expansion of a supersonic flow across a corner is termed as the *Prandtl–Meyer expansion*. M_2 is obtained by inputting $\Delta\theta$ into (7.54). From (7.44) and (7.47),

$$\frac{\mathrm{d}a}{a} = -\frac{(\gamma-1)M}{2+(\gamma-1)M^2}\mathrm{d}M$$

$$\mathrm{d}\ln a = -\frac{1}{2}\mathrm{d}\ln\{2+(\gamma-1)M^2\}$$

$$a^2\{2+(\gamma-1)M^2\} = \text{const.}$$

Then, the temperature, T, is given by

$$\frac{T}{T_{M=1}} = \frac{a^2}{a^2_{M=1}} = \frac{\frac{\gamma+1}{\gamma-1}}{M^2+\frac{2}{\gamma-1}}. \tag{7.56}$$

The pressure p is obtained from the isentropic relation.

$$\frac{p}{p_{M=1}} = \left(\frac{T}{T_{M=1}}\right)^{\frac{\gamma}{\gamma-1}} = \left(\frac{\frac{\gamma+1}{\gamma-1}}{M^2+\frac{2}{\gamma-1}}\right)^{\frac{\gamma}{\gamma-1}}. \tag{7.57}$$

Let us examine how much the flow can be refracted. From (7.49) and (7.55),

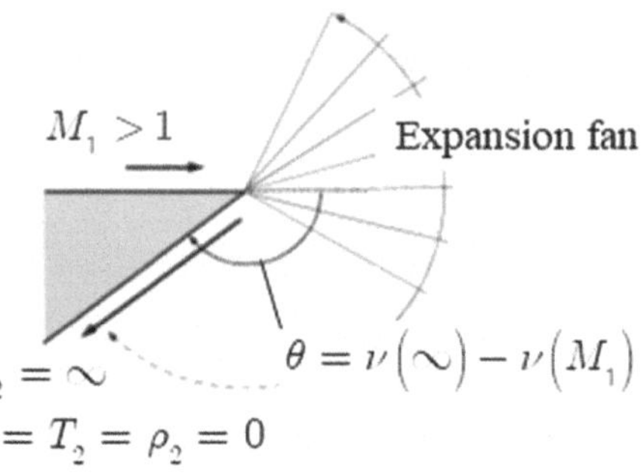

Fig. 7.5 Refraction limit of supersonic flow

$$(-\Delta\theta)_{\max} = \nu(\infty) - \nu(M_1)$$
$$= \frac{\pi}{2}\left(\sqrt{\frac{\gamma+1}{\gamma-1}} - 1\right) - \sqrt{\frac{\gamma+1}{\gamma-1}}\tan^{-1}\sqrt{\frac{\gamma-1}{\gamma+1}(M_1^2-1)} + \tan^{-1}\sqrt{M_1^2-1} \tag{7.58}$$

$$\nu(\infty) = \frac{\pi}{2}\left(\sqrt{\frac{\gamma+1}{\gamma-1}} - 1\right). \tag{7.59}$$

If the flow is refracted up to this angle, the Mach number becomes infinity, and the pressure, temperature, and density vanish. For air with $\gamma = 1.4$, $\nu(\infty) = 130.45°$. This value equals the maximum refraction angle of an incident flow of $M_1 = 1$. For $M_1 > 1$, the maximum refraction angle is given by (7.58), see Fig. 7.5.

7.3 Supersonic Flow Around a Cone

Figure 7.6 illustrates a supersonic flow over a cone. The flow is assumed to be inviscid and axisymmetric. If the cone has an infinite length, the flow becomes self-similar. The shock wave likewise has a cone shape. The solution is referred to as the *Taylor–Maccoll solution* [1].

From the mass and momentum conservation equations, (3.4) and (3.15), respectively,

$$\nabla \cdot (\rho \mathbf{u}) = 0 \tag{7.60}$$

$$\rho(\mathbf{u} \cdot \nabla)\mathbf{u} + \nabla p = 0. \tag{7.61}$$

The half-apex angles of the cone and the shock wave are designated by θ_w and β. The two-dimensional polar coordinates, (r, θ), are set as shown in Fig. 7.6. Since the

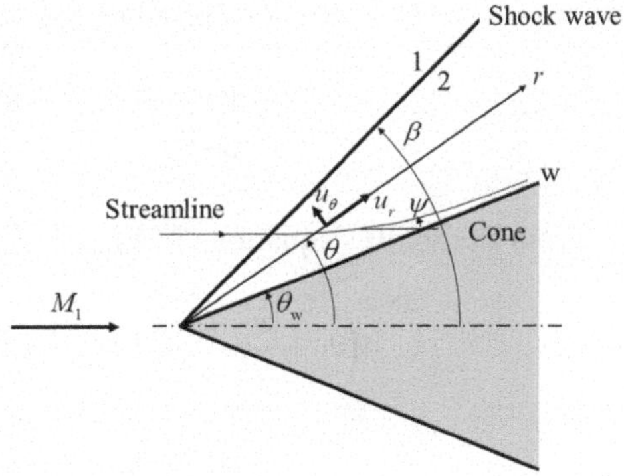

Fig. 7.6 Flow over a cone

flow is self-similar, the flow velocity vector (u_r, u_θ) and thermodynamics properties become a function only of θ, and the derivative with respect to r vanishes. With those conditions, (7.60) and (7.61) yield to

$$\frac{1}{r^2}\frac{\mathrm{d}}{\mathrm{d}r}\left(\rho r^2 u_r\right) + \frac{1}{r\sin\theta}\frac{\mathrm{d}}{\mathrm{d}\theta}(\rho u_\theta \sin\theta) = 0$$

$$2\rho u_r \sin\theta + \frac{\mathrm{d}}{\mathrm{d}\theta}(\rho u_\theta \sin\theta) = 0 \tag{7.62}$$

$$\rho\left(\frac{u_\theta}{r}\frac{\mathrm{d}u_r}{\mathrm{d}\theta} - \frac{u_\theta^2}{r}\right) = 0 \tag{7.63}$$

$$\rho\left(u_\theta\frac{\mathrm{d}u_\theta}{\mathrm{d}\theta} + u_r u_\theta\right) + \frac{\mathrm{d}p}{\mathrm{d}\theta} = 0. \tag{7.64}$$

From (7.63),

$$u_\theta = \frac{\mathrm{d}u_r}{\mathrm{d}\theta}. \tag{7.65}$$

This leads to the result that the flow is free from vorticity $\boldsymbol{\omega}$.

$$\boldsymbol{\omega} = \nabla \times \mathbf{u} = \left(\frac{u_\theta}{r} - \frac{1}{r}\frac{\partial u_r}{\partial\theta}\right)\mathbf{e}_\phi = 0. \tag{7.66}$$

From (7.62) and (7.65),

$$2\rho u_r + u_\theta\frac{\mathrm{d}\rho}{\mathrm{d}\theta} + \rho\frac{\mathrm{d}^2 u_r}{\mathrm{d}\theta^2} + \rho u_\theta \cot\theta = 0. \tag{7.67}$$

For the isentropic flow,

$$\mathrm{d}p = a^2 \mathrm{d}\rho. \tag{7.68}$$

From (7.64), (7.65), (7.67), and (7.68),

$$\frac{d^2u_r}{d\theta^2} + u_r = -\frac{u_r + u_\theta \cot\theta}{1 - \frac{u_\theta^2}{a^2}}. \tag{7.69}$$

From Bernoulli's equation (7.38),

$$d\left(h + \frac{u_r^2 + u_\theta^2}{2}\right) = d\left(\frac{a^2}{\gamma - 1} + \frac{u_r^2 + u_\theta^2}{2}\right) = 0 \tag{7.70}$$

$$\frac{1}{2}u_t^2 \equiv h_t = \frac{a^2}{\gamma - 1} + \frac{u_r^2 + u_\theta^2}{2}. \tag{7.71}$$

Here, u_t is a constant and equals the flow speed attained when it expands to vacuum in a steady-state manner. From (7.65) and (7.67),

$$\begin{pmatrix} \frac{du_r}{d\theta} \\ \frac{du_\theta}{d\theta} \end{pmatrix} = \begin{pmatrix} u_\theta \\ -u_r - \frac{u_r + u_\theta \cot\theta}{1 - \frac{u_\theta^2}{a^2}} \end{pmatrix}. \tag{7.72}$$

Integrating (7.72), the solutions $u_r = u_r(\theta)$ and $u_\theta = u_\theta(\theta)$ are obtained. The boundary conditions on the cone wall (subscript, w) are

$$u_r(\theta_w) = u_{r,w} \tag{7.73}$$

$$u_\theta(\theta_w) = 0. \tag{7.74}$$

With M_1 and θ_w given, β needs to be found. Here, we use subscripts 1 and 2 for the state upstream and immediately behind the shock wave. Defining

$$u_{r,2} \equiv u_r(\beta) \tag{7.75}$$

$$u_{\theta,2} \equiv u_\theta(\beta) < 0 \tag{7.76}$$

$$\frac{u_t^2}{2} = \frac{a^2}{\gamma - 1} + \frac{u_{r,2}^2}{2} + \frac{u_\theta^2}{2} = \frac{a_*^2}{\gamma - 1} + \frac{u_{r,2}^2}{2} + \frac{a_*^2}{2} = \frac{\gamma + 1}{2(\gamma - 1)}a_*^2 + \frac{u_{r,2}^2}{2} \tag{7.77}$$

$$a_*^2 = \frac{\gamma - 1}{\gamma + 1}\left(u_t^2 - u_{r,2}^2\right). \tag{7.78}$$

Applying the Prandtl relation (4.57),

$$M_1 a_1 \sin\beta\left(-u_{\theta,2}\right) = \frac{\gamma - 1}{\gamma + 1}\left(u_t^2 - u_{r,2}^2\right). \tag{7.79}$$

Since the flow velocity tangent to the shock wave does not change,

$$M_1 a_1 \cos\beta = u_{r,2}. \tag{7.80}$$

From (7.79) and (7.80),

$$(-u_{\theta,2}) = \frac{\gamma - 1}{\gamma + 1}\frac{u_t^2 - u_{r,2}^2}{u_{r,2}\tan\beta}. \tag{7.81}$$

Applying (7.77) to the upstream of the shock wave, and using (7.78),

$$\frac{1}{2}u_t^2 = \left(\frac{1}{\gamma - 1} + \frac{1}{2}M_1^2\right)a_1^2 = \left(\frac{1}{\gamma - 1} + \frac{1}{2}M_1^2\right)\left(\frac{u_{r,2}}{M_1\cos\beta}\right)^2$$

$$M_1^2 = \frac{2}{\gamma - 1}\frac{u_{r,2}^2}{u_t^2\cos^2\beta - u_{r,2}^2}. \tag{7.82}$$

The above relations are expressed using dimensionless quantities:

$$\widetilde{u_r} \equiv \frac{u_r}{u_t}, \quad \widetilde{u_\theta} \equiv \frac{u_\theta}{u_t}, \quad \tilde{a} \equiv \frac{a}{u_t}$$

$$\begin{pmatrix} \frac{\mathrm{d}\widetilde{u_r}}{\mathrm{d}\theta} \\ \frac{\mathrm{d}\widetilde{u_\theta}}{\mathrm{d}\theta} \end{pmatrix} = \begin{pmatrix} \widetilde{u_\theta} \\ -\widetilde{u_r} - \frac{\tilde{a}^2(\widetilde{u_r} + \widetilde{u_\theta}\cot\theta)}{\tilde{a}^2 - \widetilde{u_\theta}^2} \end{pmatrix} \tag{7.83}$$

$$\tilde{a}^2 = \frac{\gamma - 1}{2}\left\{1 - \left(\widetilde{u_r}^2 + \widetilde{u_\theta}^2\right)\right\} \tag{7.84}$$

$$\widetilde{u_r}(\theta_w) = \widetilde{u_{r,w}} \tag{7.85}$$

$$\widetilde{u_\theta}(\theta_w) = 0 \tag{7.86}$$

$$\widetilde{u_{r,2}} = \frac{M_1\cos\beta}{\sqrt{\frac{2}{\gamma-1} + M_1^2}} \tag{7.87}$$

$$(-\widetilde{u_{\theta,2}}) = \frac{\gamma - 1}{\gamma + 1}\frac{1 - \widetilde{u_{r,2}}^2}{\widetilde{u_{r,2}}\tan\beta}. \tag{7.88}$$

For a given combination of M_1 and θ_w, and assuming the value of β and $\widetilde{u_{\theta,w}}$, $\widetilde{u_r} = \widetilde{u_r}(\theta)$, $\widetilde{u_\theta} = \widetilde{u_\theta}(\theta)$ are calculated by numerically integrating (7.83). The calculation should be repeated until a solution is obtained satisfying both (7.87) and (7.88).

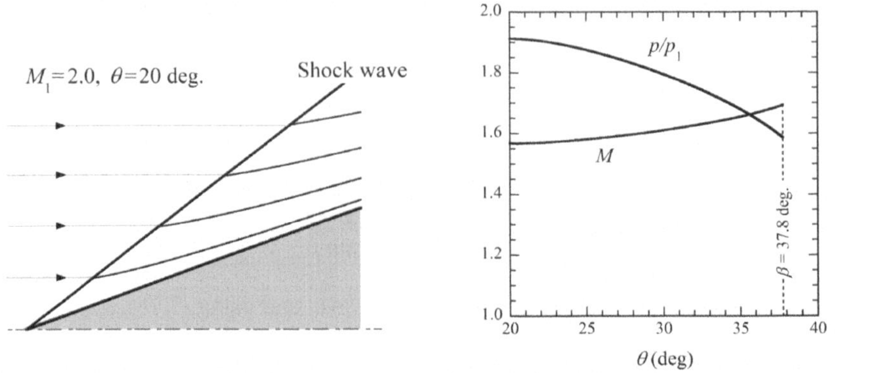

(a) Shock wave and streamlines **(b) Variation of pressure and Mach number**

Fig. 7.7 Supersonic flow around a cone, $M_1 = 2.0$, $\theta = 20°$

Once the solution is obtained, the flow quantities at the state 2 are obtained using the oblique shock-wave relations. Those between the shock wave and the cone surface are obtained using Bernoulli's and isentropic equations.

$$\frac{p_2}{p_1} = 1 + \frac{2\gamma}{\gamma+1}\left(M_1^2 \sin^2\beta - 1\right) \tag{7.89}$$

$$\frac{\gamma}{\gamma-1}\frac{p(\theta)}{\rho(\theta)} + \frac{u_r^2(\theta)}{2} + \frac{u_\theta^2(\theta)}{2} = \frac{\gamma}{\gamma-1}\frac{p_t}{\rho_t} = \frac{u_t^2}{2} \tag{7.90}$$

$$\frac{p(\theta)}{p_2} = \left(\frac{\rho(\theta)}{\rho_2}\right)^{\gamma} \tag{7.91}$$

$$\frac{p(\theta)}{p_2} = \left\{\frac{1 - \widetilde{u_r}^2(\theta) - \widetilde{u_\theta}^2(\theta)}{1 - \widetilde{u_{r,2}}^2 - \widetilde{u_{\theta,2}}^2}\right\}^{\frac{\gamma}{\gamma-1}}. \tag{7.92}$$

The deflection angle, ψ, is given by

$$\psi = \theta + \tan^{-1}\left(\frac{u_r}{u_\theta}\right) = \theta + \tan^{-1}\left(\frac{\widetilde{u_r}}{\widetilde{u_\theta}}\right) < \theta. \tag{7.93}$$

An example of a solution is shown in Fig. 7.7. In two-dimensional flow, for $M_1 = 2.0$ and $\theta = 20°$, $\beta = 53.4°$, and $p_2/p_1 = 2.84$, whereas in the conical flow $\beta = 37.8°$ and $p_2/p_1 = 1.59$ (the pressure ratio on the cone surface to the upstream equals 1.91). Thereby, in the latter the shock wave becomes much weaker.

7.4 Reflection of Shock Waves

7.4.1 Shock Reflection Patterns in Steady Flow

A shock wave is reflected on a wall. However, its reflection pattern is different from that of electromagnetic waves including light. In steady flows, the reflection pattern is subdivided to *regular reflection* (Fig. 7.8) and *Mach reflection* (Fig. 7.9). If we place a wedge with a deflection angle of θ_w in a $M_0(>1)$ flow, an oblique shock wave is generated. This shock wave propagates to the lower wall as an *incident shock wave* i. Here, it is assumed that the expansion fan originated in the lower corner of the wedge does not affect i.

In the case of regular reflection (Fig. 7.8), the incident shock wave is reflected on the point R, and subsequently, a *reflected shock wave* r is generated at the same point. In the upstream flow (state 0), the flow is along the bottom wall and has an angle β_1 from i. Behind i (state 1), the flow is bent toward the bottom wall by a reflection angle of θ_1. After the flow passes r, the flow becomes bent back by $\theta_2 = -\theta_1$, restoring the same direction as the upstream (state 2). The last process is equivalent to that occurring during the formation of an attached shock wave, as described Sect. 4.3.4.

In the Mach reflection (Fig. 7.9), the incident shock wave i is bent at the point T, at which the incident shock (i), reflected shock (r), and *Mach stem* (m) intersect. This point is referred to as a *triple point*, T. In many cases, i is connected to m with a kink. The flow above T passes through i and r; the flow direction of state 2 is not necessarily along the bottom wall. The flow below T has only m, becomes subsonic at state 3. Therefore, state 2 and 3 are interfaced with a slip line, tangent to each other

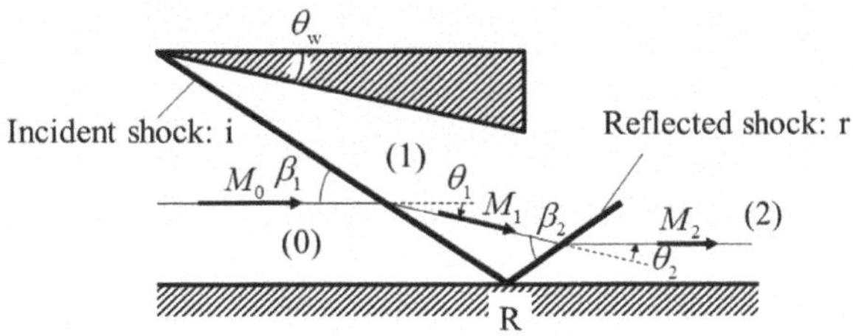

Fig. 7.8 Regular reflection

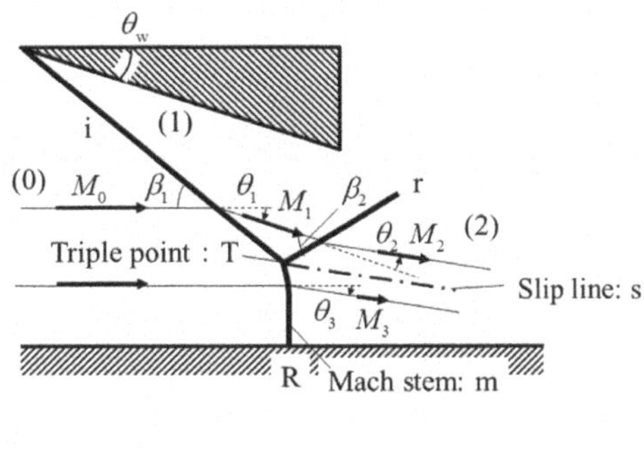

Fig. 7.9 Mach reflection

at the same pressure, but with a different flow velocity, density, and temperature. On the bottom wall, the shock wave is normal to the flow.

7.4.2 Shock Polar

In order to analyze the shock wave reflection, it is convenient to use the relation between a pressure, p, and a deflection angle, θ, that is $p-\theta$ *shock polar*. Here, we apply the oblique shock-wave relations presented in Sect. 4.3 to a supersonic flow with an upstream Mach number of M_0.

$$\frac{p_1}{p_0} = 1 + \frac{2\gamma}{\gamma+1}\left(M_0^2 \sin^2\beta - 1\right) \tag{7.94}$$

$$\tan\theta = \frac{2\cot\beta\left(M_0^2\sin^2\beta - 1\right)}{M_0^2(\gamma + \cos 2\beta) + 2}. \tag{7.95}$$

The pressures in front and behind the shock wave are designated by p_0 and p_1, respectively. From (7.94) and (7.95), the shock polar equation is obtained, see Fig. 7.10.

$$\Phi \equiv \frac{p_1}{p_0} = \Phi(M_0, \theta). \tag{7.96}$$

This equation is applicable to any shock wave appearing in the reflection with corresponding subscripts. The flow Mach number behind the shock wave is given by

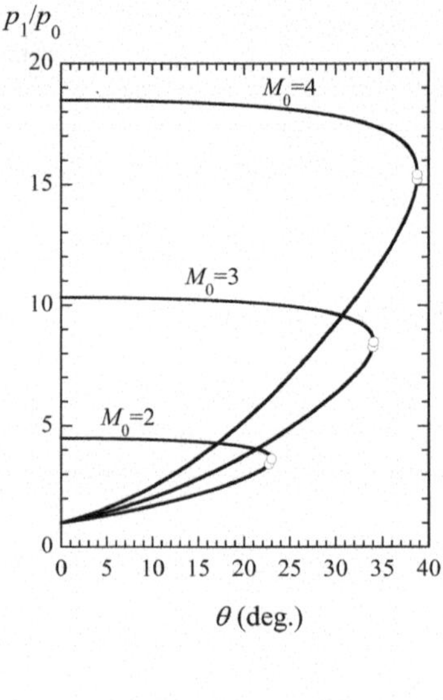

Fig. 7.10 Shock polar, $\gamma = 1.4$. The two circles on each polar correspond to $\theta = \theta_{max}$ (upper) and $M_1 = 1$ (lower), respectively

$$M_1 = \left[\frac{\{(\gamma - 1)M_0^2 \sin^2 \beta + 2\}^2 + (\gamma + 1)^2 M_0^4 \sin^2 \beta \cos^2 \beta}{\{2\gamma M_0^2 \sin^2 \beta - (\gamma - 1)\}\{(\gamma - 1)M_0^2 \sin^2 \beta + 2\}} \right]^{\frac{1}{2}}. \tag{7.97}$$

In Fig. 7.10, three polars are plotted. Tracing from $(p_1/p_0, \theta) = (1, 0°)$ along each one, the shock strength increases with increasing θ in the weak shock regime. The flow behind the shock wave is supersonic. The condition for the post-shock flow to become sonic is close to that with the maximum θ. By further increasing the shock strength, θ decreases in the strong shock regime. The strongest condition corresponds to a normal shock wave.

7.4.3 *Two Shock Theory*

Let us analyze the regular reflection, Fig. 7.8. With an upstream flow Mach number M_0, and an angle of the incident shock wave β_1 being specified, for the incident shock wave,

$$p_1 = p_0 \Phi(M_0, \theta_1). \tag{7.98}$$

Meanwhile, for the reflected shock wave,

$$p_2 = p_1 \Phi(M_1, \theta_2). \tag{7.99}$$

The boundary conditions here are

$$\theta_1 = \theta_{\mathrm{w}} \tag{7.100}$$

$$\theta_1 - \theta_2 = 0. \tag{7.101}$$

Figure 7.11 shows the shock polar for $M_0 = 3$ and $\beta_1 = 30°$. In the upstream (0), $\theta_0 = 0$ and $p/p_0 = 1$. The polar for the incident shock originates at (0) and is plotted on $\theta > 0$ side, going through the post-shock state (1). Here, β_1 is given, and thereby the corresponding deflection angle is determined as $\theta_1 = 12.8°$. Across the incident shock, the pressure becomes 2.46 times larger. The shock polar of the reflected shock originates at (1) and extends to the left. In regular reflection, the intersection of the polar with the vertical ordinate ($\theta = 0°$) corresponds to the post-shock state of the reflected shock. From the two intersections, only the lower one is the solution. In this case, the post-shock pressure becomes 5.20 times larger.

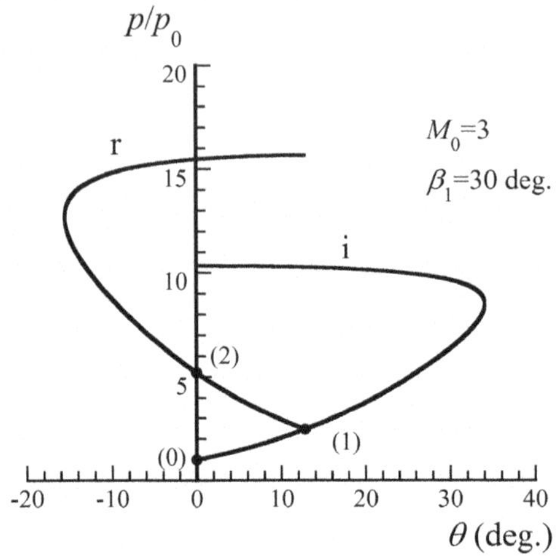

Fig. 7.11 Example of shock polar for regular reflection

7.4.4 Three-Shock Theory

Let us analyze the Mach reflection, shown in Fig. 7.9. M_0 and β_1 are specified, and thus $\theta_1 (=\theta_w)$ is obtained using (7.95). In the same way as for the regular reflection, for the incident shock wave,

$$p_1 = p_0 \Phi(M_0, \theta_1). \tag{7.102}$$

Meanwhile, for the reflected shock wave,

$$p_2 = p_1 \Phi(M_1, \theta_2). \tag{7.103}$$

To the state immediately below the Mach stem (3),

$$p_3 = p_0 \Phi(M_0, \theta_3). \tag{7.104}$$

The boundary conditions are

$$\theta_1 = \theta_w \tag{7.105}$$

$$\theta_1 - \theta_2 = \theta_3 \tag{7.106}$$

$$p_2 = p_3. \tag{7.107}$$

Figure 7.12 shows the shock polars for $M_0 = 3$ and $\theta_w = 40°$. The polar for the Mach stem is shared with the incident shock wave. The polar for the reflected shock has a unique intersection, which corresponds both to the state 2 and 3, with

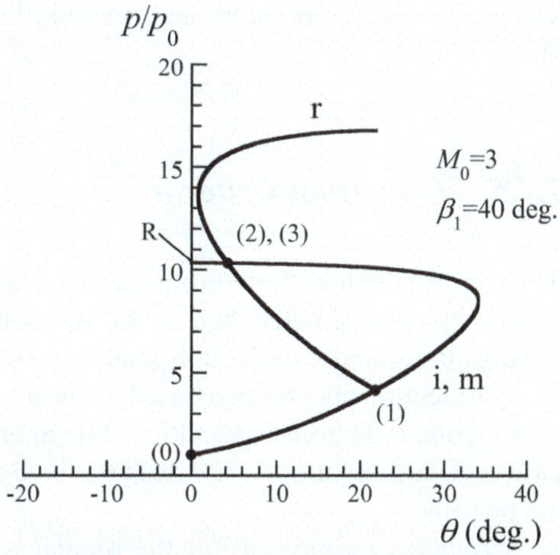

Fig. 7.12 Example of shock polar for Mach reflection

Fig. 7.13 Photograph of exhaust plume from a liquid rocket engine, LE-7A. In the middle of the plume, a conical radiation emission region is observed behind the Mach stem. Courtesy of Mitsubishi Heavy Industries Ltd.

the incident one. Behind the incident shock, the pressure becomes 4.17 times larger, and $\theta_1 = 21.8°$. Behind the reflected shock, the pressure becomes 10.3 times of the upstream value with a deflection angle toward the wall of $\theta_2 = \theta_3 = 4.21°$. The direction of the Mach stem varies toward the wall. On the reflection point R on the wall, θ vanishes, and thereby the normal shock wave condition is applied.

We can observe the Mach reflection in the exhaust plume of a liquid rocket engine, as shown in Fig. 7.13. The exhaust jet from a rocket engine is shaped like a feather, and it is called an *exhaust plume*. Usually, the rocket engine is designed such that on the ground the flow is over-expanded, meaning that the static pressure in the plume is lower than the ambient pressure.[2] In order to recover the pressure, a shock wave system where the flow repeats the compression and expansion cycle appears. In the compression processes, oblique shock waves are generated around the peripheral of the plume. At the center axis, a Mach reflection is formed. In such an axisymmetric configuration, the Mach stem is configured as a disk (*Mach disk*). Behind the

[2]For further details, refer to Sect. 11.1.

Mach disk, the temperature and pressure become high, such that they are visually observable.

7.4.5 Transition Criteria

Thus far, we dealt with two-dimensional shock wave reflection with the shock angle β, as a given parameter. Yet, in the shock polar analyses, the relation between the pressure ratio and a deflection angle, θ, is more useful. Therefore, we analyze the shock reflection phenomena based on the θ dependence. In the condition of Fig. 7.11, only regular reflection is possible, while in Fig. 7.12, only Mach reflection is possible. However, combinations of M_0 and θ_w (wedge angle) exist, with which both patterns are possible.

A necessary condition for the regular reflection is that the polar of a reflected shock wave intersects with the coordinate of $\theta = 0°$. As a critical condition, the polar of the reflected shock wave becomes tangent to the coordinate. This is termed as the *detachment criterion*, as depicted in Fig. 7.14.

$$\theta_{2,\max} = \theta_w \equiv \theta_{w,d}. \tag{7.108}$$

As described in Sect. 4.3, the condition for a maximum θ variation is effectively equivalent to the sonic condition for the post-shock flow (*sonic criterion*).

A necessary condition for the Mach reflection is that the reflected shock polar intersects with the incident shock polar in $\theta > 0°$ The critical condition is the polars having an intersection at $\theta = 0°$ (*mechanical equilibrium criterion*).

$$\theta_w - \theta_2 = \theta_3 = 0, \quad \theta_w \equiv \theta_{w,m}. \tag{7.109}$$

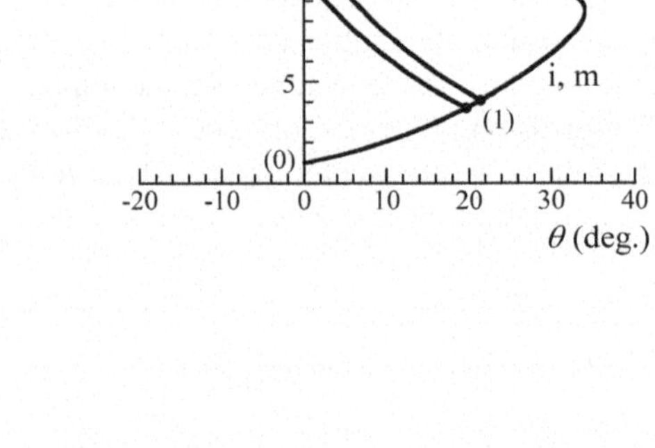

Fig. 7.14 Shock polars under critical conditions

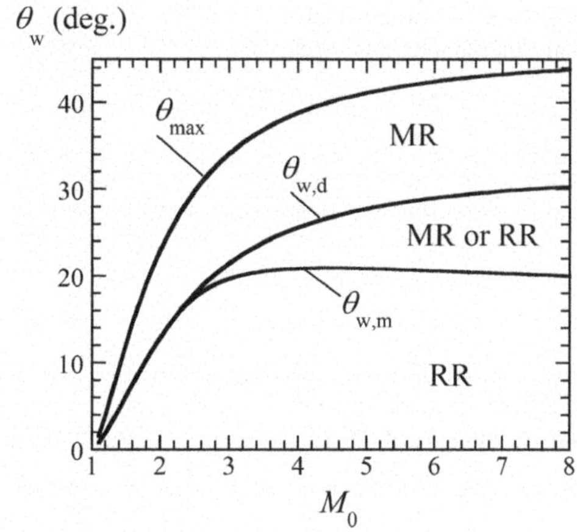

Fig. 7.15 Regimes of possible shock reflection patterns, $\gamma = 1.4$; MR, Mach reflection; RR, regular reflection

For a given flow Mach number M_0, the conditions for the respective reflection patterns are

Mach reflection:

$$\theta_{w,m} \leq \theta_w < \theta_{max} \tag{7.110}$$

Regular reflection:

$$0 \leq \theta_w < \theta_{w,d}. \tag{7.111}$$

Figure 7.15 maps the reflective domains. θ_{max} is the maximum deflection angle (see Sect. 4.3). There is no solution of an attached shock wave for $\theta_w > \theta_{max}$. With $\theta_{w,m} < \theta_{w,d}$, dual solutions exist in $\theta_{w,m} < \theta_w < \theta_{w,d}$. This is experimentally validated. In the case of the Mach reflection, the flow behind the Mach stem becomes subsonic, and thereby it is affected by the downstream condition. Which reflection pattern appears depends on boundary conditions and histories of the flow.

7.4.6 Shock Wave Reflection in Pseudo-Steady Flows

In this section, we deal with the reflection of a shock wave propagating in space over an object, e.g., a wedge. In Fig. 7.16, A planar shock wave with a shock Mach number, M_s, is reflected on a wedge of an apex angle of θ_w. In inviscid flow, if the wedge has an infinite length, a characteristic length cannot be defined, and thereby the shock reflection pattern becomes self-similar (*pseudo-steady flow*). Setting a coordinate originating at an appropriate point, a *self-similar solution* is obtained. The shock reflection that appears in pseudo-steady flows is categorized into various patterns [2].

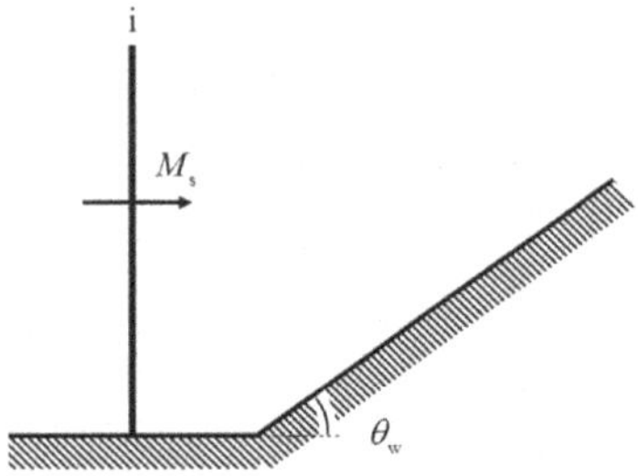

Fig. 7.16 A planar shock wave incident on a wedge

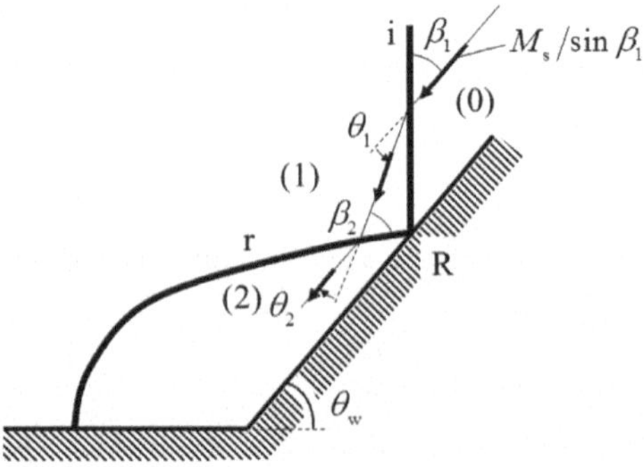

Fig. 7.17 Regular reflection in pseudo-steady flow

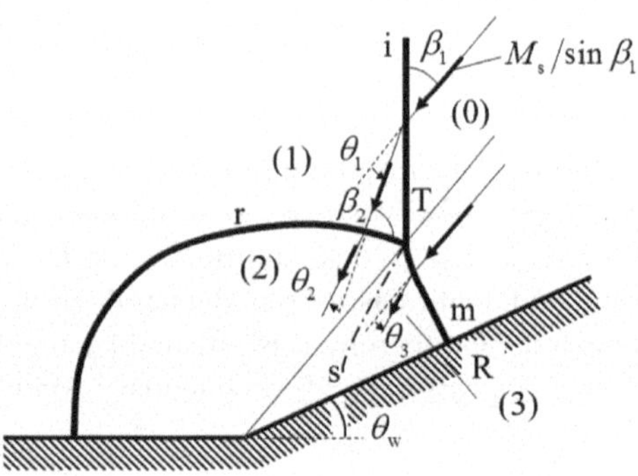

Fig. 7.18 Mach reflection in pseudo-steady flow

Figure 7.17 schematically illustrates regular reflection. At the coordinate of the origin that is set at the reflection point R, a supersonic flow with a Mach number of $M_s/\sin\beta_1$ enters with an angle of $\beta_1 = (\pi/2 - \theta_w)$ with respect to the shock wave front. The deflection angle in (2) equals that in (0).

Figure 7.18 shows the Mach reflection. A self-similar solution is obtained with an origin set at the triple point. In pseudo-steady flows, various irregular reflection patterns including the Mach reflection appear. In particular, with a shallow angle of θ_w and/or M_s close to unity, a reflection pattern that does not agree with the three-shock theory is observed. In this case, the incident shock wave is smoothly connected even at the intersection point with the reflected shock wave. The reflection pattern regimes of this unsteady flow on $M_0 - \theta_w$ coordinates do not necessarily agree with those of the steady flow.

7.5 Shock Wave—Boundary Layer Interaction

This book basically deals with inviscid flows. However, in many real flows, problems regarding the interaction between a shock wave and a boundary layer becomes important. In a boundary layer, the flow speed is low, and the pressure disturbance may propagate upstream; a shock wave over the boundary layer forms an inverse pressure gradient, which may extend upstream through the boundary layer. As a result, flow separation may be induced.

Figure 7.19 schematically illustrates the reflection of an oblique shock wave on a flat plate. Without the boundary layer (Fig. 7.19a), the incident shock wave is reflected on the wall. However, if a boundary layer exists (Fig. 7.19b), the inverse pressure gradient propagates back toward the upstream, inducing flow separation and a *separation bubble*. In front of the bubble, the flow is directed off the wall, accompanied by a *separation shock*, which is connected to the reflected shock wave. Behind the bubble, the flow reattaches with the accompanying compression waves. If the compression waves coalesce, an oblique shock wave appears (not shown in the figure). In the boundary layer, the flow Mach number varies from 0 (on the wall) to a supersonic value. Under the sonic line, the flow is subsonic, and the shock wave smears out.

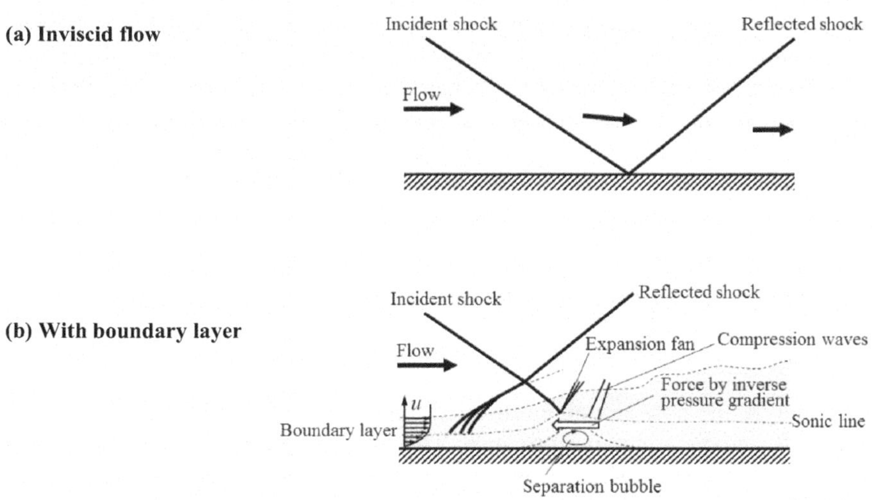

Fig. 7.19 Reflection of oblique shock wave on a plate

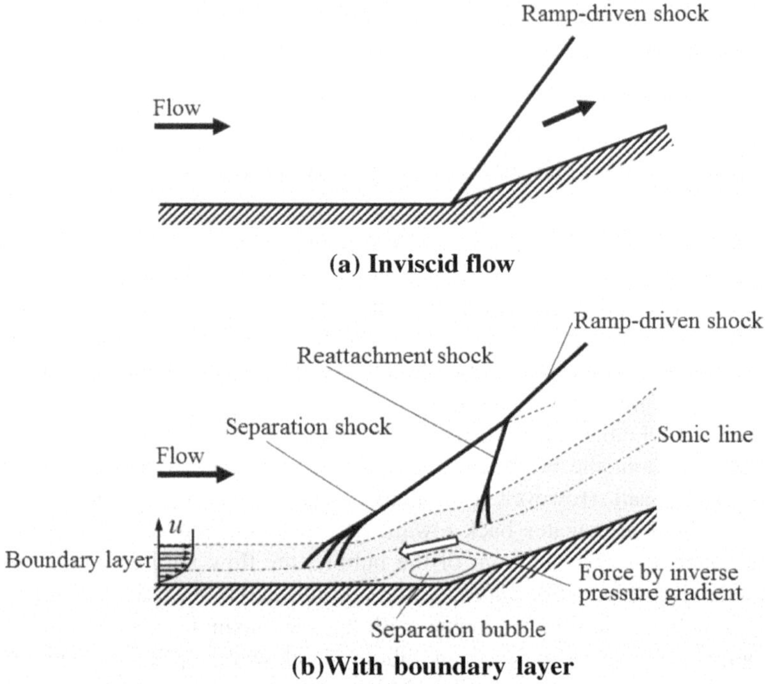

Fig. 7.20 Modification of supersonic flow over a ramp

Figure 7.20 illustrates the supersonic flows incident to a ramp. In inviscid flow (Fig. 7.20a), an oblique shock wave attaches to the corner. However, with a boundary layer (Fig. 7.20b), the flow is separated due to an effective, inverse pressure gradient being accompanied by a separation shock wave in front of the separation bubble. Downstream of the bubble, the flow reattaches, inducing a reattachment shock wave.

If such a shock wave boundary layer interaction is significant in the real flow, the effective streamline changes, possibly degrading the performance of aerodynamic devices, such as a supersonic intake. In these flows, instability often matters to a large degree. Therefore, the design should be carefully constructed.

7.6 Practice: Supersonic Flow Incident on an Inverted Triangle Wing

Let us analyze the flow around an inverted triangle wing in a supersonic flow, and evaluate its aerodynamics performance. This is good practice to deal with two-dimensional flow, where shock waves and/or expansion waves appear. The inverted isosceles triangle wing with a base angle of ϕ is in a uniform flow with a Mach number M_0 and an angle of attack α (Fig. 7.21). In the figure, under the leading wedge

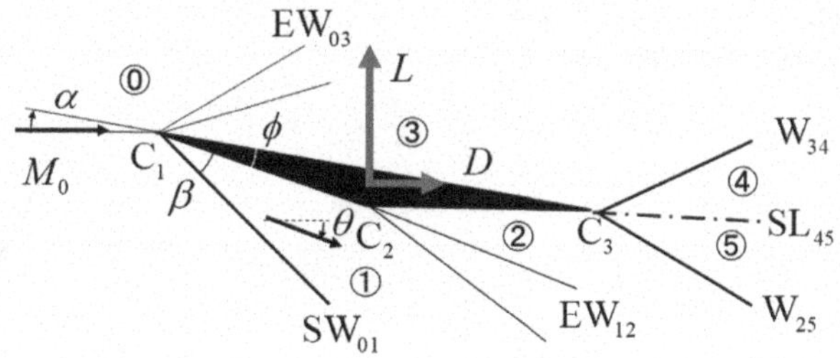

Fig. 7.21 Supersonic flow around an inverted triangle wing

(C_1), an oblique shock wave (SW_{01}) is generated. Here, the shock wave is assumed to be attached to the wing. At the apex corner (C_2), an expansion fan (EW_{12}) is formed. Above C_1, an expansion fan (EW_{03}) is formed. At the trailing edge (C_3), the two flows with a different angle of ϕ impinge on each other, forming a slip line (SL_{45}) across which the pressures are equal. Whether the waves W_{34} and W_{25} become a shock wave or an expansion fan depends on the mutual flow conditions.

The variation of the deflection angle, θ (which has a positive sign in the clockwise direction), past an expansion fan is defined as

$$\Delta\theta = \theta_{\mathrm{d}} - \theta_{\mathrm{u}} \tag{7.112}$$

where subscripts "u" and "d" indicate the upstream and downstream conditions, respectively. On the upper surface (3), an oblique shock wave is generated with $\Delta\theta < 0$, while an expansion fan is generated with $\Delta\theta > 0$ on the lower surface (1 and 2), and vice versa, respectively.

First, let us obtain the flow condition behind an oblique shock wave. The angle between the shock wave and the upstream flow, β, is obtained by implicitly solving (4.79).

$$\tan|\Delta\theta| = \frac{2\cot\beta\left(M_{\mathrm{u}}^2\sin^2\beta - 1\right)}{M_{\mathrm{u}}^2(\gamma + \cos 2\beta) + 2}. \tag{7.113}$$

Substituting (4.74) and (4.87) with β as the solution of (7.112), the downstream pressure and Mach number are obtained.

$$\frac{p_{\mathrm{d}}}{p_{\mathrm{u}}} = 1 + \frac{2\gamma}{\gamma + 1}\left(M_{\mathrm{u}}^2\sin^2\beta - 1\right) \tag{7.114}$$

$$M_{\mathrm{d}} = \frac{1}{\sin(\beta - |\Delta\theta|)}\left[\frac{(\gamma - 1)M_{\mathrm{u}}^2\sin^2\beta + 2}{2\gamma M_{\mathrm{u}}^2\sin^2\beta - (\gamma - 1)}\right]^{\frac{1}{2}}. \tag{7.115}$$

The downstream condition past an expansion fan is obtained as the solution of the implicit Eqs. (7.43) and (7.49).

$$|\Delta\theta| = \nu(M_{\mathrm{d}}) - \nu(M_{\mathrm{u}}) \tag{7.116}$$

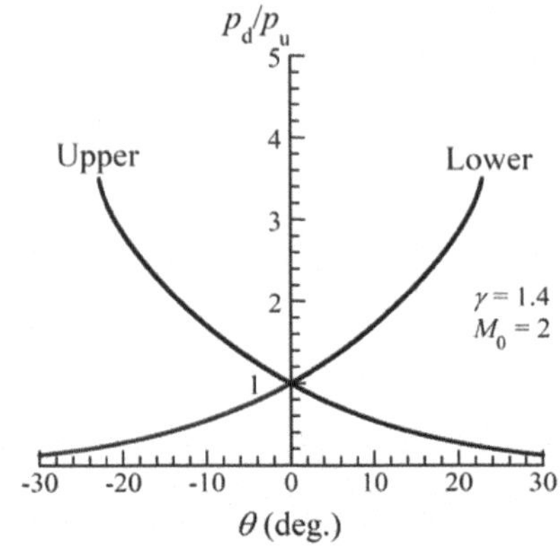

Fig. 7.22 Pressure ratio versus deflection angle, $\gamma = 1.4$, $M_u = 2.0$

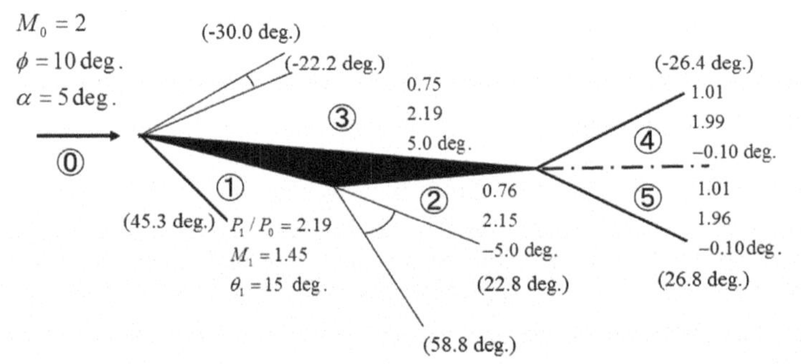

Fig. 7.23 Example of flow field, $\gamma = 1.4$, $M_0 = 2.0$, $\phi = 10°$, $\alpha = 5°$

$$\nu(M) \equiv \sqrt{\frac{\gamma+1}{\gamma-1}} \tan^{-1} \sqrt{\frac{\gamma-1}{\gamma+1}(M^2-1)} - \tan^{-1}\sqrt{M^2-1}. \tag{7.117}$$

The pressure ratio is obtained from the isentropic relation given by (7.57).

The above variations are graphically illustrated in Fig. 7.22. The regime of $p_d/p_u > 1$ corresponds to a shock wave, $p_d/p_u < 1$ to Prandtl–Meyer expansion. The curves for the upper and lower sides are symmetrical to each other, as the compression and expansion directions are opposite.

The example of the flow field is shown in Figs. 7.23 and 7.24. On the lower side of the wing, the flow is compressed by the oblique shock wave (1) and then expanded at the apex corner (2). On the upper side, the flow is expanded (3). Downstream of the trailing edge, the flow is separated by the slip line to regimes 4 (upper) and 5 (lower), which in turn are formed behind the respective oblique shock waves. Note that these flows have a small deflection angle of –0.10°, as shown in Fig. 7.24b.

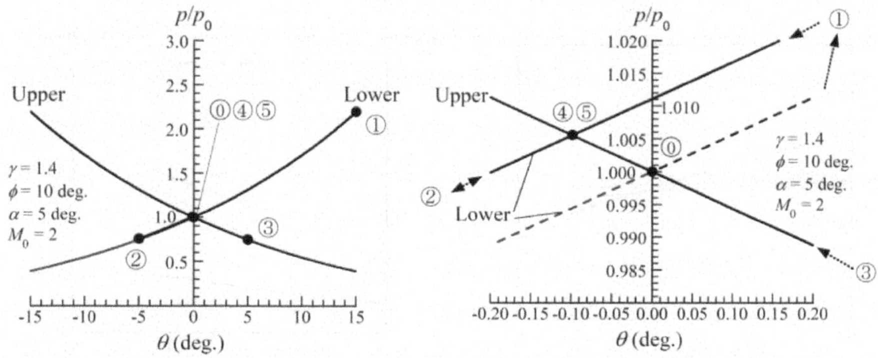

(a) Whole flow field **(b) Flow around trailing edge**

Fig. 7.24 Pressure versus deflection angle under the condition of Fig. 7.23

Fig. 7.25 Direction of forces on the respective surface

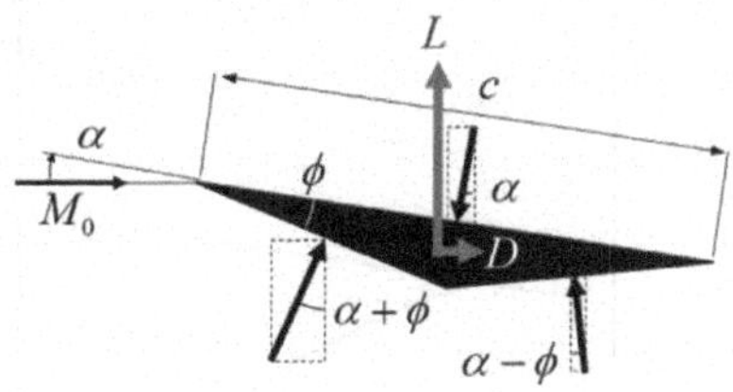

The aerodynamics performance of the wing is calculated with definitions shown in Fig. 7.25. The span and cord lengths of the wing are designated by b and c, respectively. With the forces exerted on the respective surface, the resultant force is decomposed to a drag D in the stream-wise direction and a lift L, (upward) perpendicular to it.

$$L = p_1 b \frac{c}{2\cos\phi}\cos(\alpha+\phi) + p_2 b \frac{c}{2\cos\phi}\cos(\alpha-\phi) - p_3 bc\cos\alpha \tag{7.118}$$

$$D = p_1 b \frac{c}{2\cos\phi}\sin(\alpha+\phi) + p_2 b \frac{c}{2\cos\phi}\sin(\alpha-\phi) - p_3 bc\sin\alpha \tag{7.119}$$

$$C_{\mathrm{L}} = \frac{L}{\frac{1}{2}\rho_0 U_0^2 bc} = \frac{2}{\gamma M_0^2 p_0}\left[p_1\frac{1}{2\cos\phi}\cos(\alpha+\phi) + p_2\frac{1}{2\cos\phi}\cos(\alpha-\phi) - p_3\cos\alpha\right] \tag{7.120}$$

$$C_{\mathrm{D}} = \frac{D}{\frac{1}{2}\rho_0 U_0^2 bc} = \frac{2}{\gamma M_0^2 p_0}\left[p_1\frac{1}{2\cos\phi}\sin(\alpha+\phi) + p_2\frac{1}{2\cos\phi}\sin(\alpha-\phi) - p_3\sin\alpha\right] \tag{7.121}$$

$$\frac{L}{D} = \frac{p_1\frac{1}{2\cos\phi}\cos(\alpha+\phi) + p_2\frac{1}{2\cos\phi}\cos(\alpha-\phi) - p_3\cos\alpha}{p_1\frac{1}{2\cos\phi}\sin(\alpha+\phi) + p_2\frac{1}{2\cos\phi}\sin(\alpha-\phi) - p_3\sin\alpha}. \tag{7.122}$$

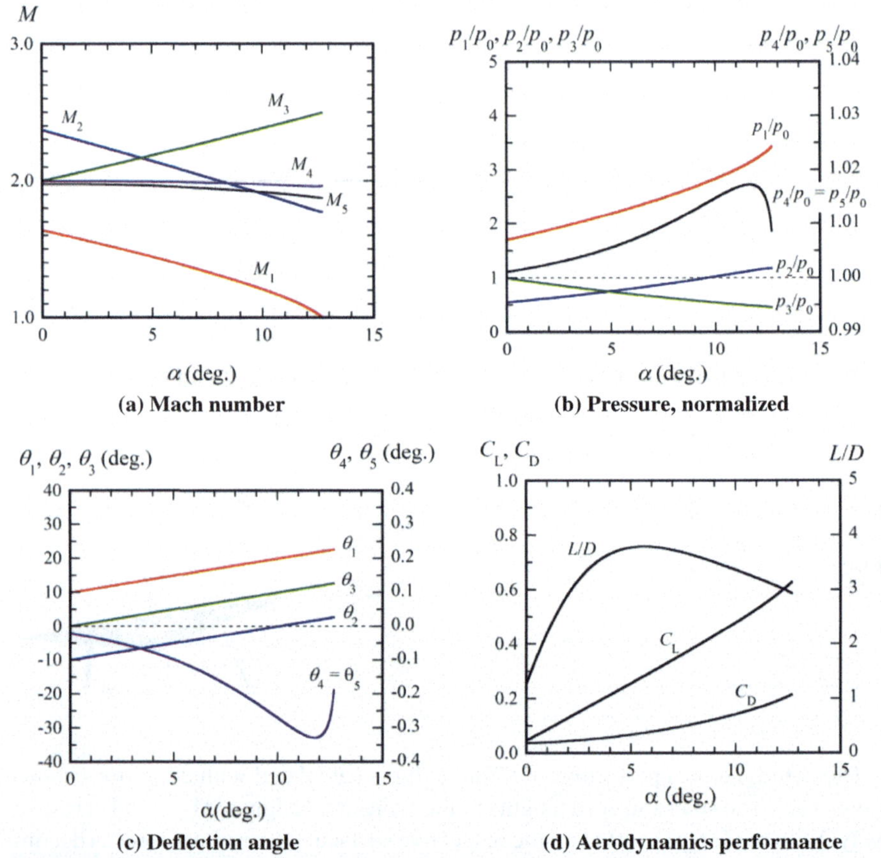

Fig. 7.26 Variation of flow and aerodynamic performance as a function of α, $\gamma = 1.4$, $M_0 = 2.0$, $\phi = 10°$

In the case of Fig. 7.23, $C_L = 0.255$, $C_D = 0.068$, and $L/D = 3.8$. The aerodynamics performance for $M_0 = 2.0$ is shown in Fig. 7.26. In the figures, only solutions where the flow is supersonic in the entire regime are shown.

References

1. Taylor GI, Maccoll JW (1933) Proc R Soc Lond A 139:278–311
2. For further details, refer to Ben-Dor G (2007) Shock wave reflection phenomena, 2nd edn., Chap. 3. Springer

Chapter 8
Unsteady, One-Dimensional Flows

Pressure waves appearing in this book are categorized in Table 8.1. The density of gas is increased by the *compression* wave, and decreased by the *expansion* wave. A shock wave is a special type of compression wave, and it is thus dealt with differently. A *sound wave* is an infinitesimally weak wave, in which the time average of the flow velocity and thermodynamic properties remain unchanged. The *speed of sound* is a fundamental quantity obtained in characteristic velocities of wave propagation, which will be formulated for unsteady, one-dimensional flow in this chapter. As described in the Appendix, this formulation can be extended to three-dimensional flow.

8.1 Sound Wave

Generally, sound signifies a pressure oscillation of air in an audible frequency band (from 20 to 20 kHz). The amplitude, frequency, and its spectra determine the sound pressure, pitch, and tone color. With the pressure oscillation, the density also oscillates while their time-averaged values vary only with much longer time scales compared to the oscillation period. As shown in Fig. 8.1, the amplitude of the oscillation is locally constant in a spatiotemporal sense, and at most on the order of one ten thousands.[1] Therefore, this variation is treated as a *disturbance*. In compressible fluid dynamics, the *sound wave* is fundamental pressure wave, which propagates pressure and density disturbances. As long as the fluctuations are treated as disturbances, there are no constraints in their frequency ranges, and they can be superimposed onto each other, such that the waves are linear.

[1]The sound pressure level is defined by $L_P = 20 \log\{(\Delta p)_{rms}/p_0\}$ [dB], where $(\Delta p)_{\rm rms}$ is the effective sound pressure, the root-mean-square of the pressure oscillation, and $p_0 = 2 \times 10^{-5}$ [Pa]. At $L_P = 120$ [dB], we cannot conduct daily conversation with $(\Delta p)_{\rm rms}$=20 [Pa], which is one five thousands of the atmospheric pressure.

A. Sasoh, *Compressible Fluid Dynamics and Shock Waves*,
https://doi.org/10.1007/978-981-15-0504-1_8

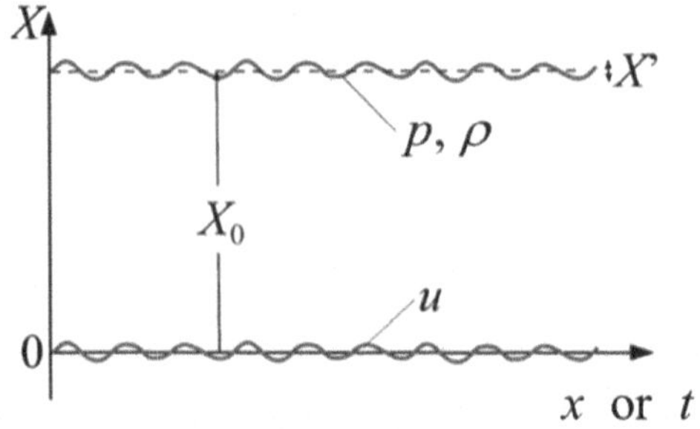

Fig. 8.1 Oscillation of X (pressure, density, or flow velocity) with respect to space, x, or time, t

Table 8.1 Categorization of pressure waves

	Expansion wave	Sound wave	Compression wave	Shock wave
Linearity	Nonlinear	Linear	Nonlinear	
Density variation	$\Delta\rho < 0$		$\Delta\rho > 0$	$\Delta\rho > 0$
Entropy variation, Δs	0			$O((\Delta p)^3) > 0$
Variation in flow parameters	Continuous	Time average, unchanged	Continuous	Discontinuous

Changes in the density of a medium induce pressure variations, which propagate toward the surroundings. In adiabatic processes, if part of a gas is compressed, its pressure is increased, thereby exerting force on the surrounding gas in order for it to expand. If expanded, the pressure is decreased, thereby being compressed by the surroundings (Fig. 8.2). These scenarios apply even to condensed matter, liquid, or solid. The speed of sound is a function of thermodynamic properties, in a calorically perfect gas only of a temperature. In a uniform gas, the speed of sound is likewise uniform. As shown in Fig. 1.2, a preceding sound wave is not caught up with the proceeding sound waves. The sound wave is an infinitesimally weak pressure wave.

Fig. 8.2 Generation and propagation of sound waves

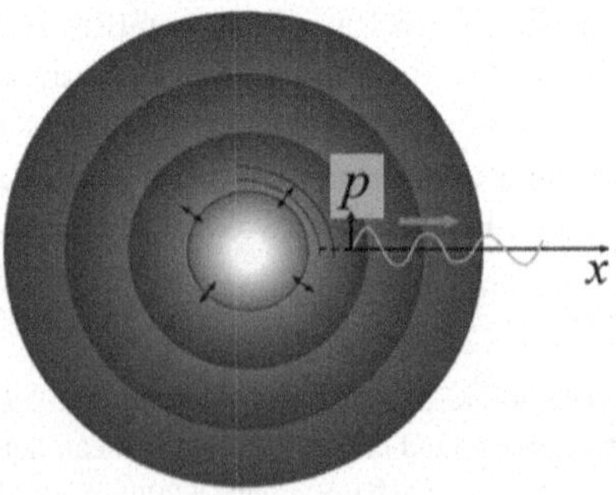

8.2 Characteristic Velocity and Invariants

Here, we derive characteristic velocity and invariants in unsteady, one-dimensional flow (in x-direction). Volume force and heat conduction are not taken into account. The flow is assumed to be isentropic. The x-components of the mass (3.4) and momentum (3.15) conservation equations are as follows:

$$\text{Mass conservation}: \quad \frac{\partial \rho}{\partial t} + u\frac{\partial \rho}{\partial x} + \rho\frac{\partial u}{\partial x} = 0 \tag{8.1}$$

$$\text{Momentum conservation}: \quad \frac{\partial u}{\partial t} + u\frac{\partial u}{\partial x} + \frac{1}{\rho}\frac{\partial p}{\partial x} = 0. \tag{8.2}$$

Energy conservation (isentropic flow):

$$T\mathrm{d}s = \mathrm{d}e + p\mathrm{d}\left(\frac{1}{\rho}\right) = \mathrm{d}h - \frac{1}{\rho}\mathrm{d}p = 0. \tag{8.3}$$

Applying the equation of state of an ideal gas (2.3) and that of a calorically perfect gas (2.97),

$$T\mathrm{d}s = C_p\mathrm{d}T - \frac{1}{\rho}\mathrm{d}p = C_p\mathrm{d}\left(\frac{p}{\rho R}\right) - \frac{1}{\rho}\mathrm{d}p = \frac{p}{\rho R}\left[(C_p - R)\frac{\mathrm{d}p}{p} - C_p\frac{\mathrm{d}\rho}{\rho}\right] = 0. \tag{8.4}$$

From (8.4),

$$\left(\frac{\partial p}{\partial \rho}\right)_s = \frac{C_p}{C_p - R}\frac{p}{\rho} = \frac{C_p}{C_v}\frac{p}{\rho} = \gamma\frac{p}{\rho} = \gamma RT. \tag{8.5}$$

We will see in the following that a which is defined by

$$a \equiv \sqrt{\left(\frac{\partial p}{\partial \rho}\right)_s} \tag{8.6}$$

has a physical meaning as the *speed of sound*. For a calorically perfect gas, this is in proportion to the square root of the temperature. For example, a monoatomic gas has a small number of degrees of freedom, such that the portion of translational motion is large. Therefore, with a large value of γ, a is high. Moreover, with a small atomic mass, for example, hydrogen and helium, a is likewise high, as illustrated in Table 8.2.

For isentropic flow, (8.6) is expressed as a substantial derivative.

$$\mathrm{d}p = a^2\mathrm{d}\rho \tag{8.7}$$

Table 8.2 Speed of sound in a calorically perfect gas

Species	Molecular mass [g/mol]	γ	a [m/s] $T = 288$ K	a [m/s] $T = 400$ K
H_2	2.0	1.4	1290	1530
He	4.0	1.67	1000	1180
Air	29	1.4	340	400
Ar	40	1.67	320	370

$$\frac{\partial p}{\partial t} + u\frac{\partial p}{\partial x} - a^2\left(\frac{\partial \rho}{\partial t} + u\frac{\partial \rho}{\partial x}\right) = 0. \tag{8.8}$$

Combining (8.1) and (8.8),

$$\frac{\partial p}{\partial t} + u\frac{\partial p}{\partial x} + \rho a^2\frac{\partial u}{\partial x} = 0. \tag{8.9}$$

The conservation Eqs. (8.1), (8.2), and (8.9) are combined as

$$\frac{\partial \mathbf{V}}{\partial t} + \mathbf{A}\frac{\partial \mathbf{V}}{\partial x} = \mathbf{0} \tag{8.10}$$

$$\mathbf{V} = \begin{pmatrix} \rho \\ u \\ p \end{pmatrix}, \quad \mathbf{A} = \begin{pmatrix} u & \rho & 0 \\ 0 & u & \frac{1}{\rho} \\ 0 & \rho a^2 & u \end{pmatrix}, \quad \mathbf{0} = \begin{pmatrix} 0 \\ 0 \\ 0 \end{pmatrix}. \tag{8.11}$$

Let us assume a wave-form solution

$$\mathbf{V} = \bar{\mathbf{V}}e^{i\phi(x,\,t)}, \qquad (i, \text{ imaginary unit}). \tag{8.12}$$

Substituting (8.10) with (8.12),

$$(\omega\mathbf{I} - k\mathbf{A})\bar{\mathbf{V}} = \mathbf{0}, \ (\mathbf{I}, \text{ unit vector}) \tag{8.13}$$

$$\omega \equiv \frac{\partial \phi}{\partial t} \tag{8.14}$$

$$k = -\frac{\partial \phi}{\partial x}. \tag{8.15}$$

Multiplying (8.13) with $1/k$, and defining the phase velocity, λ,

$$\lambda \equiv \frac{\omega}{k} \tag{8.16}$$

$$(\mathbf{A} - \lambda\mathbf{I})\bar{\mathbf{V}} = \mathbf{0}. \tag{8.17}$$

In order for $\bar{\mathbf{V}}$ to have nontrivial solutions, the following *characteristic equation* should hold.

$$|\mathbf{A} - \lambda \mathbf{I}| = 0. \tag{8.18}$$

Since

$$\mathbf{A}\bar{\mathbf{V}} = \lambda\,\bar{\mathbf{V}} \tag{8.19}$$

λ and $\bar{\mathbf{V}}$ become the eigenvalue and eigenvector of the matrix $\mathbf{A}$, respectively. As a 3×3 matrix, $\mathbf{A}$ has three eigenvalues, $\lambda_1, \lambda_2, \lambda_3$, and the corresponding eigenvectors, $\bar{\mathbf{V}}_1$, $\bar{\mathbf{V}}_2$, $\bar{\mathbf{V}}_3$. Defining

$$\mathbf{\Lambda} = \begin{pmatrix} \lambda_1 & 0 & 0 \\ 0 & \lambda_2 & 0 \\ 0 & 0 & \lambda_3 \end{pmatrix} \tag{8.20}$$

$$\mathbf{L} = \left(\bar{\mathbf{V}}_1\,,\ \bar{\mathbf{V}}_2\,,\ \bar{\mathbf{V}}_3\right). \tag{8.21}$$

Equation yields

$$\mathbf{AL} = \mathbf{L}\Lambda \quad \text{or} \quad \mathbf{\Lambda} = \mathbf{L}^{-1}\mathbf{AL}. \tag{8.22}$$

From (8.18),

$$\begin{vmatrix} u - \lambda & \rho & 0 \\ 0 & u - \lambda & \frac{1}{\rho} \\ 0 & \rho a^2 & u - \lambda \end{vmatrix} = 0$$

$$(\lambda - u - a)(\lambda - u + a)(\lambda - u) = 0. \tag{8.23}$$

Therefore, the solutions are

$$\begin{cases} \lambda_1 = u + a \\ \lambda_2 = u - a\,. \\ \lambda_3 = u \end{cases} \tag{8.24}$$

With (8.19) and

$$\bar{\mathbf{V}} \equiv \begin{pmatrix} X_1 \\ X_2 \\ X_3 \end{pmatrix} \tag{8.25}$$

$$\begin{cases} uX_1 + \rho X_2 \quad = \lambda X_1 \\ \quad uX_2 \quad + \dfrac{1}{\rho} X_3 = \lambda X_2 \\ \quad \rho a^2 X_2 + uX_3 = \lambda X_3 \end{cases}. \tag{8.26}$$

Therefore, with $\lambda = u \pm a$,

$$\begin{cases} uX_1 + \rho X_2 \qquad = (u \pm a) X_1 \\ \quad uX_2 \quad + \dfrac{1}{\rho} X_3 = (u \pm a) X_2 \\ \quad \rho a^2 X_2 + uX_3 = (u \pm a) X_3 \end{cases}$$

$$\begin{cases} \rho X_2 = \pm a X_1 \\ X_3 = \pm \rho a X_2 \end{cases} \quad \text{(double sign corresponds).}$$

The eigenvectors are

$$\text{For } \lambda_1 = u + a, \quad \bar{\mathbf{V}}_1 = \begin{pmatrix} \rho \\ a \\ \rho a^2 \end{pmatrix}, \tag{8.27}$$

$$\text{For } \lambda_2 = u - a, \quad \bar{\mathbf{V}}_2 = \begin{pmatrix} \rho \\ -a \\ \rho a^2 \end{pmatrix}. \tag{8.28}$$

For $\lambda = u$, (8.26) yields

$$\begin{cases} uX_1 + \rho X_2 \qquad = u X_1 \\ \quad uX_2 \quad + \dfrac{1}{\rho} X_3 = uX_2 \\ \quad \rho a^2 X_2 + uX_3 = u X_3 \end{cases}.$$

Therefore,

$$\begin{cases} X_1, \text{ arbitrary} \\ X_2 = 0 \\ X_3 = 0 \end{cases}.$$

Therefore, with generality,

$$\text{For } \lambda_3 = u, \quad \mathbf{V}_3 = \begin{pmatrix} 1 \\ 0 \\ 0 \end{pmatrix}. \tag{8.29}$$

Substituting (8.21) with (8.27) to (8.29),

$$\mathbf{L} = \left(\bar{\mathbf{V}}_1, \bar{\mathbf{V}}_2, \bar{\mathbf{V}}_3\right) = \begin{pmatrix} \rho & \rho & 1 \\ a & -a & 0 \\ \rho a^2 & \rho a^2 & 0 \end{pmatrix} \tag{8.30}$$

$$\mathbf{L}^{-1} = \frac{1}{2\rho a^3}\begin{pmatrix} 0 & \rho a^2 & a \\ 0 & -\rho a^2 & a \\ 2\rho a^3 & 0 & -2\rho a \end{pmatrix} = \begin{pmatrix} 0 & \frac{1}{2a} & \frac{1}{2\rho a^2} \\ 0 & -\frac{1}{2a} & \frac{1}{2\rho a^2} \\ 1 & 0 & -\frac{1}{a^2} \end{pmatrix}. \tag{8.31}$$

Multiplying (8.10) with (8.31),

$$\mathbf{L}^{-1}\frac{\partial \mathbf{V}}{\partial t} + \mathbf{L}^{-1}\mathbf{A}\frac{\partial \mathbf{V}}{\partial x} = \mathbf{0} \tag{8.32}$$

$$\mathbf{L}^{-1}\frac{\partial \mathbf{V}}{\partial t} = \begin{pmatrix} 0 & \frac{1}{2a} & \frac{1}{2\rho a^2} \\ 0 & -\frac{1}{2a} & \frac{1}{2\rho a^2} \\ 1 & 0 & -\frac{1}{a^2} \end{pmatrix}\begin{pmatrix} \frac{\partial \rho}{\partial t} \\ \frac{\partial u}{\partial t} \\ \frac{\partial p}{\partial t} \end{pmatrix} = \begin{pmatrix} \frac{1}{2a}\frac{\partial u}{\partial t} + \frac{1}{2\rho a^2}\frac{\partial p}{\partial t} \\ -\frac{1}{2a}\frac{\partial u}{\partial t} + \frac{1}{2\rho a^2}\frac{\partial p}{\partial t} \\ \frac{\partial \rho}{\partial t} - \frac{1}{a^2}\frac{\partial p}{\partial t} \end{pmatrix} \tag{8.33}$$

$$\mathbf{L}^{-1}\mathbf{A}\frac{\partial \mathbf{V}}{\partial x} = \begin{pmatrix} 0 & \frac{1}{2a} & \frac{1}{2\rho a^2} \\ 0 & -\frac{1}{2a} & \frac{1}{2\rho a^2} \\ 1 & 0 & -\frac{1}{a^2} \end{pmatrix}\begin{pmatrix} u & \rho & 0 \\ 0 & u & \frac{1}{\rho} \\ 0 & \rho a^2 & u \end{pmatrix}\begin{pmatrix} \frac{\partial \rho}{\partial x} \\ \frac{\partial u}{\partial x} \\ \frac{\partial p}{\partial x} \end{pmatrix} = \begin{pmatrix} \frac{u+a}{2a}\frac{\partial u}{\partial x} + \frac{u+a}{2\rho a^2}\frac{\partial p}{\partial x} \\ \frac{-u+a}{2a}\frac{\partial u}{\partial x} + \frac{u-a}{2\rho a^2}\frac{\partial p}{\partial x} \\ u\frac{\partial \rho}{\partial x} - \frac{u}{a^2}\frac{\partial p}{\partial x} \end{pmatrix}. \tag{8.34}$$

Therefore,

$$\begin{pmatrix} \frac{\partial u}{\partial t} + (u+a)\frac{\partial u}{\partial x} + \frac{1}{\rho a}\left\{\frac{\partial p}{\partial t} + (u+a)\frac{\partial p}{\partial x}\right\} \\ \frac{\partial u}{\partial t} + (u-a)\frac{\partial u}{\partial x} - \frac{1}{\rho a}\left\{\frac{\partial p}{\partial t} + (u-a)\frac{\partial p}{\partial x}\right\} \\ \frac{\partial \rho}{\partial t} + u\frac{\partial \rho}{\partial x} - \frac{1}{a^2}\left(\frac{\partial p}{\partial t} + u\frac{\partial p}{\partial x}\right) \end{pmatrix} = \begin{pmatrix} 0 \\ 0 \\ 0 \end{pmatrix}. \tag{8.35}$$

The variation in $Y = Y(t, x)$ by tracing the velocity c is expressed by

$$\frac{\mathrm{d}Y}{\mathrm{d}t} = \frac{\partial Y}{\partial t} + c\frac{\partial Y}{\partial x} \tag{8.36}$$

$$c \equiv \frac{\mathrm{d}x}{\mathrm{d}t}. \tag{8.37}$$

Applying this to (8.35), the following relations are obtained:

$$\text{Along } \lambda_1 \equiv c_+ = u + a \text{ (characteristic C}_+\text{)}, \ \mathrm{d}J_+ \equiv \mathrm{d}u + \frac{1}{\rho a}\mathrm{d}p = 0 \tag{8.38}$$

$$\text{Along } \lambda_2 \equiv c_- = u - a \text{ (characteristic C}_-\text{)}, \ \mathrm{d}J_- \equiv \mathrm{d}u - \frac{1}{\rho a}\mathrm{d}p = 0 \tag{8.39}$$

$$\text{Along } \lambda_3 \equiv c_0 = u \text{ (characteristic C}_0\text{)}, \quad \mathrm{d}\rho - \frac{1}{a^2}\mathrm{d}p = 0 \text{ that is } \mathrm{d}J_0 = 0, \quad J_0 \equiv s \tag{8.40}$$

c is a *characteristic velocity*, and J is a *Riemann invariant*. In (8.40), the characteristic velocity equals to the flow velocity, along which the entropy is kept constant. For calorically perfect gas,

$$e = \frac{1}{\gamma - 1}\frac{p}{\rho}. \tag{8.41}$$

From (8.3),

$$\frac{1}{\gamma - 1}\mathrm{d}\left(\frac{p}{\rho}\right) + p\mathrm{d}\left(\frac{1}{\rho}\right) = 0$$

$$\frac{\mathrm{d}p}{p} = \gamma\frac{\mathrm{d}\rho}{\rho}. \tag{8.42}$$

From (8.6) and (8.42),

$$a^2 = \gamma\frac{p}{\rho}$$

$$\mathrm{d}a^2 = \gamma\mathrm{d}\frac{p}{\rho} = \gamma\left(\frac{1}{\rho}\mathrm{d}p - \frac{p}{\rho^2}\mathrm{d}\rho\right).$$

Substituting with (8.42),

$$\frac{\mathrm{d}p}{\rho a} = \frac{2\mathrm{d}a}{\gamma - 1}. \tag{8.43}$$

Therefore, (8.38) and (8.39) are transformed to explicit forms:

$$dJ_+ = 0, \quad J_+ \equiv u + \frac{2a}{\gamma - 1} \text{ along } \lambda_1 \equiv c_+ = u + a \text{ (characteristic C}_+\text{)} \tag{8.44}$$

$$dJ_- = 0, \quad J_- \equiv u - \frac{2a}{\gamma - 1} \text{ along } \lambda_2 \equiv c_- = u - a\text{(characteristic C}_-\text{)}. \tag{8.45}$$

Appendix 8.1 Isentropic Compressibility
An *isentropic compressibility*, κ, equals to the rate of variation in a specific volume with respect to a pressure.

$$\kappa \equiv -\frac{1}{\upsilon}\left(\frac{\partial \upsilon}{\partial p}\right)_s \tag{8.46}$$

Since

$$\rho = \frac{1}{\upsilon}, \tag{8.47}$$

$$d\upsilon = -\frac{1}{\rho^2}d\rho \tag{8.48}$$

$$\kappa = -\rho\frac{d\upsilon}{d\rho}\left(\frac{\partial \rho}{\partial p}\right)_s = -\rho\left(-\frac{1}{\rho^2}\right)\left(\frac{\partial \rho}{\partial p}\right)_s = \frac{1}{\rho}\left(\frac{\partial \rho}{\partial p}\right)_s. \tag{8.49}$$

Using κ, a is given by

$$a = \sqrt{\left(\frac{\partial p}{\partial \rho}\right)_s} = \frac{1}{\sqrt{\kappa\rho}}. \tag{8.50}$$

8.3 Compression Wave

Let a piston start to move in quiescent air in the x-direction (Fig. 8.3). When the piston slowly starts moving, it compresses the air in front of it; the compression waves C_+ propagate to the right, such that a flow velocity $u(> 0)$ is induced. Although an infinite number of compression waves is induced, only small numbers are shown in Fig. 8.3. At $t = t_1$, the piston velocity is equal to u_p with the pressure distribution as shown. A compression wave C_+ propagates at a local characteristic velocity,

$$c_+ = u + a \tag{8.51}$$

Because u and a vary with respect to (x, t), c_+ varies also spatiotemporally. The leading wave "a" is generated when the piston starts to move at $t = 0$. In front of "a", the air is quiescent without information on the piston movement. The flow velocity u to the right increases while approaching the piston, and is equal to u_p at the piston (wave "c").

In contrast, C_- waves (broken lines) propagate toward the piston from the right. Since they propagate from the quiescent region (0), they have the same invariant $J_{-,\infty}$. Let us obtain the flow variables from the flow velocity, u_b, at "b." From, (8.45)

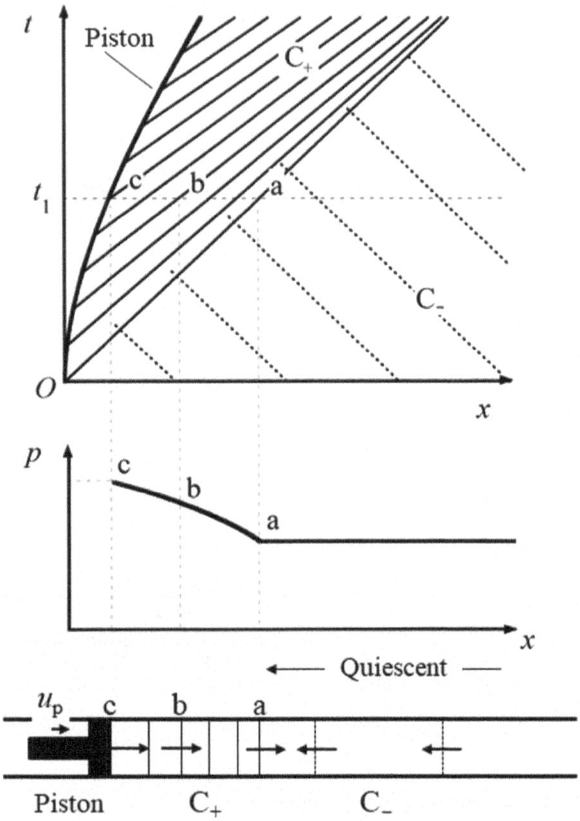

Fig. 8.3 Wave diagram and pressure distribution induced by a piston slowly accelerated to a velocity $u_p(>0)$ at $t = t_1$

$$u - \frac{2a}{\gamma - 1} = -\frac{2a_0}{\gamma - 1}$$

$$a = a_0 + \frac{\gamma - 1}{2} u. \tag{8.52}$$

Using the isentropic relation (2.91),

$$p = p_0 \left(\frac{a}{a_0} \right)^{\frac{2\gamma}{\gamma - 1}} = p_0 \left(1 + \frac{\gamma - 1}{2a_0} u \right)^{\frac{2\gamma}{\gamma - 1}}. \tag{8.53}$$

As seen in (8.53), the pressure is an increasing function of the flow velocity. Approaching the piston, the flow velocity and then the pressure gradually increase.

The propagation velocity of C_+ is given by

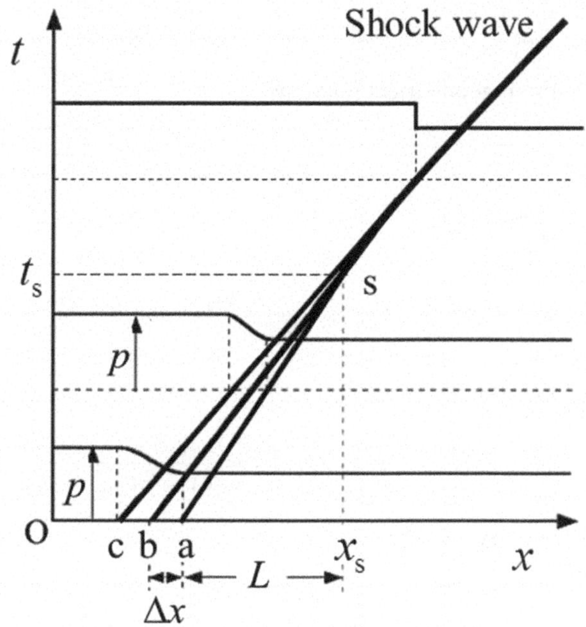

Fig. 8.4 Formation of a shock wave from coalesced compression waves

$$c_{+} = u + a_0 + \frac{\gamma - 1}{2}u = a_0 + \frac{\gamma + 1}{2}u = \left\{\frac{\gamma + 1}{2}\left(\frac{p}{p_0}\right)^{\frac{\gamma-1}{2\gamma}} - 1\right\}\frac{2}{\gamma - 1}a_0. \tag{8.54}$$

Because $(\gamma - 1)/(2\gamma) > 0$, the higher the pressure, the larger c_+ becomes. That is, *in compression waves the propagation velocity is higher at behind than at in front*. As shown in Fig. 8.4, as time elapses, the separation distance between the compression waves decreases; and the pressure gradient increases accordingly. Eventually, waves catch up with the preceding waves, and thereby invariants corresponding to the respective characteristics break up mathematically. Such inconsistencies are solved by the formation of a shock wave. In other words, if plural compression waves coalesce, the waves cannot keep their smooth variation, thereby making the transition to a shock wave. At the same time, the assumption of isentropic processes is broken, and the entropy increases behind the shock wave.

Let us obtain the *shock wave formation distance*, L, in Fig. 8.4 by considering the intersection of compression waves that originate from points "a" $(x_a, 0)$ and "b" $(x_b, 0)$ at $t = 0$. Their propagation velocities are

$$c_{+,a} = u_a + a_a = a_0 \tag{8.55}$$

$$c_{+,b} = u_b + a_b \tag{8.56}$$

$$\Delta x = x_a - x_b. \tag{8.57}$$

From geometrical relations,

$$t_s = \frac{x_s - x_a}{c_{+,a}} = \frac{x_s - x_b}{c_{+,b}} = \frac{x_s - x_a + \Delta x}{c_{+,b}}. \tag{8.58}$$

Here

$$J_{-,0} = u_a - \frac{2a_a}{\gamma - 1} = u_b - \frac{2a_b}{\gamma - 1} = -\frac{2a_0}{\gamma - 1}. \tag{8.59}$$

From (8.55) to (8.59),

$$L \equiv x_s - x_a = \frac{c_{+,a}}{c_{+,b} - c_{+,a}} \Delta x = \frac{2}{\gamma+1} \frac{a_0}{u_b} \Delta x. \tag{8.60}$$

With $\Delta x \to 0$,

$$\frac{u_a - u_b}{\Delta x} = \frac{0 - u_b}{\Delta x} = -\frac{u_b}{\Delta x} \to \left(\frac{\partial u}{\partial x}\right)_{t=0} \tag{8.61}$$

$$L = -\frac{2}{\gamma+1} \frac{a_0}{\left(\frac{\partial u}{\partial x}\right)_{t=0}} \tag{8.62}$$

$$t_s = \frac{L}{a_0} = -\frac{2}{\gamma + 1} \frac{1}{\left(\frac{\partial u}{\partial x}\right)_{t=0}}. \tag{8.63}$$

Equation (8.62) states that L is inversely proportional to the flow velocity gradient. In practice, the pressure p is measured rather than the flow velocity u. From (8.45) and (8.53),

$$\Delta u = \frac{2}{\gamma - 1} \Delta a \cong \frac{2}{\gamma - 1} \frac{\gamma - 1}{2\gamma} \frac{\Delta p}{p} a = \frac{a}{\gamma} \frac{\Delta p}{p}. \tag{8.64}$$

Combining (8.62) and (8.64),

$$L = -\frac{2\gamma}{\gamma+1} \frac{p_0}{\left(\frac{\partial p}{\partial x}\right)_{t=0}} \tag{8.65}$$

L is inversely proportional to the pressure gradient as well. In many cases, the pressure is measured at a fixed point. From (8.54), the time derivative of the pressure measured at point "b" is given by

$$\left(\frac{\partial p}{\partial t}\right)_{t=0} = -(u_b + a_b)\left(\frac{\partial p}{\partial x}\right)_{t=0} = -\left\{\frac{\gamma + 1}{2}\left(\frac{p_b}{p_0}\right)^{\frac{\gamma-1}{2\gamma}} - 1\right\} \frac{2a_0}{\gamma - 1}\left(\frac{\partial p}{\partial x}\right)_{t=0}$$

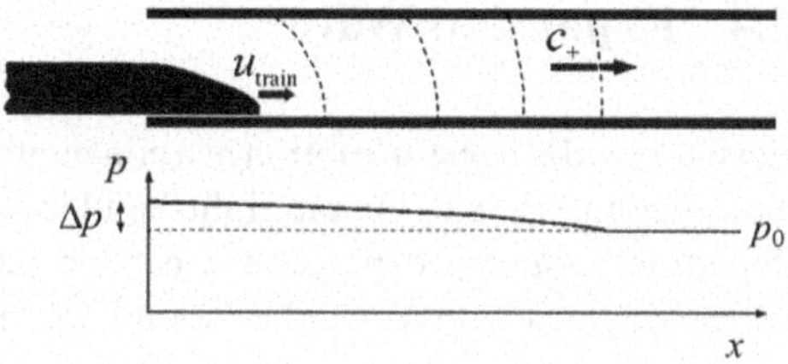

Fig. 8.5 Generation of compression waves by a high-speed train entering a tunnel

$$\rightarrow \quad -a_0\left(\frac{\partial p}{\partial x}\right)_{t=0} \quad \text{as} \quad p_b \rightarrow p_0. \tag{8.66}$$

Using (8.66), (8.65) is transformed to

$$L = \frac{2\gamma}{\gamma+1}\frac{p_0 a_0}{\left(\frac{\partial p}{\partial t}\right)_{t=0}} \tag{8.67}$$

L is in inverse proportion to the time derivative of the pressure.

When a high-speed train enters a tunnel, the train acts as a "leaky piston," generating compression waves (Fig. 8.5). Assume that a train with a nose length of $l = 20$ m enters the tunnel with a speed of $u_{\text{train}} = 100$ m/s (360 km/h). Do the generated compression waves make the transition to a shock wave? We assume that the pressure increases by 1 kPa $\Delta p/p_0 = 1\text{kPa}/100\text{kPa} = 0.01$ after the nose enters the tunnel. The period of the entry is $l/u_{\text{train}} = 20\,[\text{m}]/100\,[\text{m/s}] = 0.2\,[\text{s}]$. During that period, the leading wave advances by $a_0 l/u_{\text{train}} = 340\,[\text{m/s}] \times 0.2\,[\text{s}] = 68\,[\text{m}]$. Substituting these values with (8.65),

$$L = \frac{2\times 1.4}{1.4+1}\frac{100\,[\text{kPa}]}{\left(\frac{1\,[\text{kPa}]}{68\,[\text{m}]}\right)} \cong 7.9 \times 10^3\,[\text{m}] = 7.9\,[\text{km}]. \tag{8.68}$$

Based on this estimation, the compression waves make the transition to a shock wave in a tunnel longer than this value. If the shock wave emits from the tunnel exit, it is accompanied by a *tunnel sonic boom* [1]. However, in practice, a boundary layer grows in the long tunnel, attenuating the shock wave. To avoid the tunnel sonic boom, the effective pressure gradient upon entry should be minimized.

As will be shown in the following sections, when compression waves make the transition to a shock wave, their characteristics greatly change with a large accompanying force, impulse, or noise. To utilize such characteristics, L should be small, whereas it should be large to avoid the formation of a shock wave that can cause various hazards and noises.

8.4 Expansion Wave

Let us consider a piston motion at a constant velocity $u_p(<0)$ in the opposite direction, as shown in Fig. 8.6. On the right-hand side of the piston, the air expands due to the depletion, inducing expansion waves propagating off the piston. The leading wave "a" propagates with a speed of sound, a_0, in the quiescent air. On the left-hand side of "a," the expansion waves that originate in the origin propagate radially on the *x-t* coordinates, thereby forming an *expansion fan*. On the wave "d", which is closest to the piston, the flow velocity equals to u_p. In the regime between the wave "d" and the piston, the flow is uniform with the velocity

$$c_{+,d} = a_0 + \frac{\gamma + 1}{2} u_p.$$

The waves in the expansion fan are C_+ waves. For a wave with a propagation velocity of

$$c_+ = u + a, \tag{8.69}$$

Equations (8.53) and (8.54) yield

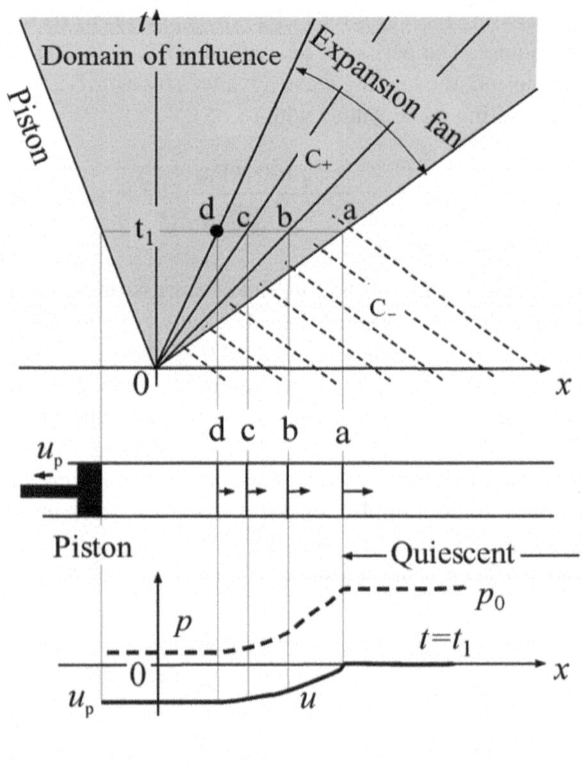

Fig. 8.6 Wave diagram, distributions of pressure and flow velocity induced by "pulling" a piston with a constant velocity $u_p(< 0)$

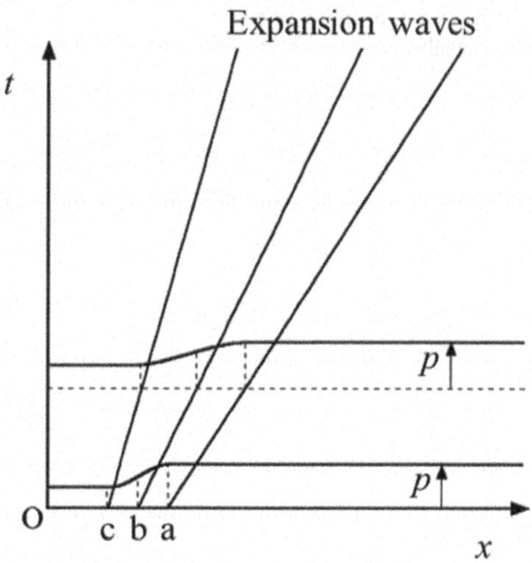

Fig. 8.7 Trajectories of expansion waves originate in a smooth pressure gradient and are associated by pressure distributions

$$u = \frac{2}{\gamma + 1}(c_{+} - a_0) \tag{8.70}$$

$$p = p_0\left(\frac{2}{\gamma + 1} + \frac{\gamma - 1}{\gamma + 1}\frac{c_{+}}{a_0}\right)^{\frac{2\gamma}{\gamma - 1}} \tag{8.71}$$

$$a_0 + \frac{\gamma + 1}{2}u_{\mathrm{p}} = c_{+,\mathrm{d}} \leq \underbrace{c_{+}}_{\text{between d and a}} \leq c_{+,\mathrm{a}} = a_0. \tag{8.72}$$

The farther the wave is from "a," the larger the absolute value of u becomes. However, in the negative direction, the speed of sound becomes lower. At the wave "d," u equals to u_{p}. Resultantly, in the case of expansion waves, the leading wave goes ahead; a preceding wave is not caught up by the waves following behind. With the elapse of time, the distance between the waves increases, and the pressure gradient becomes smoothed out. Unlike in the shock waves, the isentropic assumption does not break up. In this way, an expansion fan is formed, not necessarily centered at a single point, as shown in Fig. 8.7.

8.4.1 Exercise: Piston Falling in Tube

Problem: A straight tube with a cross-sectional area A vertically stands at the atmospheric pressure, p_0. At $t = 0$, a piston with a mass m is released for its free fall in the tube (Fig. 8.8). The tube is long enough, and perfectly seals the clearance between the piston. A friction force between the piston and the inner wall of the tube is negligible. The x coordinated is set vertically downward, and it originates in the initial location of the piston. The following steps need to be taken within the condition that compression waves do not make the transition to a shock wave.

Fig. 8.8 Piston falling in a straight tube

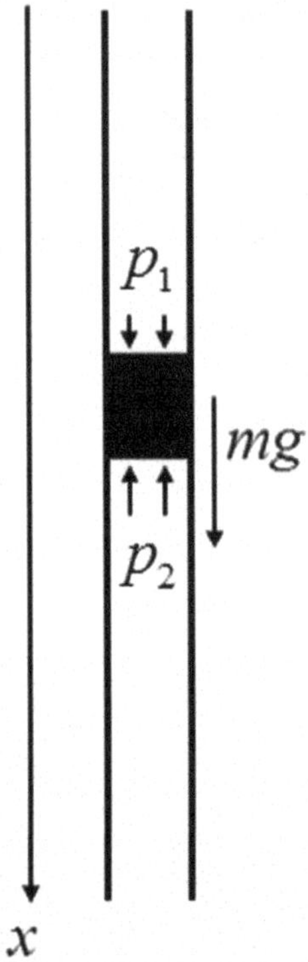

(1) The forces on the upper and lower surface of the piston need to be obtained as the function of a piston velocity, U.
(2) The shock wave formation distance needs to be determined.
(3) The piston trajectory and associated wave diagram on x-t coordinates need to be determined.

Results:

(1) The pressure on the upper and lower surface of the piston, and the gravitational acceleration are denoted by p_1, p_2 and g, respectively. The equation of motion of the piston is

$$m\frac{d^2x}{dt^2} = mg + (p_1 - p_2)A \tag{8.73}$$

$$\frac{d^2x}{dt^2} = g + \frac{\left(\frac{p_1}{p_0} - \frac{p_2}{p_0}\right)p_0 A}{m}. \tag{8.74}$$

Using the piston velocity, U,

$$U = \frac{dx}{dt}. \tag{8.75}$$

From, (8.53),

$$p_1 = p_0\left(1 - \frac{\gamma - 1}{2a_0}U\right)^{\frac{2\gamma}{\gamma-1}} \tag{8.76}$$

$$a_1 = a_0 - \frac{\gamma - 1}{2}U \tag{8.77}$$

$$p_2 = p_0\left(1 + \frac{\gamma - 1}{2a_0}U\right)^{\frac{2\gamma}{\gamma-1}} \tag{8.78}$$

$$a_2 = a_0 + \frac{\gamma - 1}{2}U. \tag{8.79}$$

Then, the respective forces exerted on the upper and lower surfaces of the piston are

$$p_1 A = p_0 A\left(1 - \frac{\gamma - 1}{2a_0}U\right)^{\frac{2\gamma}{\gamma-1}} \tag{8.80}$$

$$-p_2 A = -p_0 A\left(1 + \frac{\gamma - 1}{2a_0}U\right)^{\frac{2\gamma}{\gamma-1}} \tag{8.81}$$

(2) Since the piston is accelerated downward, expansion waves are generated on the upper surface, while compression waves are generated on the lower surface. The transition to a shock wave occurs on the lower side. From (8.74), (8.76), and (8.78), the acceleration of the piston at $t = 0$ is

$$\left(\frac{dU}{dt}\right)_{t=0} = g + \left\{\left(1 - \frac{\gamma - 1}{2a_0}U\right)^{\frac{2\gamma}{\gamma-1}} - \left(1 + \frac{\gamma - 1}{2a_0}U\right)^{\frac{2\gamma}{\gamma-1}}\right\}\frac{p_0 A}{m}. \tag{8.82}$$

Transforming (8.78) and (8.82) by using the condition of $U = 0$ at $t = 0$,

$$\left(\frac{dp_2}{dt}\right)_{t=0} = \left(\frac{\partial p_2}{\partial t}\right)_{t=0} + 0 \cdot \left(\frac{\partial p_2}{\partial x}\right)_{t=0} = \left(\frac{\partial p_2}{\partial t}\right)_{t=0} = \frac{p_0\gamma}{a_0}\left(1 + \frac{\gamma - 1}{2a_0}U\right)^{\frac{\gamma+1}{\gamma-1}}\left(\frac{dU}{dt}\right)_{t=0}. \tag{8.83}$$

The shock wave formation distance is obtained by substituting (8.67) with (8.82) and (8.83).

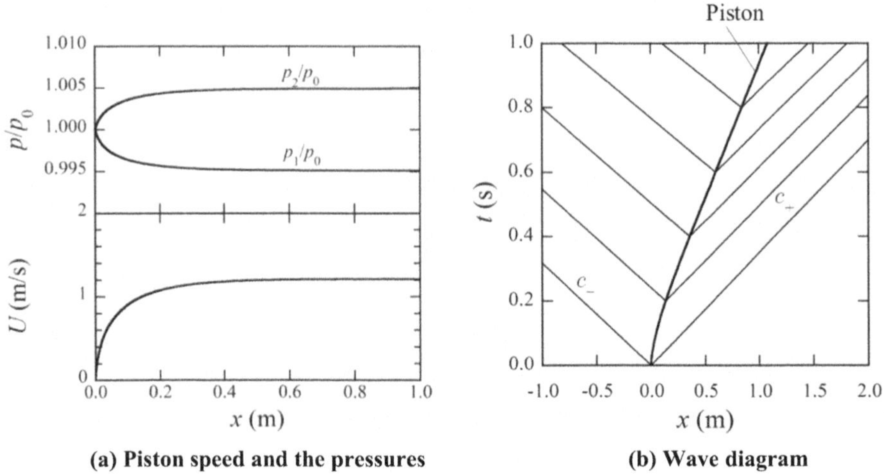

Fig. 8.9 Example of solution, $m = 1$ kg, $A = 1 \times 10^{-2}$ m^2, $p_0 = 1 \times 10^5$ Pa, $T_0 = 298$ K, $g = 9.8$ m/s^2, air (molecular mass, 29), $\gamma = 1.4$

$$
\begin{aligned}
L &= \frac{2\gamma}{\gamma+1}\frac{p_0 a_0}{\left(\frac{\partial p}{\partial t}\right)_0} \\
&= \frac{2\gamma}{\gamma+1}\frac{p_0 a_0}{\frac{p_0\gamma}{a_0}\left(1+\frac{\gamma-1}{2a_0}U\right)^{\frac{\gamma+1}{\gamma-1}}\left[g+\left\{\left(1-\frac{\gamma-1}{2a_0}U\right)^{\frac{2\gamma}{\gamma-1}}-\left(1+\frac{\gamma-1}{2a_0}U\right)^{\frac{2\gamma}{\gamma-1}}\right\}\frac{p_0A}{m}\right]} \\
&= \frac{2}{\gamma+1}\frac{a_0^2}{\left(1+\frac{\gamma-1}{2a_0}U\right)^{\frac{\gamma+1}{\gamma-1}}\left[g+\left\{\left(1-\frac{\gamma-1}{2a_0}U\right)^{\frac{2\gamma}{\gamma-1}}-\left(1+\frac{\gamma-1}{2a_0}U\right)^{\frac{2\gamma}{\gamma-1}}\right\}\frac{p_0A}{m}\right]}.
\end{aligned}
\tag{8.84}
$$

(3) By substituting (8.74) with (8.80) and (8.81), and numerically integrating it, the piston trajectory is obtained. An example solution is shown in Fig. 8.9. The acceleration of the piston decreases, because the pressure on the lower surface increases and that on the upper surface decreases. Eventually, the drag vanishes with the balance among the pressure forces and the gravitational force. The pressures asymptotically approach the respective constant values. From the piston, compression waves C_+ propagate downward, while the expansion waves propagate upward. In the regime depicted in the figure, the transition to the shock wave has not been done yet.

8.5 Pressure-Wave Propagation Around Normal Shock Wave

In Sect. 4.2.1.4, we show that the flow in front of a normal shock wave is supersonic, while it is subsonic behind it. Let us examine these relations on the laboratory frame shown in Fig. 8.10. We compare the shock-wave propagation velocity, U_s and the propagation velocities of c_+ in front of the shock wave (subscript 1) and behind it (subscript 2).

$$c_{+,1} = u_1 + a_1 \tag{8.85}$$

$$U_s = u_1 + M_s a_1. \tag{8.86}$$

Applying (4.49) and (4.51) to calorically perfect gas.

$$c_{+,2} = u_2 + a_2 = u_1 + \frac{a_1}{(\gamma+1)M_s}\left[2\left(M_s^2 - 1\right) + \left(2\gamma M_s^2 - \gamma + 1\right)^{\frac{1}{2}}\left\{(\gamma-1)M_s^2 + 2\right\}^{\frac{1}{2}}\right]. \tag{8.87}$$

From (8.85) to (8.87), because $M_s > 1$,

$$c_{+,1} < U_s < c_{+,2} \tag{8.88}$$

These relations are shown in Figs. 8.11 and 8.12. Relation (8.88) is equivalent to the condition in the frame fixed to the shock wave.

$$1 < M_s,\ M_2 < 1. \tag{8.89}$$

These imply that pressure waves can catch up with the shock wave from behind and alter the shock strength. However, the flow in front of the shock wave cannot be affected by the shock wave before its arrival. Pressure waves in front of the shock wave eventually catch up with the shock wave.

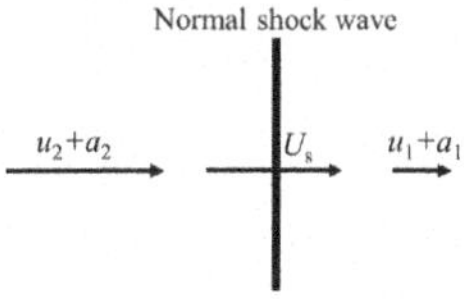

Fig. 8.10 Wave propagation relations in a laboratory frame

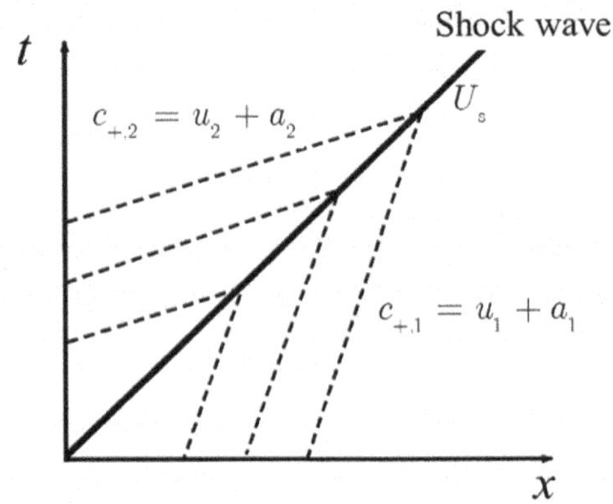

Fig. 8.11 Wave propagations in a laboratory frame

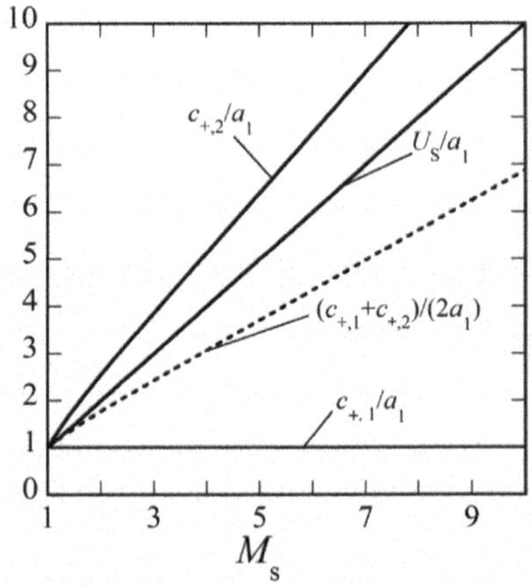

Fig. 8.12 Propagation velocity of a normal shock wave, and pressure waves in front of and behind it, in a laboratory frame. $\gamma = 1.4$ and $u_1 = 0$

$$
\begin{aligned}
\overline{c_+} &= \frac{c_{+,1} + c_{+,2}}{2} \\
&= \frac{u_1 + a_1 + u_2 + a_2}{2} \\
&= u_1 + \frac{a_1}{2(\gamma+1)M_s}\left[2\left(M_s^2 - 1\right) + \left(2\gamma M_s^2 - \gamma + 1\right)^{\frac{1}{2}}\left\{(\gamma-1)M_s^2 + 2\right\}^{\frac{1}{2}} + (\gamma+1)M_s\right] \\
&= u_1 + f(M_s)M_s a_1
\end{aligned}
$$

$$
f(M_s) \equiv \frac{2M_s^2 + (\gamma+1)M_s - 2 + \left(2\gamma M_s^2 - \gamma + 1\right)^{\frac{1}{2}}\left\{(\gamma-1)M_s^2 + 2\right\}^{\frac{1}{2}}}{2(\gamma+1)M_s^2}.
$$

If $M_s \simeq 1$,

$$
\overline{c_+} \simeq u_1 + f(1)M_s a_1 = u_1 + M_s a_1 = U_s.
$$

This means, as seen in Fig. 8.12, that the shock speed approximately equals the average of $\overline{c_+}$ across the shock wave.

8.6 Shock-Wave Propagation in Variable Area Duct

Let us analyze the shock wave behavior propagating in a duct with variable cross-sectional area [2]. The flow is assumed to be along the x-axis, and inviscid. The cross-sectional area $A(x)$ does not vary with time. As in Chap. 5, the unsteady flow conservation equations are applied to the control volume of Fig. 5.2 in a quasi-one-dimensional flow. From the mass conservation Eq. (3.2),

$$\frac{\partial}{\partial t}(\rho A \mathrm{d}x) = -\mathrm{d}(\rho u A) \tag{8.90}$$

$$\frac{\partial \rho}{\partial t} + u\frac{\partial \rho}{\partial x} + \rho\frac{\partial u}{\partial x} = -\frac{\rho u}{A}\frac{\mathrm{d}A}{\mathrm{d}x}. \tag{8.91}$$

From the momentum conservation Eq. (3.6),

$$\frac{\partial}{\partial t}(\rho u A \mathrm{d}x) = -\mathrm{d}\left(\rho u^2 A\right) - A\mathrm{d}p. \tag{8.92}$$

As explained in Sect. 5.1.1, in the second term on the right-hand side of (8.92), A is not enclosed in the derivative, because the force component exerting forces on the duct wall is balanced with the force due to the pressure on the deferential in the cross-sectional area. Transforming (8.92) with (8.90),

$$(\rho A \mathrm{d}x)\frac{\partial}{\partial t}u + \rho u A \mathrm{d}u + A\mathrm{d}p = -u\left\{\frac{\partial}{\partial t}(\rho A \mathrm{d}x) + \mathrm{d}(\rho u A)\right\} = 0$$

$$\frac{\partial u}{\partial t} + u\frac{\partial u}{\partial x} + \frac{1}{\rho}\frac{\partial p}{\partial x} = 0. \tag{8.93}$$

The momentum conservation Eq. (8.93) does not contain A, as described in Sect. 5.1.

The isentropic relation involves only the variation of thermodynamic properties.

$$\frac{\partial p}{\partial t} + u\frac{\partial p}{\partial x} - a^2\left(\frac{\partial \rho}{\partial t} + u\frac{\partial \rho}{\partial x}\right) = 0. \tag{8.94}$$

Comparing (8.90), (8.93), and (8.94) with those without cross-sectional area variation, only the mass conservation equation contains A. From these equations, the following equation for C_+ without a ρ variation is obtained.

$$\frac{\partial p}{\partial t} + (u+a)\frac{\partial p}{\partial x} + \rho a\left\{\frac{\partial u}{\partial t} + (u+a)\frac{\partial u}{\partial x}\right\} + \frac{\rho u a^2}{A}\frac{\mathrm{d}A}{\mathrm{d}x} = 0. \tag{8.95}$$

Here,

$$\frac{\partial}{\partial t} + \frac{\mathrm{d}x}{\mathrm{d}t}\frac{\partial}{\partial x} = \frac{\mathrm{d}x}{\mathrm{d}t}\frac{\mathrm{d}}{\mathrm{d}x} \tag{8.96}$$

and with

$$\frac{\mathrm{d}x}{\mathrm{d}t} = u + a \tag{8.97}$$

$$\frac{\mathrm{d}p}{\mathrm{d}x} + \rho a\frac{\mathrm{d}u}{\mathrm{d}x} + \frac{\rho u a^2}{u+a}\frac{1}{A}\frac{\mathrm{d}A}{\mathrm{d}x} = 0. \tag{8.98}$$

In (8.98), the third term due to the cross-sectional area variation is added to the first line of (8.35). Let us apply this equation to the condition behind a shock-wave propagating in the duct. The condition in front of the shock wave is designated using a subscript "0". From the normal shock relations (4.47)–(4.49),

$$\frac{p}{p_0} = 1 + \frac{2\gamma}{\gamma+1}\left(M_s^2 - 1\right) \tag{8.99}$$

$$\frac{\rho}{\rho_0} = \frac{(\gamma+1)M_s^2}{(\gamma-1)M_s^2 + 2} \tag{8.100}$$

$$u = \frac{2a_0}{\gamma+1}\left(M_s - \frac{1}{M_s}\right). \tag{8.101}$$

From (8.99) and (8.100),

$$\frac{a}{a_0} = \sqrt{\frac{p}{\rho}\frac{\rho_0}{p_0}} = \frac{\sqrt{(2\gamma M_s^2 - \gamma + 1)\{(\gamma-1)M_s^2 + 2\}}}{(\gamma+1)M_s}. \tag{8.102}$$

Here, u is the flow velocity in the laboratory frame, which is equal to zero in front of the shock wave. Substituting (8.98) with (8.99) to (8.102), and using $a_0^2 = \gamma p_0/\rho_0$, the equation with respect to M_s variation is obtained.

$$g(M_s)\frac{\mathrm{d}M_s}{\mathrm{d}x} + \frac{1}{A}\frac{\mathrm{d}A}{\mathrm{d}x} = 0 \tag{8.103}$$

$$g(M_s) \equiv \frac{M_s}{M_s^2 - 1}\left(2\mu + 1 + \frac{1}{M_s^2}\right)\left(1 + \frac{2}{\gamma+1}\frac{1-\mu^2}{\mu}\right) \tag{8.104}$$

$$\mu^2 \equiv \frac{(\gamma-1)M_s^2 + 2}{2\gamma M_s^2 - \gamma + 1}. \tag{8.105}$$

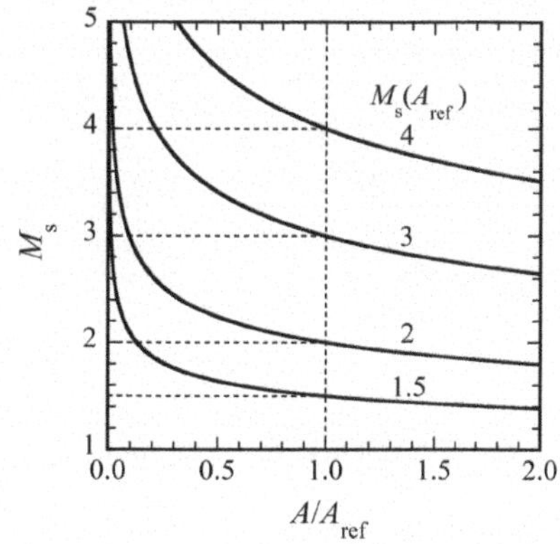

Fig. 8.13 M_S as a function of A

Integrating,

$$\frac{A}{A_{ref}} = \exp\left[-\int_{M_{ref}} g(M_s) \mathrm{d}M_s \right]. \tag{8.106}$$

Equation (8.98) depicts a characteristic quantity along C_+ propagating behind the shock wave. Without the cross-sectional area variation, the quantity is kept constant as long as the shock condition does not change. However, with the cross-sectional area variation, C_- waves from the shock wave are reflected against the duct wall, thereby interacting with the shock wave as C_+ waves. Since (8.106) is obtained by neglecting this effect, it is not strictly correct. However, this equation is often used, because the simple relation between A and M_s reproduces real phenomena with a reasonable accuracy.

Figure 8.13 shows the examples of M_s as a function of A. The larger the A, the lower M_s becomes, and vice versa. The extent of M_s variation depends on the value of M_s at a reference area A. Figure 4.21 explains that a planar shock wave is stable owing to this relation.

8.7 Blast Wave

When energy is suddenly released from a confined medium, such as explosives, volcanoes, or heating by a laser pulse, a shock wave is generated, and followed by an expansion zone, causing large impact or hazards. Such a wave is referred to as a *blast wave* (Fig. 8.14). The blast wave is induced by a rapid expansion of an energetic medium through an interface that pushes the surrounding gas. Usually, in the explosion of an energetic medium, the contact surface experiences the Rayleigh–Taylor instability (Chap. 4), becoming disturbed. However, because the shock wave is stable, it shapes almost as a sphere centered at the point of the explosion. The leading wave propagates as a shock wave. The pressure behind the shock wave decreases,

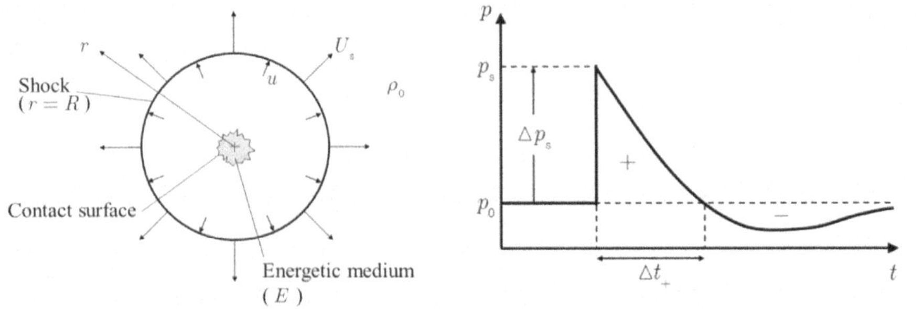

(a) Shock wave and flow **(b) Pressure variation measured at a fixed point**

Fig. 8.14 Blast wave

because the volume of the energetic medium is finite. During its expansion, the pressure surplus from the surroundings gradually decreases. The contact surface is decelerated with generating expansion waves propagating outward. As shown in Fig. 8.14b, the measured pressure p at a fixed point reaches the peak value p_s, and then decreases. During a period Δt_+, the impulse has a positive sign with respect to the atmospheric value, p_0. Then, the pressure becomes lower than p_0, resulting in a negative impulse.[2]

It is important to estimate the pressure and the impulse induced by an explosion as a function of a distance from the center. The theoretical model is applicable to a limited extent. Usually, practical estimation needs to be done by combining measurements and numerical simulations, as well as theoretical considerations.

Here, from *dimensional analyses*, let us obtain a similarity in blast waves induced by a point explosion. In order to ease the analyses, approximations for a strong shock wave are applied. The energetic medium expands from an infinitesimal to an infinite volume. Two independent parameters, the energy of the medium, E, and the density of the surrounding gas, ρ_0, determine the phenomena, meaning that the elapsed time, t, shock speed, U_s, pressure immediately behind the shock wave, p_s, etc., are determined as a function of the shock wave radius, R. E and ρ_0 have a dimension of $[ML^2\,T^{-2}]$ and $[ML^{-3}]$, respectively. They cannot define a characteristic time. Therefore, a dimensionless quantity, η, is defined together with r[L] and t [T].

$$\eta \equiv E^{-\frac{1}{5}} \rho_0^{\frac{1}{5}} r t^{-\frac{2}{5}} = \frac{r}{\left(\frac{E}{\rho_0}\right)^{\frac{1}{5}} t^{\frac{2}{5}}}. \tag{8.107}$$

Defining η_0 as the value of η at a shock location, $r = R$. From (8.107),

---

[2]When the contact surface is decelerated, compression waves are generated, propagating to the center. These compression waves are reflected from the center, propagating as the *secondary shock wave*, which in Fig. 8.14 is not drawn for simplicity.

$$R = \eta_0 \left(\frac{E}{\rho_0} \right)^{\frac{1}{5}} t^{\frac{2}{5}}. \tag{8.108}$$

Therefore, the shock wave location is proportional to $t^{\frac{2}{5}}$. Differentiating (8.108),

$$U_s = \frac{dR}{dt} = \frac{2}{5}\eta_0 \left(\frac{E}{\rho_0} \right)^{\frac{1}{5}} t^{-\frac{3}{5}} = \frac{2}{5}\frac{R}{t} = \frac{2}{5}\eta_0^{\frac{5}{2}} R^{-\frac{3}{2}} \left(\frac{E}{\rho_0} \right)^{\frac{1}{2}}. \tag{8.109}$$

Applying the strong shock relation (4.59),

$$p_s \approx \frac{2\gamma}{\gamma+1} p_0 M_s^2 = \frac{2\gamma p_0}{\gamma+1}\frac{U_s^2}{a_0^2} = \frac{2\gamma p_0}{\gamma+1}\frac{U_s^2}{\frac{\gamma p_0}{\rho_0}} = \frac{2}{\gamma+1}\rho_0 U_s^2. \tag{8.110}$$

Substituting with (8.109),

$$p_s \approx \frac{2}{\gamma+1}\rho_0 U_s^2 \sim R^{-3}\left(\frac{E}{\rho_0} \right) \tag{8.111}$$

$$R \sim \left(\frac{E/\rho_0}{p_s} \right)^{\frac{1}{3}}. \tag{8.112}$$

Equation (8.112) provides a scaling law in the overpressure due to the explosion. In order to reproduce the same magnitude of overpressure, the pressure should be measured at a distance proportional to the cubic root of the explosion energy. This relation is useful in estimating a safety distance from an explosive by scaled experiments.

Appendix: Characteristics and Invariants in Three-Dimensional Flow

Let us derive the characteristics and invariants in the three-dimensional Euler equation. Transforming the equation to the form shown below, characteristics are obtained as eigenvalues. Here, we neglect the body force and heat transfer. The conservation equations shown in Chap. 3 are as follows:

$$\text{Mass conservation}: \quad \frac{\partial \rho}{\partial t} + \nabla \cdot \rho \mathbf{u} = 0 \tag{8.113}$$

$$\text{Momentum conservation}: \quad \rho \frac{\partial \mathbf{u}}{\partial t} + \rho(\mathbf{u} \cdot \nabla)\mathbf{u} = -\nabla p \tag{8.114}$$

$$\text{Energy conservation}: \quad \rho \frac{\mathrm{D}e}{\mathrm{D}t} = -p\nabla \cdot \mathbf{u}. \tag{8.115}$$

In three-dimensional flows, the momentum conservation (8.114) has three components. Then, we have five equations in total. Now, we will transform those equations to differential equations with respect only to the following five parameters, ρ, $\mathbf{u} = (u, v, w)$ and p.

From the first law of thermodynamics,

$$T\mathrm{d}s = \mathrm{d}e + p\mathrm{d}\left(\frac{1}{\rho}\right) = \mathrm{d}e - \frac{p}{\rho^2}\mathrm{d}\rho. \tag{8.116}$$

Assuming isentropic flow,

$$\left(\frac{\partial e}{\partial \rho}\right)_s = \frac{p}{\rho^2}. \tag{8.117}$$

Using the definition of the speed of sound from (8.6),

$$\left(\frac{\partial e}{\partial p}\right)_s = \frac{\left(\frac{\partial e}{\partial \rho}\right)_s}{\left(\frac{\partial p}{\partial \rho}\right)_s} = \frac{\frac{p}{\rho^2}}{a^2} = \frac{p}{\rho^2 a^2}. \tag{8.118}$$

Substituting it with (8.116),

$$\rho\frac{\mathrm{D}e}{\mathrm{D}t} = \rho\left[\left(\frac{\partial e}{\partial p}\right)_s \frac{\mathrm{D}p}{\mathrm{D}t}\right] = \frac{p}{\rho a^2}\frac{\mathrm{D}p}{\mathrm{D}t} = -p\nabla\cdot\mathbf{u}$$

$$\frac{\partial p}{\partial t} + \mathbf{u}\cdot\nabla p + \rho a^2 \nabla\cdot\mathbf{u} = 0. \tag{8.119}$$

Expressing (8.113), (8.114), and (8.119) using the velocity components,

$$\frac{\partial \rho}{\partial t} + \left(u\frac{\partial \rho}{\partial x} + \rho\frac{\partial u}{\partial x}\right) + \left(v\frac{\partial \rho}{\partial y} + \rho\frac{\partial v}{\partial y}\right) + \left(w\frac{\partial \rho}{\partial z} + \rho\frac{\partial w}{\partial z}\right) = 0 \tag{8.120}$$

$$\begin{pmatrix} \frac{\partial u}{\partial t} \\ \frac{\partial v}{\partial t} \\ \frac{\partial w}{\partial t} \end{pmatrix} + \begin{pmatrix} u\frac{\partial u}{\partial x} + \frac{1}{\rho}\frac{\partial p}{\partial x} + v\frac{\partial u}{\partial y} + w\frac{\partial u}{\partial z} \\ u\frac{\partial v}{\partial x} + v\frac{\partial v}{\partial y} + \frac{1}{\rho}\frac{\partial p}{\partial y} + w\frac{\partial v}{\partial z} \\ u\frac{\partial w}{\partial x} + v\frac{\partial w}{\partial y} + w\frac{\partial w}{\partial z} + \frac{1}{\rho}\frac{\partial p}{\partial z} \end{pmatrix} = \begin{pmatrix} 0 \\ 0 \\ 0 \end{pmatrix} \tag{8.121}$$

$$\frac{\partial p}{\partial t} + \left(u\frac{\partial p}{\partial x} + \rho a^2\frac{\partial u}{\partial x}\right) + \left(v\frac{\partial p}{\partial y} + \rho a^2\frac{\partial v}{\partial y}\right) + \left(w\frac{\partial p}{\partial z} + \rho a^2\frac{\partial w}{\partial z}\right) = 0. \tag{8.122}$$

These are written in a vector form.

$$\frac{\partial \mathbf{V}}{\partial t} + (\tilde{\mathbf{A}}\cdot\nabla)\mathbf{V} = \mathbf{0},\ (\tilde{\mathbf{A}}\cdot\nabla)\mathbf{V} = \left(\mathbf{A}\frac{\partial}{\partial x} + \mathbf{B}\frac{\partial}{\partial y} + \mathbf{C}\frac{\partial}{\partial z}\right)\mathbf{V} \tag{8.123}$$

$$\mathbf{V} = \begin{pmatrix} \rho \\ \mathbf{u} \\ P \end{pmatrix}, \quad \tilde{\mathbf{A}} = \begin{pmatrix} \mathbf{A} \\ \mathbf{B} \\ \mathbf{C} \end{pmatrix}, \quad \nabla = \begin{pmatrix} \dfrac{\partial}{\partial x} \\ \dfrac{\partial}{\partial y} \\ \dfrac{\partial}{\partial z} \end{pmatrix} \tag{8.124}$$

$$\mathbf{A} = \begin{pmatrix} u & \rho & 0 & 0 & 0 \\ 0 & u & 0 & 0 & \frac{1}{\rho} \\ 0 & 0 & u & 0 & 0 \\ 0 & 0 & 0 & u & 0 \\ 0 & \rho a^2 & 0 & 0 & u \end{pmatrix} \tag{8.125}$$

$$\mathbf{B} = \begin{pmatrix} v & 0 & \rho & 0 & 0 \\ 0 & v & 0 & 0 & 0 \\ 0 & 0 & v & 0 & \frac{1}{\rho} \\ 0 & 0 & 0 & v & 0 \\ 0 & 0 & \rho a^2 & 0 & v \end{pmatrix} \tag{8.126}$$

$$\mathbf{C} = \begin{pmatrix} w & 0 & 0 & \rho & 0 \\ 0 & w & 0 & 0 & 0 \\ 0 & 0 & w & 0 & 0 \\ 0 & 0 & 0 & w & \frac{1}{\rho} \\ 0 & 0 & 0 & \rho a^2 & w \end{pmatrix} \tag{8.127}$$

$$\mathbf{0} = \begin{pmatrix} 0 \\ 0 \\ 0 \end{pmatrix}. \tag{8.128}$$

Assuming a wave solution to (8.123),

$$\mathbf{V} = \bar{\mathbf{V}} e^{i\varphi(\mathbf{x},t)}, \ (i, \ \text{imaginary unit}) \tag{8.129}$$

Substituting (8.129) to (8.123),

$$\frac{\partial \varphi}{\partial t}\bar{\mathbf{V}} + \tilde{\mathbf{A}} \cdot (\nabla\varphi)\, \bar{\mathbf{V}} = \mathbf{0} \text{ or } \left(\frac{\partial \varphi}{\partial t}\mathbf{I} + \mathbf{A}\frac{\partial \varphi}{\partial x} + \mathbf{B}\frac{\partial \varphi}{\partial y} + \mathbf{C}\frac{\partial \varphi}{\partial z}\right)\bar{\mathbf{V}} = \mathbf{0}. \tag{8.130}$$

We define the followings:

$$\lambda \equiv -\frac{\partial \varphi}{\partial t} \tag{8.131}$$

$$\mathbf{k} \equiv \nabla\varphi \tag{8.132}$$

$$\mathbf{K} \equiv \tilde{\mathbf{A}} \cdot \mathbf{k} \tag{8.133}$$

$\mathbf{k}$ is a unit normal vector ($|\mathbf{k}| = 1$) to a characteristic, where $\varphi = \text{const.}$ Substituting (8.130) with (8.131)–(8.133),

$$(\mathbf{K} - \lambda\mathbf{I})\bar{\mathbf{V}} = \mathbf{0} \quad \mathbf{K}\bar{\mathbf{V}} = \lambda\bar{\mathbf{V}} \tag{8.134}$$

$\mathbf{I}$ is a unit matrix. In order for $\bar{\mathbf{V}}$ to have a solution other than $\mathbf{0}$, the following characteristic equation should be satisfied.

$$|\mathbf{K} - \lambda\mathbf{I}| = 0 \tag{8.135}$$

The 5×5 matrix, $\mathbf{K}$, has five sets of an eigenvalue, λ, and eigenvector, $\bar{\mathbf{V}}$. Let $\mathbf{\Lambda}$ be defined as a matrix that has only diagonal components of the eigenvalues.

$$\mathbf{KL} = \mathbf{L}\Lambda \quad \text{or} \quad \mathbf{\Lambda} = \mathbf{L}^{-1}\mathbf{KL} \tag{8.136}$$

$$\mathbf{L} = \left(\bar{\mathbf{V}}_1, \bar{\mathbf{V}}_2, \ldots \bar{\mathbf{V}}_5\right) \tag{8.137}$$

$$\mathbf{\Lambda} = \begin{pmatrix} \lambda_1 & & & \\ & \cdot & & \\ & & \cdot & \\ & & & \lambda_5 \end{pmatrix} \tag{8.138}$$

$$\mathbf{K} = \begin{pmatrix} \mathbf{u}\cdot\mathbf{k} & \rho\,k_x & \rho\,k_y & \rho\,k_z & 0 \\ 0 & \mathbf{u}\cdot\mathbf{k} & 0 & 0 & \frac{1}{\rho}k_x \\ 0 & 0 & \mathbf{u}\cdot\mathbf{k} & 0 & \frac{1}{\rho}k_y \\ 0 & 0 & 0 & \mathbf{u}\cdot\mathbf{k} & \frac{1}{\rho}k_z \\ 0 & \rho\,a^2k_x & \rho\,a^2k_y & \rho\,a^2k_z & \mathbf{u}\cdot\mathbf{k} \end{pmatrix}. \tag{8.139}$$

Therefore, (8.135) yields

$$\begin{vmatrix} \mathbf{u}\cdot\mathbf{k}-\lambda & \rho\,k_x & \rho\,k_y & \rho\,k_z & 0 \\ 0 & \mathbf{u}\cdot\mathbf{k}-\lambda & 0 & 0 & \frac{1}{\rho}k_x \\ 0 & 0 & \mathbf{u}\cdot\mathbf{k}-\lambda & 0 & \frac{1}{\rho}k_y \\ 0 & 0 & 0 & \mathbf{u}\cdot\mathbf{k}-\lambda & \frac{1}{\rho}k_z \\ 0 & \rho\,a^2k_x & \rho\,a^2k_y & \rho\,a^2k_z & \mathbf{u}\cdot\mathbf{k}-\lambda \end{vmatrix} = 0 \tag{8.140}$$

$$(\,\mathbf{u}\cdot\mathbf{k}-\lambda\,)^3\left[(\,\mathbf{u}\cdot\mathbf{k}-\lambda)^2 - a^2\right] = 0. \tag{8.141}$$

Using the solutions

$$\begin{cases} \lambda_1 = \lambda_2 = \lambda_3 = \mathbf{u} \cdot \mathbf{k} \\ \lambda_4 = \mathbf{u} \cdot \mathbf{k} + a = (\mathbf{u} + a\mathbf{k}) \cdot \mathbf{k} \\ \lambda_5 = \mathbf{u} \cdot \mathbf{k} - a = (\mathbf{u} - a\mathbf{k}) \cdot \mathbf{k} \end{cases} \tag{8.142}$$

$$\bar{\mathbf{V}} = \begin{pmatrix} X_1 \\ \vdots \\ X_5 \end{pmatrix}. \tag{8.143}$$

Equation (8.134) is transformed to

$$\begin{cases} \mathbf{u} \cdot \mathbf{k} X_1 + \rho k_x X_2 + \rho k_y X_3 + \rho k_z X_4 = \lambda X_1 \\ \mathbf{u} \cdot \mathbf{k} X_2 + \frac{1}{\rho} k_x X_5 = \lambda X_2 \\ \mathbf{u} \cdot \mathbf{k} X_3 + \frac{1}{\rho} k_y X_5 = \lambda X_3 \\ \mathbf{u} \cdot \mathbf{k} X_4 + \frac{1}{\rho} k_z X_5 = \lambda X_4 \\ \rho a^2 k_x X_2 + \rho a^2 k_y X_3 + \rho a^2 k_z X_4 + \mathbf{u} \cdot \mathbf{k} X_5 = \lambda X_5 \end{cases}. \tag{8.144}$$

For $\lambda = \mathbf{u} \cdot \mathbf{k}$,

$$\begin{cases} X_1 = \text{arbitrary} \\ \begin{pmatrix} X_2 \\ X_3 \\ X_4 \end{pmatrix} \cdot \begin{pmatrix} k_x \\ k_y \\ k_z \end{pmatrix} = 0. \\ X_5 = 0 \end{cases} \tag{8.145}$$

For example, the following unit vectors satisfy (8.145).

$$\bar{\mathbf{V}}_1 = \begin{pmatrix} k_x \\ 0 \\ k_z \\ -k_y \\ 0 \end{pmatrix}, \quad \bar{\mathbf{V}}_2 = \begin{pmatrix} k_y \\ -k_z \\ 0 \\ k_x \\ 0 \end{pmatrix}, \quad \bar{\mathbf{V}}_3 = \begin{pmatrix} k_z \\ k_y \\ -k_x \\ 0 \\ 0 \end{pmatrix}. \tag{8.146}$$

For $\lambda = \mathbf{u} \cdot \mathbf{k} \pm a$ (double-sign corresponds),

$$\begin{cases} X_1 = \dfrac{1}{a^2} X_5 \\ X_2 = \pm \dfrac{k_x}{\rho a} X_5 \\ X_3 = \pm \dfrac{k_y}{\rho a} X_5 \\ X_4 = \pm \dfrac{k_z}{\rho a} X_5 \end{cases}. \tag{8.147}$$

For example,

$$\overline{\mathbf{V}}_4 = \begin{pmatrix} \frac{\rho}{2a} \\ \frac{k_x}{2} \\ \frac{k_y}{2} \\ \frac{k_z}{2} \\ \frac{\rho a}{2} \end{pmatrix} \text{ for } \lambda = \mathbf{u} \cdot \mathbf{k} + a, \tag{8.148}$$

$$\bar{\mathbf{V}}_5 = \begin{pmatrix} \frac{\rho}{2a} \\ -\frac{k_x}{2} \\ -\frac{k_y}{2} \\ -\frac{k_z}{2} \\ \frac{\rho a}{2} \end{pmatrix} \text{ for } \lambda = \mathbf{u} \cdot \mathbf{k} - a. \tag{8.149}$$

The inverse matrix of (8.137) is

$$\mathbf{L}^{-1} = \begin{pmatrix} k_x & 0 & k_z & -k_y & -\frac{k_x}{a^2} \\ k_y & -k_z & 0 & k_x & -\frac{k_y}{a^2} \\ k_z & k_y & -k_x & 0 & -\frac{k_z}{a^2} \\ 0 & -k_x & k_y & k_z & \frac{1}{\rho a} \\ 0 & k_x & -k_y & -k_z & \frac{1}{\rho a} \end{pmatrix}. \tag{8.150}$$

Multiplying (8.150) to the left of (8.123)

$$\mathbf{L}^{-1} \frac{\partial \mathbf{V}}{\partial t} + \mathbf{L}^{-1} (\tilde{\mathbf{A}} \cdot \nabla) \mathbf{V} = \mathbf{0} \tag{8.151}$$

$$\mathbf{L}^{-1} \frac{\partial \mathbf{V}}{\partial t} = \begin{pmatrix} k_x \frac{\partial \rho}{\partial t} & & +k_z \frac{\partial v}{\partial t} & -k_y \frac{\partial w}{\partial t} & -\frac{k_x}{a^2} \frac{\partial p}{\partial t} \\ k_y \frac{\partial \rho}{\partial t} & -k_z \frac{\partial u}{\partial t} & & +k_x \frac{\partial w}{\partial t} & -\frac{k_y}{a^2} \frac{\partial p}{\partial t} \\ k_z \frac{\partial \rho}{\partial t} & +k_y \frac{\partial u}{\partial t} & -k_x \frac{\partial v}{\partial t} & & -\frac{k_z}{a^2} \frac{\partial p}{\partial t} \\ & k_x \frac{\partial u}{\partial t} & +k_y \frac{\partial v}{\partial t} & +k_z \frac{\partial w}{\partial t} & +\frac{1}{\rho a} \frac{\partial p}{\partial t} \\ & -k_x \frac{\partial u}{\partial t} & -k_y \frac{\partial v}{\partial t} & -k_z \frac{\partial w}{\partial t} & +\frac{1}{\rho a} \frac{\partial p}{\partial t} \end{pmatrix} \tag{8.152}$$

$$\tilde{\mathbf{A}} \cdot \nabla = \begin{pmatrix} \mathbf{u} \cdot \nabla & \rho \frac{\partial}{\partial x} & \rho \frac{\partial}{\partial y} & \rho \frac{\partial}{\partial z} & 0 \\ 0 & \mathbf{u} \cdot \nabla & 0 & 0 & \frac{1}{\rho} \frac{\partial}{\partial x} \\ 0 & 0 & \mathbf{u} \cdot \nabla & 0 & \frac{1}{\rho} \frac{\partial}{\partial y} \\ 0 & 0 & 0 & \mathbf{u} \cdot \nabla & \frac{1}{\rho} \frac{\partial}{\partial z} \\ 0 & \rho a^2 \frac{\partial}{\partial x} & \rho a^2 \frac{\partial}{\partial y} & \rho a^2 \frac{\partial}{\partial z} & \mathbf{u} \cdot \nabla \end{pmatrix} \tag{8.153}$$

$$\mathbf{L}^{-1}(\tilde{\mathbf{A}} \cdot \nabla) =$$

$$\begin{pmatrix} k_x \mathbf{u} \cdot \nabla & 0 & k_z \mathbf{u} \cdot \nabla & -k_y \mathbf{u} \cdot \nabla & k_z \frac{1}{\rho} \frac{\partial}{\partial y} - k_y \frac{1}{\rho} \frac{\partial}{\partial z} - \frac{k_x}{a^2} \mathbf{u} \cdot \nabla \\ k_y \mathbf{u} \cdot \nabla & -k_z \mathbf{u} \cdot \nabla & 0 & k_x \mathbf{u} \cdot \nabla & -k_z \frac{1}{\rho} \frac{\partial}{\partial x} + k_x \frac{1}{\rho} \frac{\partial}{\partial z} - \frac{k_y}{a^2} \mathbf{u} \cdot \nabla \\ k_z \mathbf{u} \cdot \nabla & k_y \mathbf{u} \cdot \nabla & -k_x \mathbf{u} \cdot \nabla & 0 & k_y \frac{1}{\rho} \frac{\partial}{\partial x} - k_x \frac{1}{\rho} \frac{\partial}{\partial y} - \frac{k_z}{a^2} \mathbf{u} \cdot \nabla \\ 0 & k_x \mathbf{u} \cdot \nabla + a \frac{\partial}{\partial x} & k_y \mathbf{u} \cdot \nabla + a \frac{\partial}{\partial y} & k_z \mathbf{u} \cdot \nabla + a \frac{\partial}{\partial z} & k_x \frac{1}{\rho} \frac{\partial}{\partial x} + k_y \frac{1}{\rho} \frac{\partial}{\partial y} + k_z \frac{1}{\rho} \frac{\partial}{\partial z} + \frac{1}{\rho a} \mathbf{u} \cdot \nabla \\ 0 & -k_x \mathbf{u} \cdot \nabla + a \frac{\partial}{\partial x} & -k_y \mathbf{u} \cdot \nabla + a \frac{\partial}{\partial y} & -k_z \mathbf{u} \cdot \nabla + a \frac{\partial}{\partial z} & -k_x \frac{1}{\rho} \frac{\partial}{\partial x} - k_y \frac{1}{\rho} \frac{\partial}{\partial y} - k_z \frac{1}{\rho} \frac{\partial}{\partial z} + \frac{1}{\rho a} \mathbf{u} \cdot \nabla \end{pmatrix} \tag{8.154}$$

Substituting (8.151) with (8.152) and (8.154), the first line yields

$$k_x \underbrace{\left[\left(\frac{\partial}{\partial t} + \mathbf{u} \cdot \nabla \right) \rho - \frac{1}{a^2} \left(\frac{\partial}{\partial t} + \mathbf{u} \cdot \nabla \right) p \right]}_{\text{(A)}} - k_y \underbrace{\left[\left(\frac{\partial}{\partial t} + \mathbf{u} \cdot \nabla \right) w + \frac{1}{\rho} \frac{\partial p}{\partial z} \right]}_{\text{(B)}} + k_z \underbrace{\left[\left(\frac{\partial}{\partial t} + \mathbf{u} \cdot \nabla \right) v + \frac{1}{\rho} \frac{\partial p}{\partial y} \right]}_{\text{(C)}} = 0.. \tag{8.155}$$

For this equation to be satisfied for arbitrary wave number components,

$$\text{(A) Along } \frac{\mathrm{d}\mathbf{x}}{\mathrm{d}t} = \mathbf{u}, \mathrm{d}\rho - \frac{1}{a^2} \mathrm{d}p = 0 \text{ that is } \mathrm{d}s = 0 \tag{8.156}$$

(B) is equivalent to the momentum conservation equation in the z-direction.
(C) is equivalent to the momentum conservation equation in the y-direction.

The same results are obtained from the second and third lines of (8.151). From the fourth line,

$$\mathbf{k} \cdot \left[\left(\frac{\partial}{\partial t} + \mathbf{u} \cdot \nabla \right) \mathbf{u} \right] + a \nabla \cdot \mathbf{u} + \frac{1}{\rho a} \left[\frac{\partial}{\partial t} + (\mathbf{u} + a\mathbf{k}) \cdot \nabla \right] p = 0. \tag{8.157}$$

Here, we assume two unit vectors, **l** and **m**, which are normal to each other and to **k**.

$$\mathbf{k} \cdot \mathbf{l} = \mathbf{k} \cdot \mathbf{m} = \mathbf{l} \cdot \mathbf{m} = 0, \qquad |\mathbf{l}| = |\mathbf{m}| = 1. \tag{8.158}$$

Since

$$\nabla_{\mathbf{k}} = \mathbf{k} \cdot \nabla, \qquad \nabla_{\mathbf{l}} = \mathbf{l} \cdot \nabla, \qquad \nabla_{\mathbf{m}} = \mathbf{m} \cdot \nabla \tag{8.159}$$

$$\nabla \cdot \mathbf{u} = \mathbf{k} \cdot \{(\mathbf{k} \cdot \nabla)\mathbf{u}\} + \mathbf{l} \cdot \{(\mathbf{l} \cdot \nabla)\mathbf{u}\} + \mathbf{m} \cdot \{(\mathbf{m} \cdot \nabla)\mathbf{u}\} \tag{8.160}$$

k is a normal vector, which does not change along **l** and **m**, that is. $\mathbf{l}\cdot\nabla = \mathbf{m}\cdot\nabla = 0$,

$$\nabla \cdot \mathbf{u} = \mathbf{k} \cdot \{(\mathbf{k} \cdot \nabla)\mathbf{u}\}. \tag{8.161}$$

From (8.157) and (8.161),

$$\mathbf{k} \cdot \left[\left\{\frac{\partial}{\partial t} + (\mathbf{u} + a\mathbf{k}) \cdot \nabla\right\}\mathbf{u}\right] + \frac{1}{\rho a}\left[\frac{\partial}{\partial t} + (\mathbf{u} + a\mathbf{k}) \cdot \nabla\right]p = 0. \tag{8.162}$$

Defining

$$\mathrm{d}_+ \equiv \frac{\partial}{\partial t} + (\mathbf{u} + a\mathbf{k}) \cdot \nabla \tag{8.163}$$

$$\mathbf{k} \cdot \mathrm{d}_+\mathbf{u} + \frac{1}{\rho a}\mathrm{d}_+ p = 0. \tag{8.164}$$

In the same way, from the fifth line of (8.151),

$$\mathrm{d}_- \equiv \frac{\partial}{\partial t} + (\mathbf{u} - a\mathbf{k}) \cdot \nabla \tag{8.165}$$

$$\mathbf{k} \cdot \left[\left\{\frac{\partial}{\partial t} + (\mathbf{u} - a\mathbf{k}) \cdot \nabla\right\}\mathbf{u}\right] - \frac{1}{\rho a}\left[\frac{\partial}{\partial t} + (\mathbf{u} - a\mathbf{k}) \cdot \nabla\right]p = 0 \tag{8.166}$$

$$\mathbf{k} \cdot \mathrm{d}_-\mathbf{u} - \frac{1}{\rho a}\mathrm{d}_- p = 0. \tag{8.167}$$

Along the *characteristics*, therefore,

$$\mathbf{c}_\pm = \mathbf{u} \pm a\mathbf{k} \tag{8.168}$$

$$\mathrm{d}J_\pm \equiv \mathbf{k} \cdot \mathrm{d}_\pm\mathbf{u} \pm \frac{1}{\rho a}\mathrm{d}_\pm p = 0 \tag{8.169}$$

J_+ and J_- are Riemann invariants.

In summary of the above results, taking **k** to the direction of the wave propagation,

$$\text{Along } \frac{\mathrm{d}\mathbf{x}}{\mathrm{d}t} = \mathbf{c}_+ = \mathbf{u} + a\mathbf{k},\ \mathrm{d}J_+ = \mathbf{k} \cdot \mathrm{d}_+\mathbf{u} + \frac{1}{\rho a}\mathrm{d}_+ p = 0 \tag{8.170}$$

$$\text{Along } \frac{\mathrm{d}\mathbf{x}}{\mathrm{d}t} = \mathbf{c}_- = \mathbf{u} - a\mathbf{k}, \ \mathrm{d}J_- = \mathbf{k} \cdot \mathrm{d}_-\mathbf{u} - \frac{1}{\rho a}\mathrm{d}_- p = 0 \tag{8.171}$$

$$\text{Along } \frac{\mathrm{d}\mathbf{x}}{\mathrm{d}t} = \mathbf{c}_0 = \mathbf{u}, \ \mathrm{d}s = 0. \tag{8.172}$$

References

1. Takayama K, Sasoh A, Onodera O, Kaneko R, Matsui Y (1995) Shock Waves 5:127–138
2. Chester W (1954) Phil Mag 45:1293 (Chisnel RF (1957) J Fluid Mech 2:286–298; Whitham GB (1958) J Fluid Mech 4:337–360; Whitham GB (1974) Linear and nonlinear waves. Chap. 8. Wiley)

Chapter 9
Riemann Problem

Using the characteristics and invariants dealt with in Chap. 8, an initial value problem with two one-dimensional elements can be solved. This is termed as the *Riemann problem*, which gives a solution to Euler's equation that may include a discontinuity. In this chapter, we study Riemann's solutions in detail and apply them to practical problems.

9.1 Definition and Solution

Here, we deal with the interaction between two pressure waves or the interaction between a pressure wave and a contact surface appearing in two one-dimensional elements in contact. As shown in Fig. 9.1, at a time $t = 0$, two uniform states L and R in which thermodynamic properties and/or a flow velocity may differ, start to interact at the origin O. We do not consider the history before the interaction ($t < 0$) and solve this initial value problem.

The boundary between these two states at O is equivalent to a contact surface. Let us denote the states after the interaction as L* on the left and R* on the right. A *left-running wave* linked with the motion of the interface propagates in L, and subsequently a *right-running* wave propagates in R. Here, the direction of propagation refers to the interface, which is not necessarily the same as in the laboratory frame. Each wave is either compression or expansion. In the case of the compression, we assume that the compression waves coalesce right after the interaction, and the transition forms into a shock wave. In Fig. 9.1, each wave is represented by two straight lines. In the case of the shock wave, they should merge to a single line. In the case of expansion, they represent the leading wave and the tail of an expansion fan between which the flow experiences a smooth variation.

The solution of the Riemann problem is categorized depending on the type of waves. In extreme situations, vacuum regions may appear. Figure 9.2 shows all possible patterns without the vacuum region. Since a possible wave is either a shock

A. Sasoh, *Compressible Fluid Dynamics and Shock Waves*,
https://doi.org/10.1007/978-981-15-0504-1_9

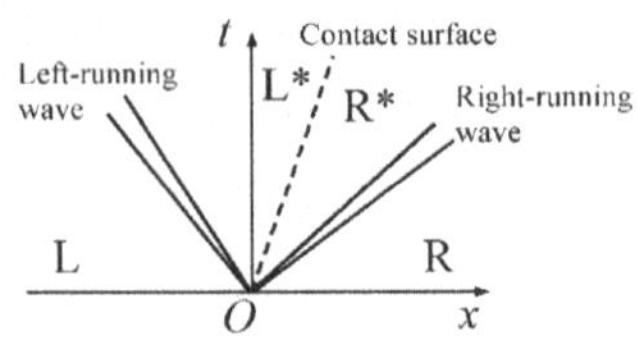

Fig. 9.1 Wave diagram of the Riemann problem in general form

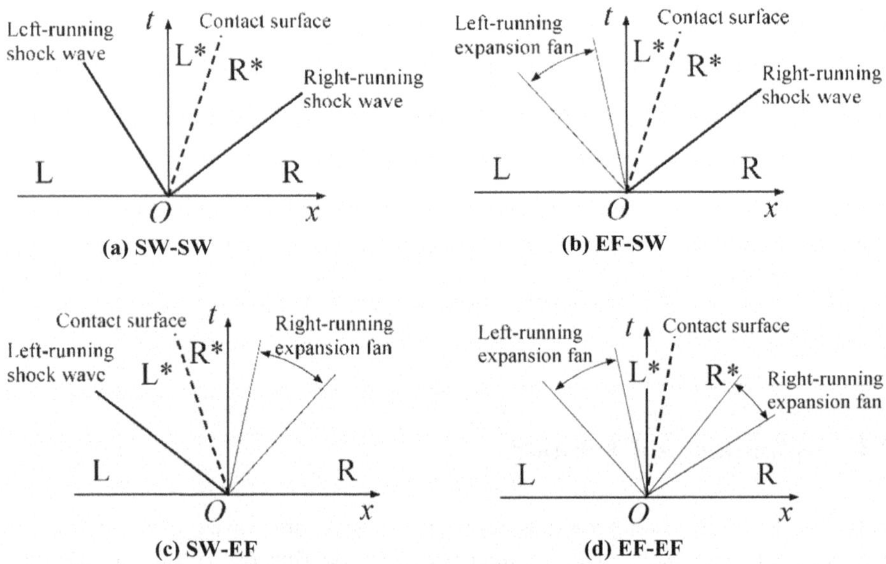

Fig. 9.2 Solution patterns without a vacuum region

wave (SW) or an expansion fan (EF), four patterns exist: (a) is a shock–shock pattern, (b) and (c) are the combinations of SW and EF, and (d) is an EF-EF pattern.

Figure 9.3 shows solution patterns with vacuum regions. In all the patterns, the gas only expands, it is not compressed. Neither a shock wave nor a contact surface is formed. Including the right and left difference, there are three patterns: (a) the gases on both sides expand outward, and a vacuum region is formed in the middle; (b) and (c) one of the initial states is in vacuum; the gas on the other side expands toward the vacuum.

In the following analyses, the gas is assumed to be ideal and calorically perfect. With the given states L and R, let us obtain states L* and R* after the interaction with accompanying wave conditions. Each state is defined by a flow velocity and two thermodynamic properties. Here, we employ the pressure, p, and density, ρ as independent parameters.

When the left-running wave is a shock wave with a shock Mach number of $M_{\mathrm{s,L}}$ and $p_{\mathrm{L}*}/p_{\mathrm{L}} > 1$, the states L and L* are related using (4.47) to (4.50),

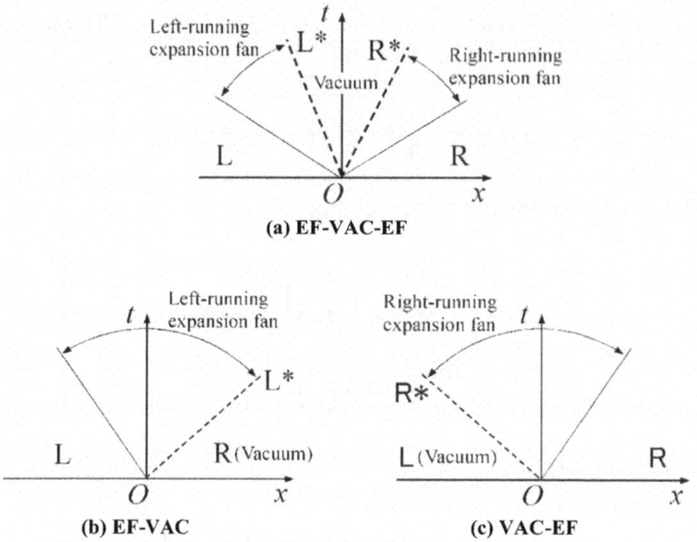

Fig. 9.3 Solution patterns with a vacuum region

$$\frac{p_{\mathrm{L}*}}{p_{\mathrm{L}}} = 1 + \frac{2\gamma}{\gamma+1}\left(M_{\mathrm{s,L}}^2 - 1\right) \tag{9.1}$$

$$\frac{\rho_{\mathrm{L}*}}{\rho_{\mathrm{L}}} = \frac{(\gamma+1)M_{\mathrm{s,L}}^2}{(\gamma-1)M_{\mathrm{s,L}}^2 + 2} \tag{9.2}$$

$$-u_{\mathrm{L}*} + u_{\mathrm{L}} = \frac{2a_{\mathrm{L}}}{\gamma+1}\left(M_{\mathrm{s,L}} - \frac{1}{M_{\mathrm{s,L}}}\right) \tag{9.3}$$

$$M_{\mathrm{s,L}} = \frac{-U_{\mathrm{s,L}} + u_{\mathrm{L}}}{a_{\mathrm{L}}}. \tag{9.4}$$

Here, the flow velocity u is based in a laboratory frame and has a positive value to the right. The shock Mach number, $M_{\mathrm{s,L}}$, refers to the upstream flow in the frame fixed on the shock wave, always having a positive value. From (9.1) and (9.3),

$$u_{\mathrm{L}*} - u_{\mathrm{L}} = -a_{\mathrm{L}}\left(\frac{p_{\mathrm{L}*}}{p_{\mathrm{L}}} - 1\right)\left(\frac{\frac{2}{\gamma(\gamma+1)}}{\frac{p_{\mathrm{L}*}}{p_{\mathrm{L}}} + \frac{\gamma-1}{\gamma+1}}\right)^{\frac{1}{2}}. \tag{9.5}$$

From (9.1) and (9.2),

$$\frac{\rho_{\mathrm{L}*}}{\rho_{\mathrm{L}}} = \frac{\frac{p_{\mathrm{L}*}}{p_{\mathrm{L}}} + \frac{\gamma-1}{\gamma+1}}{\frac{\gamma-1}{\gamma+1}\frac{p_{\mathrm{L}*}}{p_{\mathrm{L}}} + 1}. \tag{9.6}$$

If the left-running wave is an expansion fan ($p_{L*}/p_L < 1$), the Riemann invariant, J_+, is kept constant across it. From (8.44),

$$u_{L*} + \frac{2}{\gamma - 1} a_{L*} = u_L + \frac{2}{\gamma - 1} a_L. \tag{9.7}$$

The isentropic equations of (2.91) yield to

$$\frac{\rho_{L*}}{\rho_L} = \left(\frac{p_{L*}}{p_L}\right)^{\frac{1}{\gamma}} \tag{9.8}$$

$$\frac{a_{L*}}{a_L} = \left(\frac{p_{L*}}{p_L}\right)^{\frac{\gamma-1}{2\gamma}}. \tag{9.9}$$

From (9.7) and (9.9),

$$u_{L*} - u_L = \frac{2a_L}{\gamma - 1}\left\{1 - \left(\frac{p_{L*}}{p_L}\right)^{\frac{\gamma-1}{2\gamma}}\right\}. \tag{9.10}$$

To integrate the results, the pressure ratio is given as an implicit function of the flow velocity difference.

$$\frac{p_{L*}}{p_L} \equiv \Phi_L\left(\frac{u_{L*} - u_L}{a_L}\right)$$

$$\frac{u_{L*} - u_L}{a_L} = \begin{cases} -\left(\frac{p_{L*}}{p_L} - 1\right)\left\{\frac{2}{\gamma\left\{(\gamma+1)\frac{p_{L*}}{p_L} + \gamma - 1\right\}}\right\}^{\frac{1}{2}}, & \left(\frac{p_{L*}}{p_L} > 1\right) \\ \frac{2}{\gamma-1}\left\{1 - \left(\frac{p_{L*}}{p_L}\right)^{\frac{\gamma-1}{2\gamma}}\right\}, & \left(\frac{p_{L*}}{p_L} \le 1\right) \end{cases} \tag{9.11}$$

$$\frac{\rho_{L*}}{\rho_L} \begin{cases} = \frac{\frac{p_{L*}}{p_L} + \frac{\gamma-1}{\gamma+1}}{\frac{\gamma-1}{\gamma+1}\frac{p_{L*}}{p_L} + 1}, & \left(\frac{p_{L*}}{p_L} > 1\right) \\ = \left(\frac{p_{L*}}{p_L}\right)^{\frac{1}{\gamma}}, & \left(\frac{p_{L*}}{p_L} \le 1\right) \end{cases}. \tag{9.12}$$

Next, let us obtain the relation between R and R*. These states are separated by a wave propagating on the right-hand side of the contact surface. By carefully noting the direction and the invariant of the wave, the following equations are obtained. For a shock wave ($p_{R*}/p_R > 1$),

$$\frac{p_{R*}}{p_R} = 1 + \frac{2\gamma}{\gamma + 1}(M_{s,R}^2 - 1) \tag{9.13}$$

$$\frac{\rho_{R*}}{\rho_R} = \frac{(\gamma + 1)M_{s,R}^2}{(\gamma - 1)M_{s,R}^2 + 2} \tag{9.14}$$

$$u_{R*} - u_R = \frac{2a_R}{\gamma + 1}\left(M_{s,R} - \frac{1}{M_{s,R}}\right) \tag{9.15}$$

$$M_{s,R} = \frac{U_{s,R} - u_R}{a_R}. \tag{9.16}$$

Therefore,

$$u_{R*} - u_R = a_R\left(\frac{p_{R*}}{p_R} - 1\right)\left(\frac{\frac{2}{\gamma(\gamma+1)}}{\frac{p_{R*}}{p_R} + \frac{\gamma-1}{\gamma+1}}\right)^{\frac{1}{2}} \tag{9.17}$$

$$\frac{\rho_{R*}}{\rho_R} = \frac{\frac{p_{R*}}{p_R} + \frac{\gamma-1}{\gamma+1}}{\frac{\gamma-1}{\gamma+1}\frac{p_{R*}}{p_R} + 1}. \tag{9.18}$$

In the case of an expansion fan ($p_{R*}/p_R < 1$), a Riemann invariant, J_-, is kept constant. From (8.45),

$$u_{R*} - \frac{2}{\gamma - 1}a_{R*} = u_R - \frac{2}{\gamma - 1}a_R. \tag{9.19}$$

Because isentropic equations equivalent to (9.8) and (9.9) are applicable,

$$u_{R*} - u_R = \frac{2a_R}{\gamma - 1}\left\{\left(\frac{p_{R*}}{p_R}\right)^{\frac{\gamma-1}{2\gamma}} - 1\right\}. \tag{9.20}$$

Those relations are integrated as follows:

$$\frac{p_{R*}}{p_R} \equiv \Phi_R\left(\frac{u_{R*} - u_R}{a_R}\right)$$
$$\frac{u_{R*} - u_R}{a_R} = \begin{cases} \left(\frac{p_{R*}}{p_R} - 1\right)\left\{\frac{2}{\gamma\left\{(\gamma+1)\frac{p_{R*}}{p_R} + \gamma - 1\right\}}\right\}^{\frac{1}{2}}, & \left(\frac{p_{R*}}{p_R} > 1\right) \\ \frac{2}{\gamma-1}\left\{\left(\frac{p_{R*}}{p_R}\right)^{\frac{\gamma-1}{2\gamma}} - 1\right\}, & \left(\frac{p_{R*}}{p_R} \le 1\right) \end{cases} \tag{9.21}$$

$$\frac{\rho_{R*}}{\rho_R} = \begin{cases} = \frac{\frac{p_{R*}}{p_R} + \frac{\gamma-1}{\gamma+1}}{\frac{\gamma-1}{\gamma+1}\frac{p_{R*}}{p_R} + 1}, & \left(\frac{p_{R*}}{p_R} > 1\right) \\ = \left(\frac{p_{R*}}{p_R}\right)^{\frac{1}{\gamma}}, & \left(\frac{p_{R*}}{p_R} \le 1\right) \end{cases}. \tag{9.22}$$

In the extreme condition where the gas expands such that the pressure and the temperature vanish, the flow velocity has the following limiting value corresponding to its total enthalpy. Substituting (9.10) with this condition,

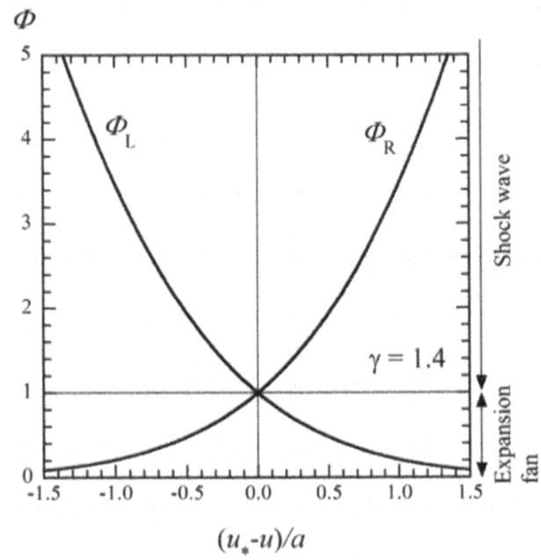

Fig. 9.4 Pressure ratios, Φ_L and Φ_R with respect to flow velocity (dimensionless)

$$u_e \equiv [u_{L^*} - u_L]_{p_{L*}/p_L \to 0} = \frac{2a_L}{\gamma - 1} \tag{9.23}$$

u_e is the highest velocity achievable by unsteady expansion, which is referred to as the *escape velocity*. A similar result in the opposite direction is obtained from (9.20).

As solutions with the appearance of a vacuum region by the left-running expansion fan, two patterns of Fig. 9.3a, b are possible. In the pattern of Fig. 9.3a, a vacuum region appears in the middle of the two states L and R. A left-running expansion fan propagates between L and L*, while a right-running expansion fan propagates between R and R*. In the pattern of Fig. 9.3b, L* and R are in vacuum. In either case, the wave-propagation velocity on the right-hand side is given by

$$c_{-,L*} = u_{L^*} - a_{L*} = u_L + u_e = u_L + \frac{2a_L}{\gamma - 1}. \tag{9.24}$$

Similar results are obtained for the right-running expansion fan, as shown in Fig. 9.3a, c. The propagation velocity of the left boundary is given by

$$c_{+,R^*} = u_{R^*} + a_{R^*} = u_R - u_e = u_R - \frac{2a_R}{\gamma - 1}. \tag{9.25}$$

Figure 9.4 depicts (9.11) and (9.21). A symbol without and with * corresponds to a quantity before and after the interaction. For the right-running wave (Φ_R), when a positive flow velocity is induced, the pressure increases, corresponding to a shock wave solution. When a negative flow velocity is induced, the pressure decreases, corresponding to an expansion fan. Φ_L is symmetrical to Φ_R, and thus opposite relations hold.

Figures 9.5, 9.6 and 9.7 show possible solution patterns in dimensional coordinates. After the interaction, a contact surface is generated between L* and R*. Across the contact surface, u and p are constant, yet ρ and other thermodynamic properties are not necessarily kept constant, as shown in Figs. 9.5b, 9.6b, and 9.7b.

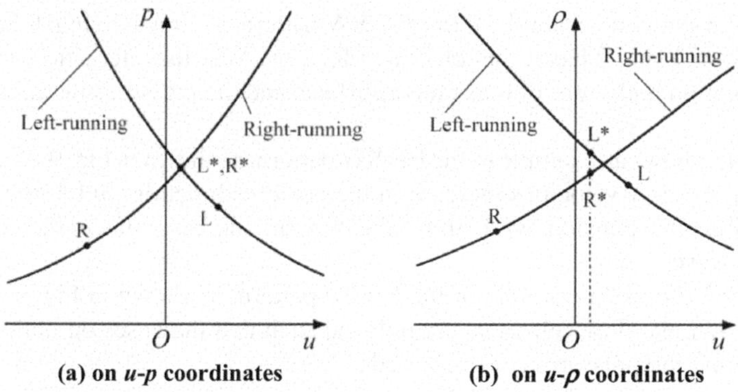

Fig. 9.5 Example of SW-SW solution (Fig. 9.2a)

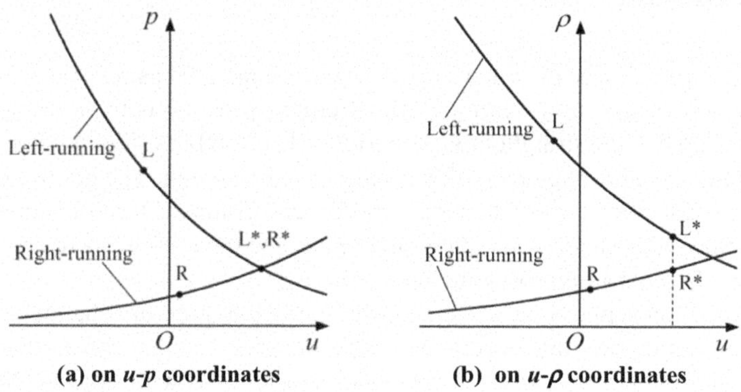

Fig. 9.6 Example of EF-SW solution (Fig. 9.2b)

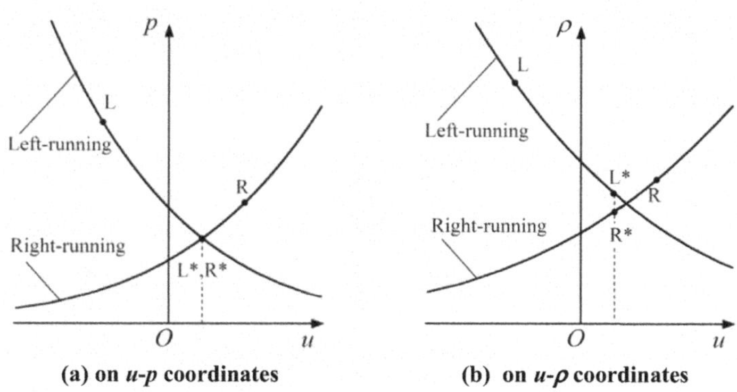

Fig. 9.7 Example of EF-EF solution (Fig. 9.2d)

Figure 9.5 shows an example of the SW-SW pattern, as shown in Fig. 9.2a. Before the interaction, L and R have counter-flow velocities. After the interaction, two shock waves appear on both sides of the contact surface, and the pressures increase on both sides.

Figure 9.6 shows an example of the EF-SW pattern, as shown in Fig. 9.2b. After the interaction, the flow velocities increase in the positive direction. In L*, the pressure decreases due to expansion, while in R* the pressure increases due to the generation of a shock wave.

Figure 9.7 shows an example of the EF-EF pattern, as shown in Fig. 9.2d. After the interaction, the flow velocities are induced such that the gases on both sides are expanded, and the pressures are decreased.

9.2 Shock Tube

The *shock tube* is a device in which a shock wave is generated using a pressure difference between two gases as a solution of the Riemann problem with $u_L = u_R = 0$. As observed in Fig. 9.8, the straight tube with a closed end on both sides is separated into high- and low- pressure channels using a sheet of a diaphragm. The gas in the respective channels can be of different species and/or temperature. Once the diaphragm is suddenly removed, a shock wave starts propagating in the low-pressure channel, and an expansion fan in the high-pressure channel.

The test section is placed near the end wall in the low-pressure channel, equipped with windows, pressure transducers, and other sensors. Using a shock tube, we can investigate the interaction between a normal shock wave with objects. Moreover, as will be seen later, a shock tube is useful for investigating characteristics of a high-temperature gas generated behind the reflected shock wave.

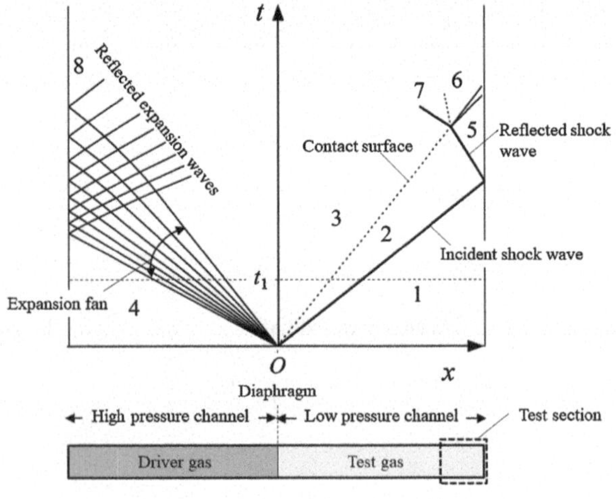

Fig. 9.8 Example of wave diagram in shock tube operation, with air in the low-pressure channel, and helium in the high-pressure channel, $p_1 = 1.0 \times 10^4$[Pa], $T_1 = 290$ [K], $p_4 = 1.0 \times 10^5$[Pa], $T_4 = 290$ [K]

As shown in Fig. 9.8, the initial states at $t = 0$ in the low- and high-pressure channels are labeled "1" and "4" ($p_4 > p_1$), respectively. Once the diaphragm is removed, an *incident shock wave* starts propagating in the low-pressure channel. The state behind the shock wave is labeled "2". After the removal of the diaphragm, the two gases that are separated by the diaphragm directly contact with each other at the contact surface move at a velocity u_2 to the right. The state left to the contact surface is labeled "3" ($u_3 = u_2$). In the high-pressure channel, an expansion fan is formed. In the expansion fan, the gas experiences unsteady expansion, and the flow state varies smoothly. The leading wave of the expansion fan is a sound wave propagating with a velocity $-a_4$. The tail of the expansion fan corresponds to the state 3.

The initial conditions 4 and 1 correspond to the states L and R in the Riemann problem described in Sect. 9.1. The respective gas can have different species, temperature and/or value of γ. Applying (9.11) and (9.21),

$$u_3 = \frac{2a_4}{\gamma_4 - 1}\left\{1 - \left(\frac{p_3}{p_4}\right)^{\frac{\gamma_4-1}{2\gamma_4}}\right\} \tag{9.26}$$

$$u_2 = a_1\left(\frac{p_2}{p_1} - 1\right)\left\{\frac{\frac{2}{\gamma_1(\gamma_1+1)}}{\frac{p_2}{p_1} + \frac{\gamma_1-1}{\gamma_1+1}}\right\}^{\frac{1}{2}}. \tag{9.27}$$

For the contact surface,

$$p_2 = p_3 \tag{9.28}$$

$$u_2 = u_3 \tag{9.29}$$

From (9.26) to (2.29),

$$\frac{p_4}{p_1} = \frac{p_2}{p_1}\left\{1 - \frac{(\gamma_4 - 1)\frac{a_1}{a_4}\left(\frac{p_2}{p_1} - 1\right)}{\sqrt{2\gamma_1\left[2\gamma_1 + (\gamma_1 + 1)\left(\frac{p_2}{p_1} - 1\right)\right]}}\right\}^{-\frac{2\gamma_4}{\gamma_4-1}}. \tag{9.30}$$

Using (9.30), the pressure ratio p_2/p_1 is obtained as an implicit function of the initial operation parameters ($p_4/p_1, a_1/a_4, \gamma_1, \gamma_4$). The shock Mach number, M_s, is obtained from (9.13).

$$M_s = \sqrt{\frac{\gamma_1 + 1}{2\gamma_1}\frac{p_2}{p_1} + \frac{\gamma_1 - 1}{2\gamma_1}}. \tag{9.31}$$

The densities of "2" and "3" are obtained from (9.22) and (9.12), respectively.

$$\frac{\rho_2}{\rho_1} = \frac{\frac{p_2}{p_1} + \frac{\gamma_1 - 1}{\gamma_1 + 1}}{\frac{\gamma_1 - 1}{\gamma_1 + 1}\frac{p_2}{p_1} + 1} \tag{9.32}$$

$$\frac{\rho_3}{\rho_4} = \left(\frac{p_3}{p_4}\right)^{\frac{1}{\gamma_4}} = \left(\frac{p_2}{p_4}\right)^{\frac{1}{\gamma_4}}. \tag{9.33}$$

The speeds of sound are as follows.

$$\frac{a_2}{a_1} = \left(\frac{p_2}{p_1}\right)^{\frac{1}{2}}\left(\frac{\rho_2}{\rho_1}\right)^{-\frac{1}{2}} = \frac{\{2\gamma_1 M_S^2 - (\gamma_1 - 1)\}^{\frac{1}{2}}\{(\gamma_1 - 1)M_s^2 + 2\}^{\frac{1}{2}}}{(\gamma_1 + 1)M_s} \tag{9.34}$$

$$\frac{a_3}{a_4} = \left(\frac{p_3}{p_4}\right)^{\frac{\gamma_4 - 1}{2\gamma_4}} = \left(\frac{p_2}{p_4}\right)^{\frac{\gamma_4 - 1}{2\gamma_4}}. \tag{9.35}$$

Let us obtain the pressure distribution in the expansion fan. Here, the origin is set to O in Fig. 9.8. From (8.44), (8.45), and (2.91),

$$u + \frac{2a}{\gamma - 1} = \frac{2a_4}{\gamma - 1} \tag{9.36}$$

$$\left(\frac{\mathrm{d}x}{\mathrm{d}t}\right) = c_- = u - a \tag{9.37}$$

$$\frac{a}{a_4} = \left(\frac{p}{p_4}\right)^{\frac{\gamma_4 - 1}{2\gamma_4}}. \tag{9.38}$$

From these equations, the relation between the lay slope and the pressure is given by

$$\left(\frac{\mathrm{d}x}{\mathrm{d}t}\right) = \left\{1 - \frac{\gamma + 1}{2}\left(\frac{p}{p_4}\right)^{\frac{\gamma_4 - 1}{2\gamma_4}}\right\}\frac{2a_4}{\gamma - 1}$$
$$\frac{p}{p_4} = \left[\frac{\gamma - 1}{\gamma + 1}\left\{\frac{2}{\gamma - 1} - \frac{1}{a_4}\left(\frac{\mathrm{d}x}{\mathrm{d}t}\right)\right\}\right]^{\frac{2\gamma_4}{\gamma_4 - 1}}. \tag{9.39}$$

Herewith, all conditions of "2" and "3" are obtained. Figure 9.9 shows the distributions of the pressure, density, and flow velocity, and Fig. 9.10 shows the solution with respect to dimensionless pressure-flow velocity coordinates.

Figure 9.11 shows the shock tube operation performance as the function of the initial pressure ratio. All the quantities shown in Fig. 9.11a are an increasing function of p_4/p_1. Only ρ_2/ρ_1 becomes saturated with $p_4/p_1 \to \infty$. As seen in (9.30) and Fig. 9.11b, the higher a_4/a_1, the stronger the shock wave becomes, that is the higher p_2/p_1 becomes. To do so, a light gas, hydrogen or helium, is used. As calculated using (8.5) and (8.6), the speed of sound of molecular hydrogen gases is 3.8 times

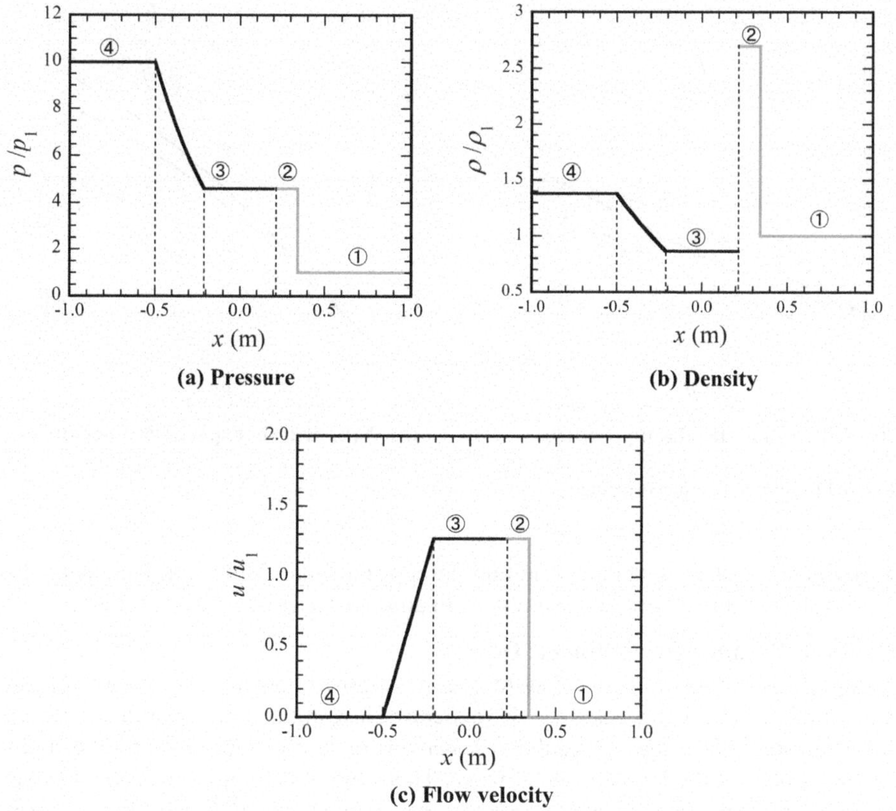

Fig. 9.9 Pressure, density, and flow velocity distributions at $t = t_1$ in Fig. 9.8

Fig. 9.10 Shock tube operation on dimensionless u-p coordinates

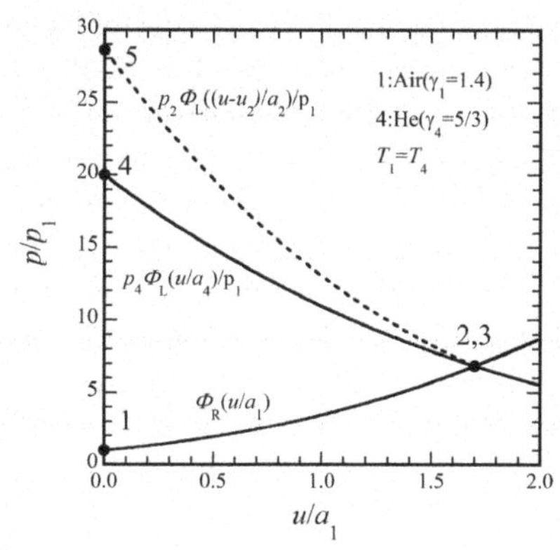

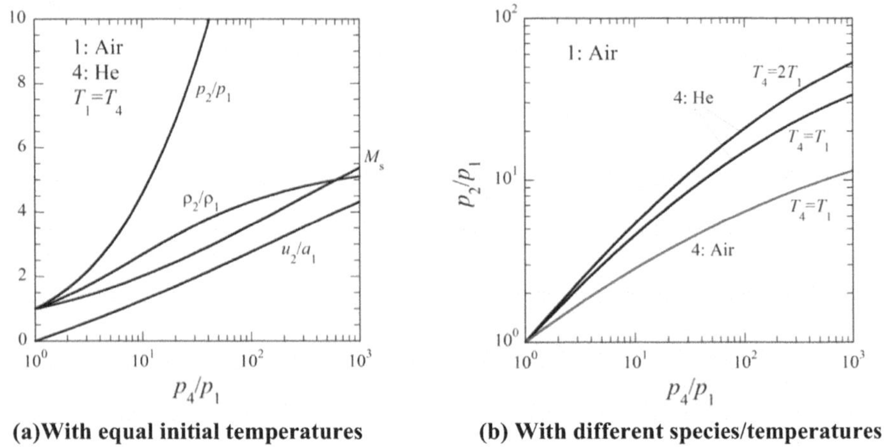

Fig. 9.11 Shock tube performance

as high as that of air, and that of helium 2.9 times as high. Moreover, increasing the temperature of the driver gas also results in an increase in the shock strength.

Column: Counter-Driver Shock Tube

Using a shock tube in the usual style of Fig. 9.8, the interaction between the post-shock flow "2" and the reflected shock wave can be investigated. However, as explained later, the strength of the reflected shock wave is uniquely determined by the shock Mach number of the incident shock wave. Therefore, their conditions cannot be set in an independent manner, and the counter-driver shock tube (CD-ST, Fig C9.1) was developed to perform this function [1]. It has two high-pressure channels ("drivers") on both sides. By setting the relative timing of diaphragm rupture, the interaction of the interest can be placed at a desired location. In Fig. C9.1a, for example, the interaction between the post-shock flow R-PSF behind the incident shock wave R-SW(i) driven by the right driver (R-Driver) and the transmitted shock wave L-SW(t) driven by the left driver (L-Driver) is set to occur at the test section. Each diaphragm is made of cellophane and ruptured using the needle (Fig. C9.1c) driven by a pneumatic cylinder, which in turn is driven by the high-pressure air supplied through an electromagnetic valve. The mutual diaphragm rupture timing is controlled with electrical trigger signals to the respective electromagnetic valves, as shown in Fig. C9.1d.

In the CD-ST, the conditions of a shock wave and a counter flow can be controlled in an independent manner. Naturally, experiments of the interaction between two shock waves with different strengths are also possible.

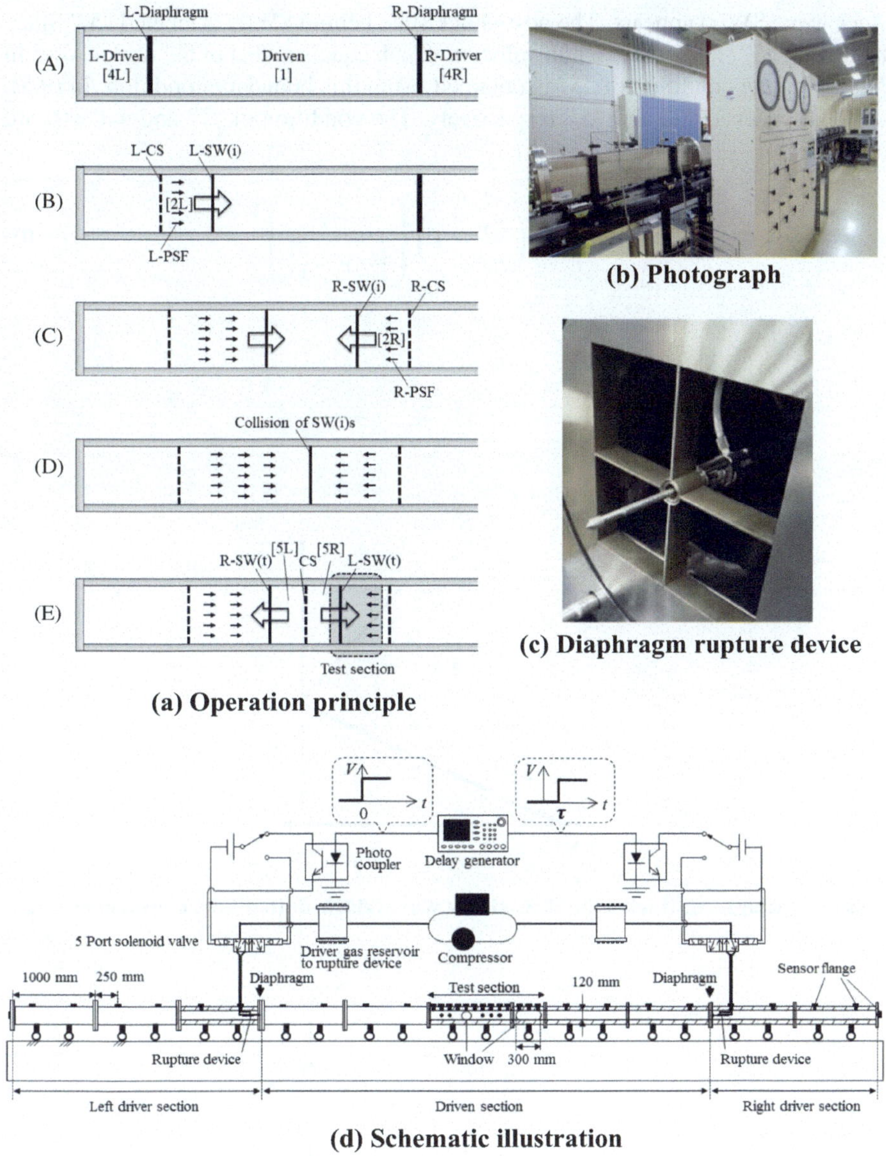

Fig. C9.1 Counter-driver shock tube at Nagoya University

9.3 Reflection of Normal Shock Wave

Let us consider a normal shock wave that is reflected on a solid wall as an application of the Riemann problem. This occurs on the right end of the wall in the shock tube in Fig. 9.8. After the incident shock wave SW(i) is reflected on the wall, the *reflected* *shock wave* SW(r) appears. The post-shock state behind SW(r) is labeled "5." Since the flow in "5" should have a null velocity, which equals to that of the wall, shown in Fig. 9.13a, the condition of "5" is obtained using this boundary condition. In (9.5), L and L* refer to "2" and "5," respectively. The condition of "2" and $u_5 (= 0)$ are known.

$$-u_2 = -a_2\left(\frac{p_5}{p_2} - 1\right)\left\{\frac{\frac{2}{\gamma_1(\gamma_1+1)}}{\frac{p_5}{p_2} + \frac{\gamma_1-1}{\gamma_1+1}}\right\}^{\frac{1}{2}}. \tag{9.40}$$

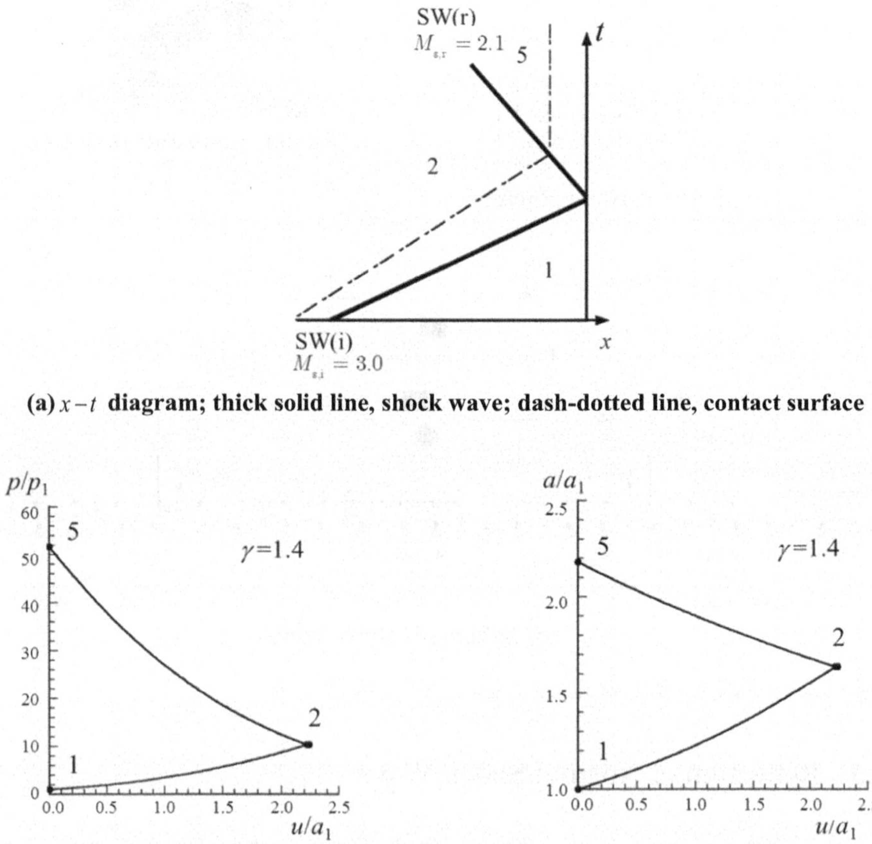

(a) $x-t$ **diagram; thick solid line, shock wave; dash-dotted line, contact surface**

(b) $p-u$ **curves**

(c) $a-u$ **curve**

Fig. 9.13 Reflection of normal shock wave on a solid wall

Solving (9.40),

$$\frac{p_5}{p_2} = 1 + \gamma_1 \frac{u_2}{a_2} \left\{ \frac{\gamma_1 + 1}{4} \frac{u_2}{a_2} + \sqrt{\left(\frac{\gamma_1 + 1}{4} \frac{u_2}{a_2} \right)^2 + 1} \right\}. \tag{9.41}$$

Using (9.31)–(9.33), the shock Mach number of SW(r), $M_{s,r}$, the density ratio, and the speed of sound ratio are obtained.

$$M_{s,r} = \sqrt{\frac{\gamma_1 + 1}{2\gamma_1} \frac{p_5}{p_2} + \frac{\gamma_1 - 1}{2\gamma_1}} \tag{9.42}$$

$$\frac{\rho_5}{\rho_2} = \frac{\frac{p_5}{p_2} + \frac{\gamma_1 - 1}{\gamma_1 + 1}}{\frac{\gamma_1 - 1}{\gamma_1 + 1} \frac{p_5}{p_2} + 1} \tag{9.43}$$

$$\frac{a_5}{a_2} = \left(\frac{T_5}{T_2} \right)^{\frac{1}{2}} = \frac{\{2\gamma_1 M_{s,r}^2 - (\gamma_1 - 1)\}^{\frac{1}{2}} \{(\gamma_1 - 1) M_{s,r}^2 + 2\}^{\frac{1}{2}}}{(\gamma_1 + 1) M_{s,r}}. \tag{9.44}$$

Note here that $M_{s,r}$ is the shock Mach number referring to "2", which has a positive flow velocity. For SW(i), (9.17) and (9.18) are applied.

$$u_2 = a_1 \left(\frac{p_2}{p_1} - 1 \right) \left(\frac{\frac{2}{\gamma_1(\gamma_1 + 1)}}{\frac{p_2}{p_1} + \frac{\gamma_1 - 1}{\gamma_1 + 1}} \right)^{\frac{1}{2}} \tag{9.45}$$

$$\frac{\rho_2}{\rho_1} = \frac{\frac{p_2}{p_1} + \frac{\gamma_1 - 1}{\gamma_1 + 1}}{\frac{\gamma_1 - 1}{\gamma_1 + 1} \frac{p_2}{p_1} + 1}. \tag{9.46}$$

From (9.40), (9.45), and (9.46),

$$\left\{ \left(\frac{\gamma_1 - 1}{\gamma_1 + 1} \frac{p_2}{p_1} + 1 \right) \frac{p_5}{p_2} - \frac{3\gamma_1 - 1}{\gamma_1 + 1} \frac{p_2}{p_1} + \frac{\gamma_1 - 1}{\gamma_1 + 1} \right\} \left(\frac{p_2}{p_1} \frac{p_5}{p_2} - 1 \right) = 0. \tag{9.47}$$

With $p_5 \neq p_1$,

$$\frac{p_5}{p_2} = \frac{\left(\frac{3\gamma_1 - 1}{\gamma_1 - 1} \right) \frac{p_2}{p_1} - 1}{\frac{p_2}{p_1} + \frac{\gamma_1 + 1}{\gamma_1 - 1}}. \tag{9.48}$$

Substituting (9.42)–(9.44) with (9.48), the post-shock condition of SW(r) is related to the condition of SW(i). For example, applying (9.13) and (9.1) to SW(i) and SW(r), respectively, the following equation is obtained.

$$M_{s,r} = \sqrt{\frac{2\gamma_1 M_{s,i}^2 - (\gamma_1 - 1)}{(\gamma_1 - 1)M_{s,i}^2 + 2}}. \tag{9.49}$$

Figure 9.14 shows the variation in the pressure and temperature ratios and $M_{s,r}$ as a function of $M_{s,i}$. For linear waves, including sound waves, pressures of an incident and reflected waves can be superimposed on each other. The pressure amplitude equals to the sum of the corresponding amplitude. However, a shock wave is a nonlinear wave, and thus such superimposition is not applicable. Instead, the pressure of the post-shock state "5" is obtained by multiplying the pressure ratios across the incident and reflected shock waves, as shown in Fig. 9.14a. The same scenario applies to the

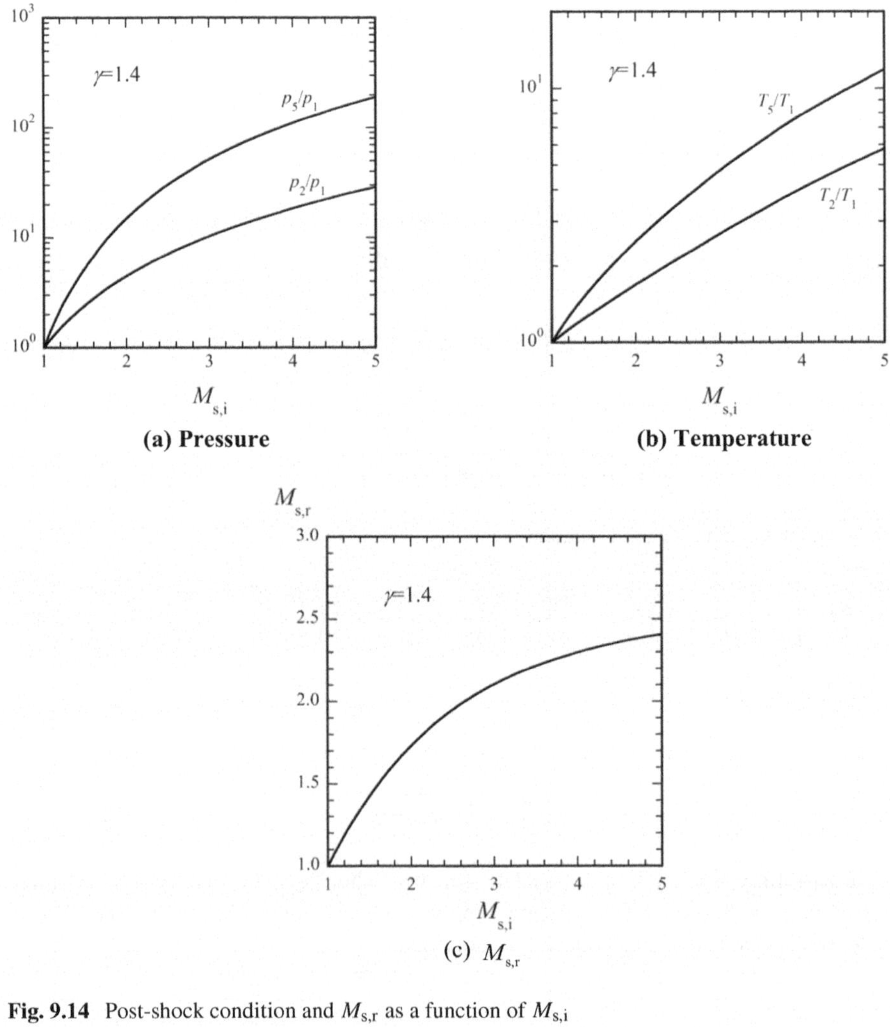

Fig. 9.14 Post-shock condition and $M_{s,r}$ as a function of $M_{s,i}$

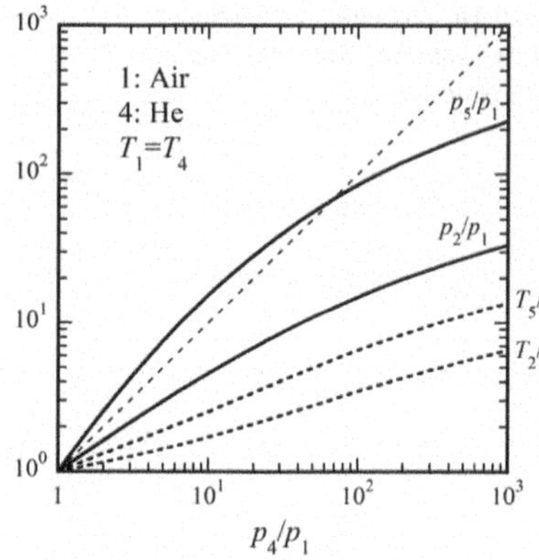

Fig. 9.15 Pressure and temperature ratios generated in a shock tube, as a function of the initial pressure ratio

post-shock temperature (Fig. 9.14b) As seen in Fig. 9.14c, $M_{s,r}$ is an increasing function of $M_{s,i}$. The higher $M_{s,i}$, the higher the pressure and temperature ratios.

In this way, the shock tube is a useful tool to generate the high-pressure/temperature state by utilizing the principle of unsteady, compressible fluid dynamics. It is often used to measure chemical reaction rates at high temperature. The pressure and temperature ratios obtained as a function of the initial pressure ratio, by a shock tube is shown in Fig. 9.15. Here, a regime where $p_5 > p_4$ exists. This implies that when using the unsteady fluid dynamics principle, a pressure higher than the initial value can be generated, which is not possible in steady compression. However, this does not mean that the total energy of the whole system is increased, but that part of the system can become more energetic during a limited period.

9.4 Reflection of Expansion Fan

In the high-pressure channel of the shock tube in Fig. 9.8, after the diaphragm is ruptured, the expansion fan propagates to the left end wall and is reflected there. Unlike the shock wave, it is composed of expansion waves, which form a fan in the x-t diagram. Accordingly, the reflected waves also span a finite width.

Let us obtain the condition of "8," the state behind the reflected expansion fan in Fig. 9.8, with the condition of "3" and "4" being known. The reflected waves are right-running expansion waves, to which (9.20) applies. Moreover, because $u_8 = 0$,

$$-u_3 = \frac{2a_3}{\gamma_4 - 1}\left\{\left(\frac{p_8}{p_3}\right)^{\frac{\gamma_4-1}{2\gamma_4}} - 1\right\} \tag{9.50}$$

$$\frac{p_8}{p_3} = \left(1 - \frac{\gamma_4 - 1}{2}M_3\right)^{\frac{2\gamma_4}{\gamma_4-1}}. \tag{9.51}$$

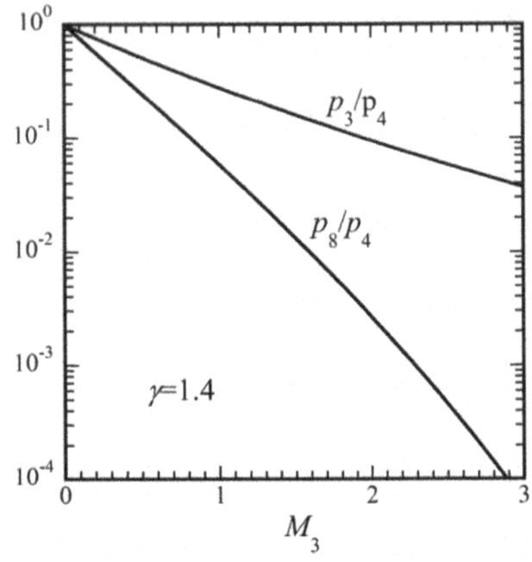

Fig. 9.16 Pressure variation behind incident and reflected expansion fan

The relation between "3" and "4" are obtained using (9.9) and (9.10) and $u_4 = 0$.

$$\frac{p_4}{p_3} = \left(1 + \frac{\gamma_4 - 1}{2} M_3\right)^{\frac{2\gamma_4}{\gamma_4 - 1}}. \tag{9.52}$$

Therefore,

$$\frac{p_8}{p_4} = \left(\frac{1 - \frac{\gamma_4 - 1}{2} M_3}{1 + \frac{\gamma_4 - 1}{2} M_3}\right)^{\frac{2\gamma_4}{\gamma_4 - 1}} \tag{9.53}$$

$$\frac{\rho_8}{\rho_4} = \left(\frac{p_8}{p_4}\right)^{\frac{1}{\gamma_4}}. \tag{9.54}$$

Figure 9.16 shows the pressure variation behind the incident and reflected expansion fan as a function of M_3. Regarding the expansion waves, the pressure variation is nonlinear, lowering to a large extent with increasing M_3.

In the shock tube operation, if the length of the high-pressure channel is not sufficiently long, the incident shock wave is contaminated by the reflected expansion fan, and thereby the effective test time may be shortened or even vanish. Figure 9.17 shows the trajectory of the leading wave of the reflected expansion fan. The effective test flow that is not contaminated by the reflected expansion fan is colored gray. As long as the reflected expansion waves do not arrive ($0 < x < x_m$), the effective test time behind the (incident) shock wave increases in proportion to the distance from the origin. However, for $x > x_m$ the reflected expansion wave propagates in "2" before the contact surface arrives, thereby the test time becomes shortened. As shown in Fig. 9.17, there is an optimum distance, x_m, to maximize the effective test time to τ_m. In order to obtain this condition, let us trace the trajectory of the leading wave of the reflected expansion wave.

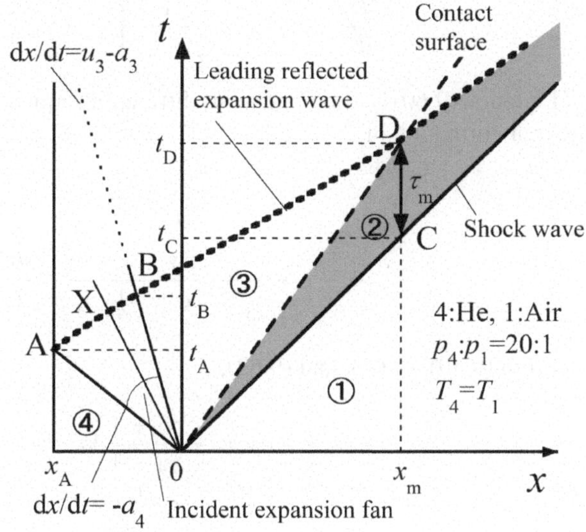

Fig. 9.17 Reflection of reflected expansion fan on the left end wall. Only the leading wave is plotted

For a calorically perfect gas, x_m is obtained analytically. In Fig. 9.17, the reflection point of the leading wave of the reflected expansion fan is labeled "A", and the point where this wave terminates the interaction with the incident expansion fan is labeled "B." X(x, t) is an arbitrary point in between A and B on the trajectory. In region 3, the flow condition is uniform. If we can obtain t_B, at which the leading reflected wave terminates the interaction, we can readily trace the trajectory to the contact surface. At A, the following equation holds.

$$u_4 - a_4 = -a_4 = \frac{x_A}{t_A}. \tag{9.55}$$

At X, from the geometrical relation and Riemann invariant,

$$u - a = \frac{x}{t} \tag{9.56}$$

$$u + \frac{2a}{\gamma_4 - 1} = u_4 + \frac{2a_4}{\gamma_4 - 1} = \frac{2a_4}{\gamma_4 - 1}. \tag{9.57}$$

Solving (9.55) to (9.57)

$$u = a + \frac{x}{t} \tag{9.58}$$

$$a = \frac{2a_4}{\gamma_4 + 1} - \frac{\gamma_4 - 1}{\gamma_4 + 1}\frac{x}{t}. \tag{9.59}$$

Because the reflected wave propagates at a characteristic velocity c_+,

$$\frac{dx}{dt} = u + a = \frac{1}{\gamma_4 + 1}\left\{(3 - \gamma_4)\frac{x}{t} + 4a_4\right\}. \tag{9.60}$$

Equation (9.60) is a homogeneous differential equation, which can be solved using the transformation of

$$y = \frac{x}{t} \tag{9.61}$$

$$\frac{dy}{-\frac{2(\gamma_4-1)}{\gamma_4+1}y + \frac{4}{\gamma_4+1}a_4} = \frac{dt}{t}. \tag{9.62}$$

From (9.59), (9.61), and (9.62),

$$\mathrm{d}y = -\frac{\gamma_4 + 1}{\gamma_4 - 1}\mathrm{d}a$$

$$\frac{-\frac{\gamma_4+1}{2(\gamma_4-1)}\mathrm{d}a}{a} = \frac{dt}{t}. \tag{9.63}$$

Integrating (9.63) with the condition that A is equivalent to "4,"

$$\frac{t_\mathrm{X}}{t_\mathrm{A}} = \left(\frac{a_\mathrm{X}}{a_\mathrm{A}}\right)^{-\frac{\gamma_4+1}{2(\gamma_4-1)}}. \tag{9.64}$$

Applying (9.64) to B, the condition that is equivalent to that of "3,"

$$\frac{t_\mathrm{B}}{t_\mathrm{A}} = \left(\frac{a_3}{a_4}\right)^{-\frac{\gamma_4+1}{2(\gamma_4-1)}}. \tag{9.65}$$

The duration of the wave passage is expressed by such a simple equation. Using this, and from Fig. 9.17,

$$x_\mathrm{m} = U_{\mathrm{s,i}} t_\mathrm{C} = u_3 t_\mathrm{D} = (u_3 + a_3)(t_\mathrm{D} - t_\mathrm{B}) + x_\mathrm{B}$$
$$x_\mathrm{B} = (u_3 - a_3)t_\mathrm{B}.$$

Therefore,

$$t_\mathrm{D} = 2t_\mathrm{B}$$
$$t_\mathrm{C} = \frac{2u_3}{U_{\mathrm{s,i}}}t_\mathrm{B}.$$

From the above equations and substituting (9.9) and (9.26) with (9.65),

$$t_B = \left(\frac{a_3}{a_4}\right)^{-\frac{\gamma_4+1}{2(\gamma_4-1)}} t_A = \left(\frac{a_3}{a_4}\right)^{-\frac{\gamma_4+1}{2(\gamma_4-1)}} \frac{|x_A|}{a_4} = \left(\frac{p_3}{p_4}\right)^{-\frac{\gamma_4+1}{4\gamma_4}} \frac{|x_A|}{a_4}$$

$$x_m = \frac{4}{\gamma_4 - 1}\left\{1 - \left(\frac{p_3}{p_4}\right)^{\frac{\gamma_4-1}{2\gamma_4}}\right\}\left(\frac{p_3}{p_4}\right)^{-\frac{\gamma_4+1}{4\gamma_4}} |x_A| \tag{9.66}$$

$$\tau_m = t_D - t_C = 2\left[1 - \frac{2}{\gamma_4 - 1}\frac{a_4}{U_{s,i}}\left\{1 - \left(\frac{p_3}{p_4}\right)^{\frac{\gamma_4-1}{2\gamma_4}}\right\}\right]\left(\frac{p_3}{p_4}\right)^{-\frac{\gamma_4+1}{4\gamma_4}} \frac{|x_A|}{a_4}. \tag{9.67}$$

In (9.66) and (9.67), all the variables on the right-hand side including $U_{s,i}$ and p_3/p_4 are determined from the shock tube operation condition, and thereby x_m and τ_m are obtained from the operation parameters. They are proportional to the length of the high-pressure channel, $|x_A|$.

9.5 Shock–Shock Interactions

9.5.1 Head-on Collision

Figure 9.18 shows the head-on collision of SW(a) and SW(b). SW(a) propagates with $M_{s,a} = 3$ to the right, in the positive direction of the x coordinate, SW(b) with $M_{s,b} = 2$ to the left. The state ahead of the shock waves is labeled as "1"; those behind SW(a) and SW(b) are labeled "2" and "3," respectively. After the collision, the shock waves change their strength and direction of propagation as SW(c) and SW(d). Between them, a contact surface that separates "4" and "5" appears. The condition of "4" and "5" is obtained in Fig. 9.18c as the intersection between the left-running shock polar $\Phi_{-,2}$ and the right-running shock polar $\Phi_{+,3}$. The pressures in "4" and "5" are equal to each other as the flow velocities. However, across the contact surface, the speed of sound (temperature and density) is different. This is because the shock compression processes and the entropy productions are different on the respective sides, although the pressures become equilibrated. The appearance of the contact surface is an important difference from the interaction between isentropic waves dealt with in Chap. 8.

In "4" and "5," the pressure rises to a very high value. For example, in the case of Fig. 9.18, the pressure increment behind SW(a) and SW(b) equals $(p_2-p_1)/p_1 = 9.3$ and $(p_3 - p_1)/p_1 = 3.5$. If the linear relation is applied, both $(p_4 - p_1)/p_1$ and $(p_5 - p_1)/p_1$ would equal to 12.8. However, the solution is nonlinear, and thus $(p_4 - p_1)/p_1 = (p_5 - p_1)/p_1 = 28.3$. As observed in this example, the shock wave is a highly nonlinear wave; the stronger the shock wave, the more significant its nonlinearity, i.e., the increment in the pressure ratio, will become. Utilizing this characteristic, we can generate high-pressure/temperature states.

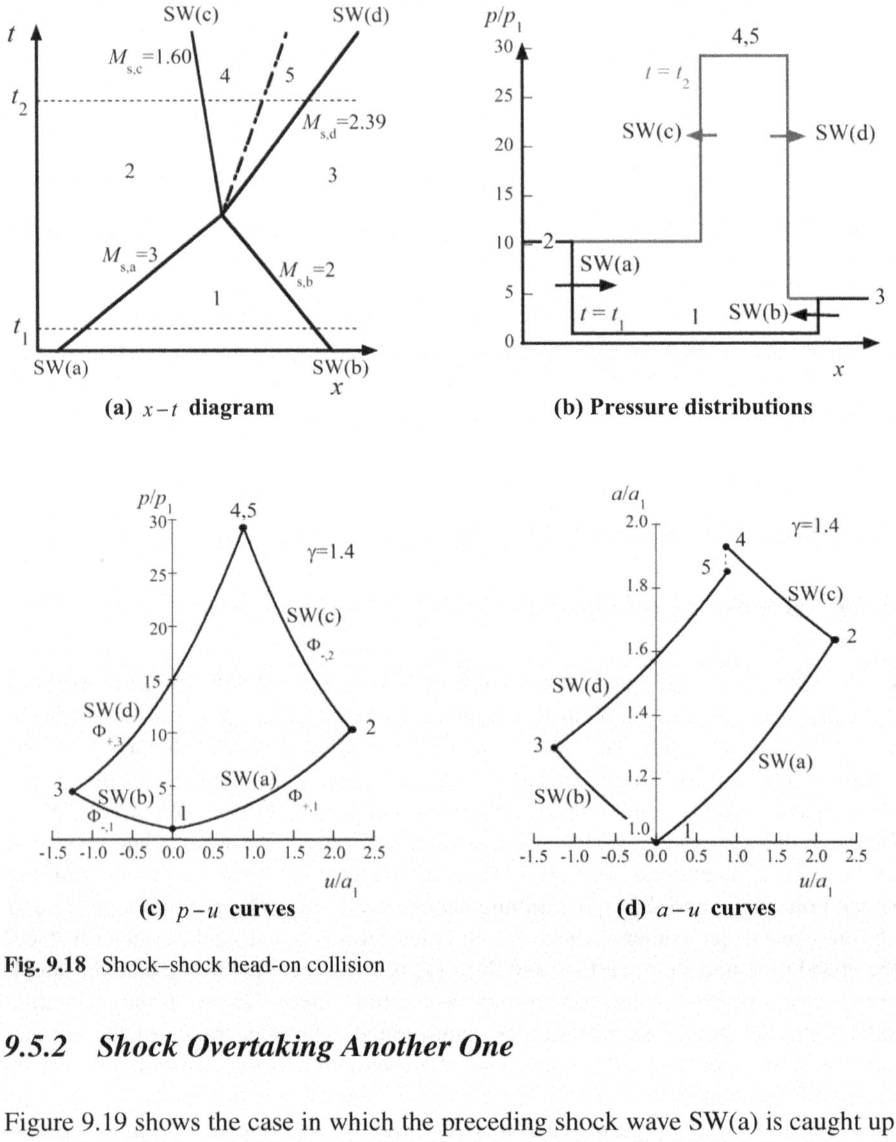

Fig. 9.18 Shock–shock head-on collision

9.5.2 Shock Overtaking Another One

Figure 9.19 shows the case in which the preceding shock wave SW(a) is caught up with by SW(b) from behind. After catching up, the strength of SW(a) is enhanced; yet behind the contact surfaces an expansion fan is generated. The pressure across the CS is constant, whereas the speeds of sound are different.

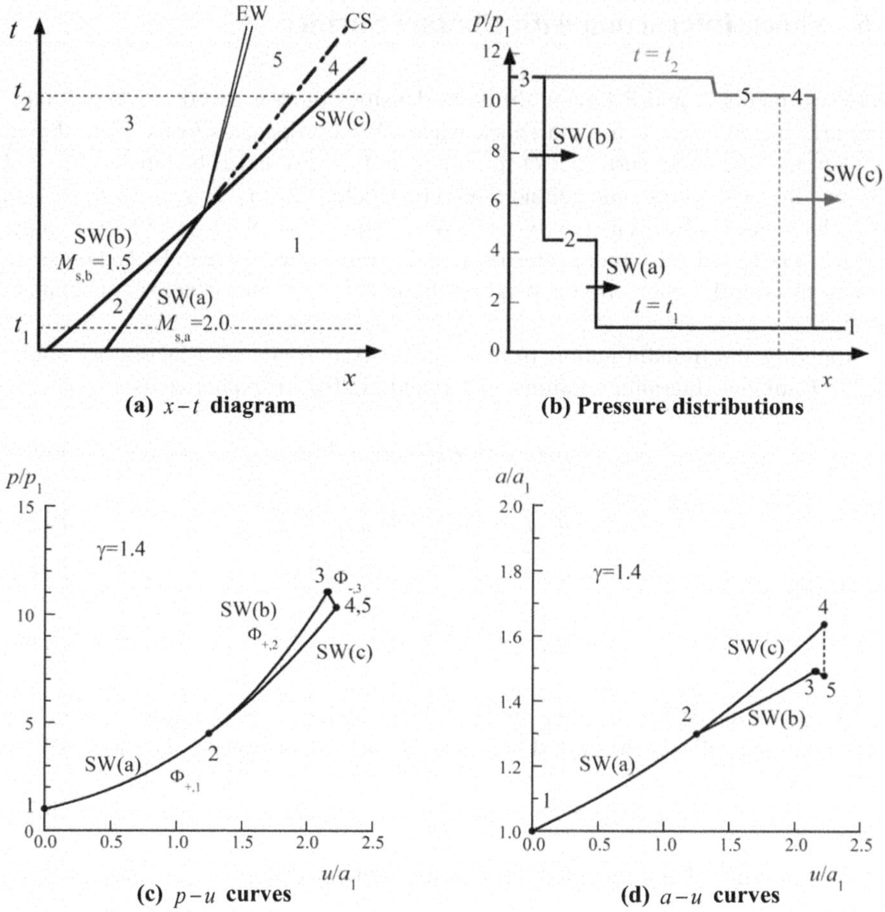

Fig. 9.19 A shock wave overtaking another one

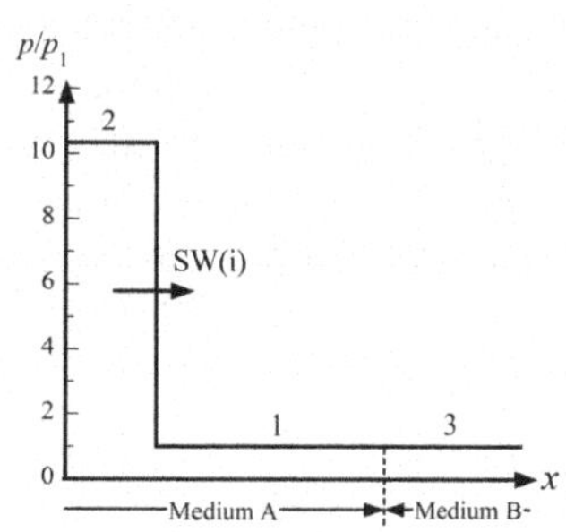

Fig. 9.20 Shock interaction with contact surface

9.6 Shock Interaction with Contact Surface

Here two media A and B having different densities (and/or speeds of sound, temperature) are in contact. In A, a shock wave SW(i) propagates to the right, being incident to the contact surface. Let the initial states of A and B be labeled "1" and "3," and the post-shock state behind SW(i) be labeled "2" ($u_1 = u_3 = 0$, $p_1 = p_3$). In A, the pressure always increases past SW(i). After the interaction with the contact surface, a reflected wave propagates in A, and a transmitted wave in B. The pressure variations after the interaction depend on the interface condition across the contact surface.

Applying the transformation from a shock-fixed frame to a laboratory frame (4.25), Rankine–Hugoniot relations, (4.19) and (4.20), are expressed as

$$\rho_1 U_{\mathrm{s,i}} = \rho_2 \left(U_{\mathrm{s,i}} - u_2 \right) \tag{9.68}$$

$$\rho_1 U_{\mathrm{s,i}}^2 + p_1 = \rho_2 \left(U_{\mathrm{s,i}} - u_2 \right)^2 + p_2. \tag{9.69}$$

Therefore,

$$p_2 - p_1 = \rho_1 U_{\mathrm{s,i}} u_2. \tag{9.70}$$

Note here that (9.70) is equivalent to (1.23), which gives an impulsive force. The same relation applies to the post-shock state "4" behind the transmitted shock SW(t).

$$p_4 - p_3 = p_4 - p_1 = \rho_3 U_{\mathrm{s,t}} u_4. \tag{9.71}$$

The condition that a reflected wave is not generated is $p_4 = p_2$ and $u_4 = u_2$, yielding

$$\rho_1 U_{\mathrm{s,i}} = \rho_3 U_{\mathrm{s,t}} \tag{9.72}$$

ρU is referred to as the *shock impedance*. From (9.72), the condition of no reflected wave is equivalent to the equality in the shock impedance. However, (9.72) is not explicitly set using the initial conditions, because in general U is not known a priori.

In the limiting case of an infinitesimally weak shock wave, U approximately equals to a, and (9.72) to

$$\rho_1 a_1 = \rho_3 a_3 \tag{9.73}$$

ρa is known as *acoustic impedance*, and it is a more convenient parameter, because it is determined only by thermodynamic properties of the medium. On the interface between media with an equal acoustic impedance, sound waves are not reflected. Recalling that in Sect. 4.2.1.3, a shock wave solution is equivalent to an isentropic wave to the third order in its pressure, the above reflection criterion is effectively

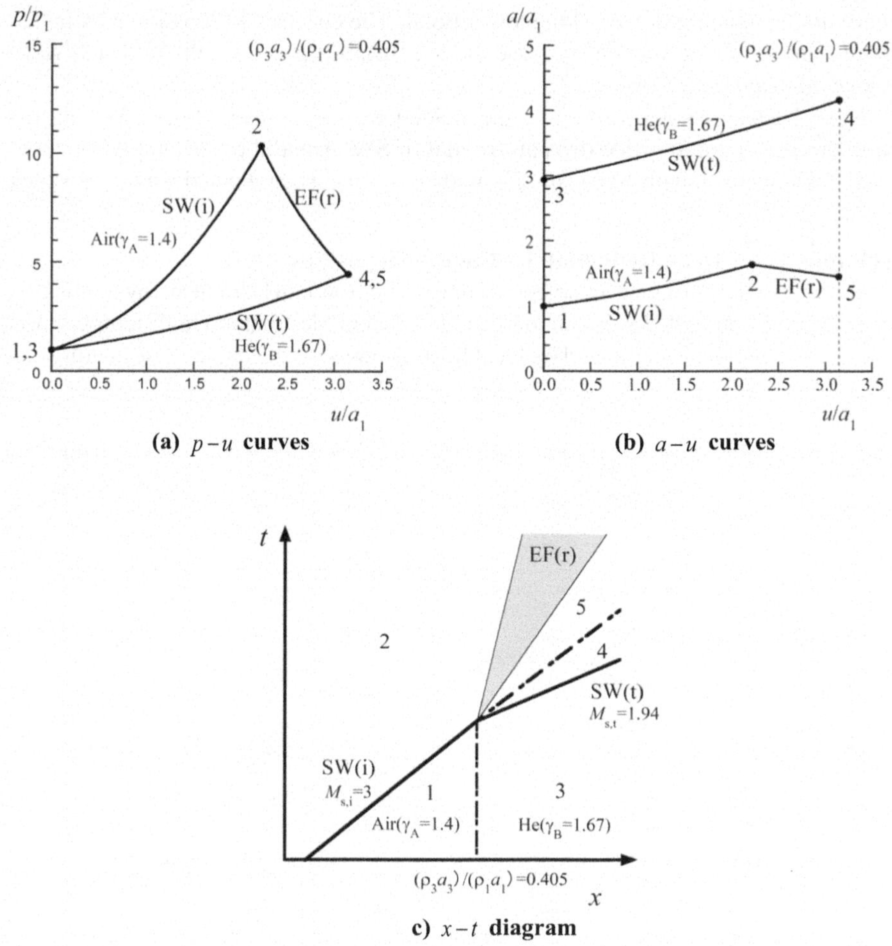

Fig. 9.21 Shock interaction with contact surface from high-to-low acoustic impedance ($\rho_1 a_1 > \rho_3 a_3$). Gray zone marks the expansion fan

applicable to a weak shock wave. The pressure increment, Δp, approximately equals to

$$\Delta p = \rho a \Delta u \tag{9.74}$$

This implies that the slope of $p - u$ equals the acoustic impedance. The type of the reflected wave, either the shock SW(r) or the expansion fan EF(r), is determined from the relative magnitude of their acoustic impedances.

Let us consider the case where SW(i) is incident on the interface from a high-to-low acoustic impedance, $\rho_1 a_1 > \rho_3 a_3$ (Fig. 9.21). In $p - u$ coordinates, the respective shock polars are plotted from the common initial state "1" and "3". In this case, SW(i) curve marks upper, and SW(3) the lower polar. The curve of SW(t) drawn from "2" should intersect with the SW(t) curve at "5". Because $p_5 < p_2$, the reflected wave is an expansion fan EF(r).

Next, let us consider a low-to-high incidence $\rho_1 a_1 < \rho_3 a_3$ (Fig. 9.22). In this case, the $p - u$ curve of SW(t) is above that of SW(i), and the reflected wave curve from "2" intersects with SW(t) at "5", with $p_5 > p_2$. The reflected wave is a shock wave.

Column: Wave Drag Reduction by Energy Deposition

It is known that the drag on an object in supersonic flow can be reduced by depositing an energy in front of it. As shown in Fig. C9.2, A "bubble" that is a high-temperature, low-density region, is generated by local heating ahead of the object. The heating can

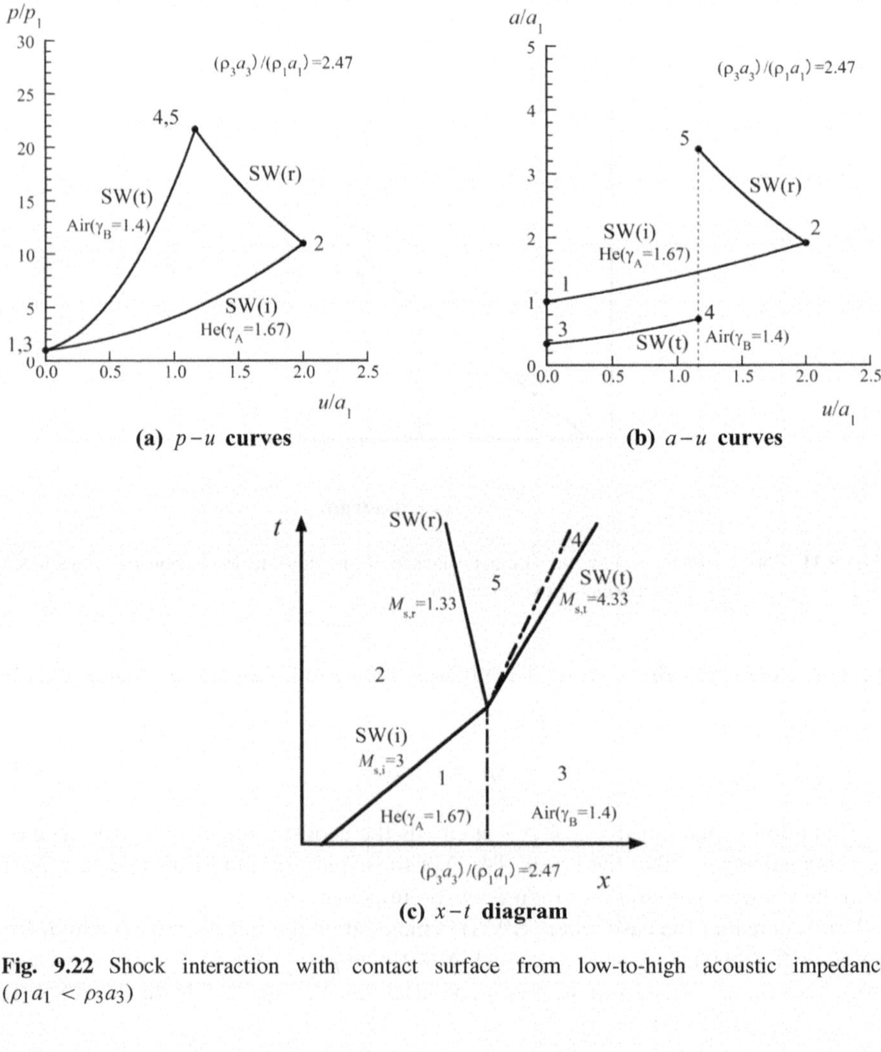

Fig. 9.22 Shock interaction with contact surface from low-to-high acoustic impedance ($\rho_1 a_1 < \rho_3 a_3$)

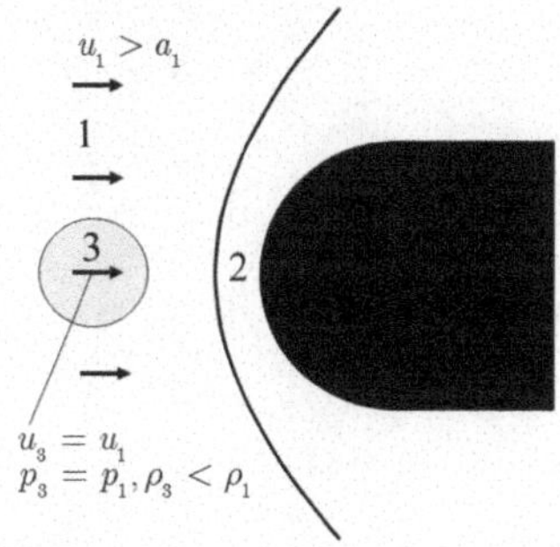

Fig. C9.2 Schematic illustration of energy deposition; 1, upstream without energy deposition; 2, immediately behind the bow shock wave; 3, thermal bubble generated by energy deposition

be performed using a laser pulse, electrical discharge, etc. After the bubble enters the shock layer in front of the object, the wall pressure and subsequently the drag is reduced. This scheme is referred to as *energy deposition*, and it is reported to be effective particularly for high-Mach number flows [2].

Figure C9.3 shows experimental results with a flathead, circular cylinder (diameter, 20 mm) set in Mach 1.94 flow. The laser pulse energy is repetitively sent to the focal point at 40 mm upstream from the cylinder head. Without energy deposition, a bow shock wave is formed ahead of the cylinder. When thermal bubbles are repetitively generated using a pulse laser, the thermal bubbles transitioned to a toroidal vortex ring are piled up in front of the cylinder, which in turn act as a virtual spike to reduce the wave drag.

Although the actual mechanisms of the drag reduction are complicated, here we analyze the drag reduction as a Riemann problem. The initial condition shown in Fig. C9.2 is equivalent to the Riemann problem of Fig. 9.21. Here, they are in different coordinates, as the direction of the wave propagation is opposite.

A part of the region "1" is heated at constant volume by an energy q per unit mass (state 1'). The state after expansion to the equalized pressure is labeled "3." Here, the expansion process is assumed to be isentropic. For a calorically perfect gas, the following equations hold [3].

$$p_3 = p_1 \tag{C9.1}$$

$$\frac{1}{\gamma - 1} R T_{1'} = \frac{1}{\gamma - 1} R T_1 + q \tag{C9.2}$$

$$\frac{p_{1'}}{T_{1'}} = \frac{p_1}{T_1} \tag{C9.3}$$

$$\frac{p_3}{p_{1'}} = \frac{p_1}{p_{1'}} = \left(\frac{\rho_3}{\rho_{1'}}\right)^{\gamma} = \left(\frac{\rho_3}{\rho_1}\right)^{\gamma} \tag{C9.4}$$

$$\frac{a_3}{a_{1'}} = \left(\frac{T_3}{T_{1'}}\right)^{\frac{1}{2}} = \left(\frac{\rho_3}{\rho_1}\right)^{\frac{\gamma-1}{2}}. \tag{C9.5}$$

(a)Schlieren image without energy deposition **(b) Schlieren image with *f* = 80 kHz.**

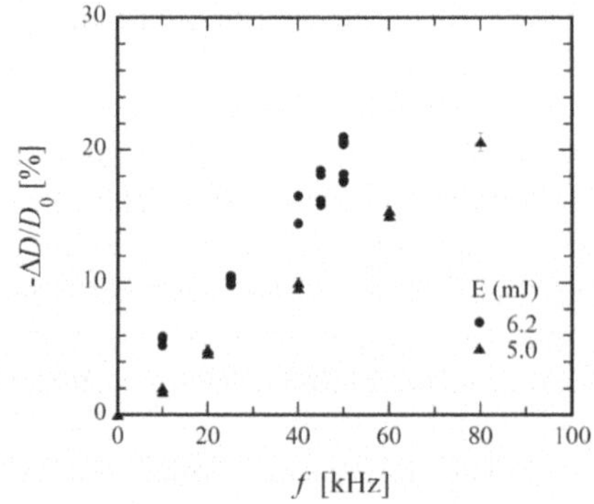

(c) Normalized drag reduction vs. *f*

Fig. C9.3 Example of supersonic drag reduction with energy deposition; *f*, repetition frequency of laser pulse; *E*, laser pulse energy; D_0, drag without energy deposition

Therefore,

$$\frac{\rho_3}{\rho_1} = (1 + Q)^{-\frac{1}{\gamma}} \tag{C9.6}$$

$$\frac{a_3}{a_1} = (1 + Q)^{\frac{1}{2\gamma}} \tag{C9.7}$$

$$\frac{\rho_3 a_3}{\rho_1 a_1} = (1 + Q)^{-\frac{1}{2\gamma}} \; (< 1) \tag{C9.8}$$

$$Q \equiv \frac{q}{\frac{RT_1}{\gamma - 1}} = \frac{q}{e_1}. \tag{C9.9}$$

From (C9.8), the acoustic impedance is decreased with the energy deposition. Using the normal shock relation, the L and R states in a Riemann problem apply to "3" and "2," respectively.

$$\frac{p_{\mathrm{R}}}{p_{\mathrm{L}}} = \frac{p_2}{p_3} = \frac{p_2}{p_1} = 1 + \frac{2\gamma}{\gamma+1}\left(M_1^2 - 1\right) \tag{C9.10}$$

$$\begin{cases} \dfrac{u_{\mathrm{L}}}{a_1} = \dfrac{u_3}{a_1} = M_1 \\ \dfrac{\rho_{\mathrm{L}}}{\rho_1} = \dfrac{\rho_3}{\rho_1} = (1+Q)^{-\frac{1}{\gamma}} \\ \dfrac{a_{\mathrm{L}}}{a_1} = \dfrac{a_3}{a_1} = (1+Q)^{\frac{1}{2\gamma}} \end{cases} \tag{C9.11}$$

$$\begin{cases} \dfrac{u_{\mathrm{R}}}{a_1} = \dfrac{u_2}{a_1} = \dfrac{u_1}{a_1}\dfrac{u_2}{u_1} = M_1\dfrac{\rho_1}{\rho_2} = \dfrac{(\gamma-1)M_1^2+2}{(\gamma+1)M_1} \\ \dfrac{\rho_{\mathrm{R}}}{\rho_1} = \dfrac{\rho_2}{\rho_1} = \dfrac{(\gamma+1)M_1^2}{(\gamma-1)M_1^2+2} \\ \dfrac{a_{\mathrm{R}}}{a_1} = \dfrac{a_2}{a_1} = \dfrac{(2\gamma M_1^2-\gamma+1)^{\frac{1}{2}}\{(\gamma-1)M_1^2+2\}^{\frac{1}{2}}}{(\gamma+1)M_1} \end{cases}. \tag{C9.12}$$

Since for a normal shock wave,

$$\frac{\rho_2 a_2}{\rho_1 a_1} > 1$$

$$\frac{\rho_{\mathrm{L}} a_{\mathrm{L}}}{\rho_{\mathrm{R}} a_{\mathrm{R}}} = \frac{\rho_3 a_3}{\rho_2 a_2} = \frac{\rho_3 a_3}{\rho_1 a_1}\frac{\rho_1 a_1}{\rho_2 a_2} < 1.$$

After the bubble enters the shock layer, a transmitted shock wave propagates in the bubble, whereas behind the bow shock the pressure is decreased due to the expansion fan. Applying the relation to (9.11) and (9.21), the solution of this Riemann problem is obtained by combining the following equations.

$$\frac{u_*}{a_1} - \frac{u_3}{a_1} = -\frac{a_3}{a_1}\left(\frac{p_*}{p_1} - 1\right)\left\{\frac{\frac{2}{\gamma(\gamma+1)}}{\frac{p_*}{p_1} + \frac{\gamma-1}{\gamma+1}}\right\}^{\frac{1}{2}} \tag{C9.13}$$

$$\frac{u_*}{a_1} - \frac{u_2}{a_1} = \frac{2}{\gamma-1}\left\{\left(\frac{p_*}{p_2}\right)^{\frac{\gamma-1}{2\gamma}} - 1\right\}\frac{a_2}{a_1} \tag{C9.14}$$

$$\frac{p_*}{p_1} = f(Q, M_1)$$

$$- (1+Q)^{\frac{1}{2\gamma}}\left(\frac{p_*}{p_1} - 1\right)\left\{\frac{\frac{2}{\gamma}}{(\gamma+1)\frac{p_*}{p_1} + \gamma - 1}\right\}^{\frac{1}{2}}\frac{\gamma+1}{2}M_1 = -\left(M_1^2 - 1\right)$$

$$+ \frac{1}{\gamma-1}\left\{\left(\frac{p_*}{p_1}\right)^{\frac{\gamma-1}{2\gamma}}\left(\frac{2\gamma M_1^2-\gamma+1}{\gamma+1}\right)^{-\frac{\gamma-1}{2\gamma}} - 1\right\}\left(2\gamma M_1^2-\gamma+1\right)^{\frac{1}{2}}\left\{(\gamma-1)M_1^2+2\right\}^{\frac{1}{2}}. \tag{C9.15}$$

Substituting (C9.13) with (C9.15), the flow velocity after the interaction, u_*, is obtained. From normal shock wave relations,

$$\frac{p_*}{p_1} = 1 + \frac{2\gamma}{\gamma+1}\left(M_{s,3*}^2 - 1\right) \tag{C9.16}$$

$$M_{s,3*} = \sqrt{\frac{\gamma+1}{2\gamma}\left(\frac{p_{3*}}{p_1} + \frac{\gamma-1}{\gamma+1}\right)} \tag{C9.17}$$

$$\frac{a_{3*}}{a_1} = \frac{a_3}{a_1}\frac{a_{3*}}{a_3} = (1+Q)^{\frac{1}{2\gamma}}\frac{\left(2\gamma M_{s,3*}^2 - \gamma + 1\right)^{\frac{1}{2}}\left\{(\gamma-1)M_{s,3*}^2 + 2\right\}^{\frac{1}{2}}}{(\gamma+1)M_{s,3*}}. \tag{C9.18}$$

The velocity of the transmitted shock wave in the bubble, $U_{s,3*}$, is given by

$$\frac{U_{s,3*}}{a_1} = \frac{u_3 - M_{s,3*}a_{3*}}{a_1} = M_1 - \frac{a_{3*}}{a_1}M_{s,3*}. \tag{C9.19}$$

As observed in Fig. C9.4, by the energy deposition, the pressure behind the bow shock wave is decreased; the flow velocity in the opposite direction to the upstream flow is induced. The transmitted shock wave propagates toward the upstream in the thermal bubble. This is experimentally observed and termed as the *lens effect* [4].

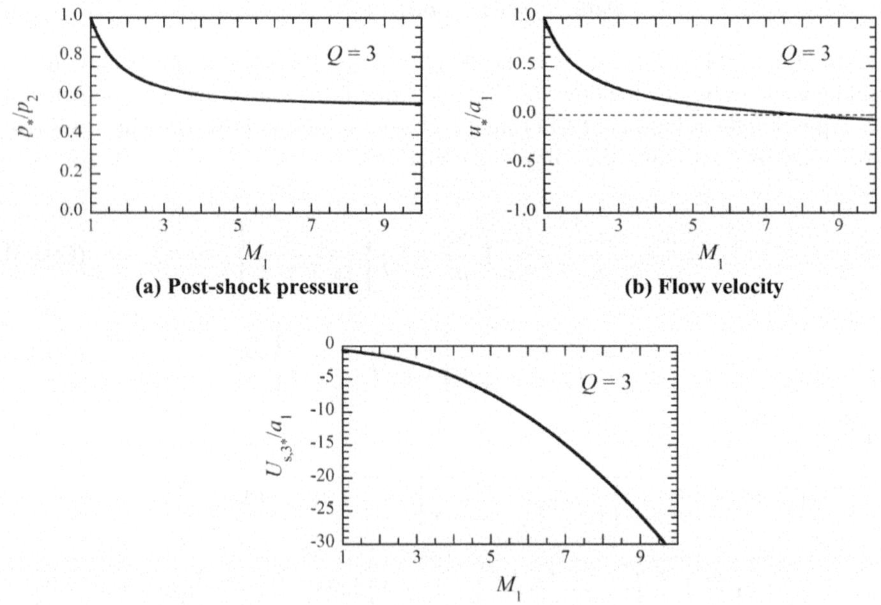

Fig. C9.4 Post-shock pressure, flow velocity and propagation velocity of transmitted shock wave due to interaction with the thermal bubble

With a low flow Mach number, $1 < M_1 < 2$, the effects of the pressure decrease and the induction of the counter flow are large. The direction of the flow becomes reversed with $M_1 \geq 8$. With a high flow Mach number, $5 < M_1$, those effects become saturated, yet the speed of the transmitted shock wave in the opposite direction, $|U_{s,3*}|$, continues to increase. Thus, the extrusion of the bow shock wave due to the lens effect becomes significant.

References

1. Tamba T et al (2015) Shock Waves 25:667–674
2. Knight D (2008) J Prop Power 24:1153–1167 (2008)/Tret'yakov PK et al (1996) Phys-Doklady 41:566–567/Kim J-H et al (2011) AIAA J 49:2076–2078
3. Iwakawa A et al (2017) Transactions of the JSASS. Aerosp Technol Jpn 60(5):303–311
4. Knight D (2008) J Prop Power 24:1153–1167

Chapter 10
Method of Characteristics

In a steady, two-dimensional flow (Chap. 7) and an unsteady, one-dimensional flow (Chap. 8), the flow velocity and thermodynamic properties are expressed as functions of two independent variables by applying Euler's equation. In those cases, the differential equation is *hyperbolic* and can be solved by sequentially solving interactions between characteristics with an appropriate boundary or initial conditions. This method, called the *method of characteristics*, is useful in such simple problems and will be dealt with in this chapter.

10.1 Design of Supersonic Nozzle

Let us design a supersonic nozzle as an example of steady, two-dimensional flow. Here, the gas is assumed to flow along the nozzle wall without separation. In the one-dimensional flow in Chap. 5, the flow Mach number is a function of the cross-sectional area uniquely. Yet, in reality, the flow is multidimensional.

10.1.1 Characteristics and Flow Variation

If supersonic flow turns such that the flow passage area increases, its flow Mach number is increased due to the Prandtl–Meyer expansion. Transforming a smooth curve of a nozzle wall profile to a polygonal line, the flow is accelerated through an expansion fan at each corner, as shown in Fig. 10.1a. The expansion fan is sandwiched by the upstream (1) and downstream (2) regions by the respective Mach waves. However, such flow variations with a finite width complicate the design procedure of the flow passage. Instead, in the method of characteristics an expansion fan is replaced by a "representative" characteristic, as shown in Fig. 10.1b, which is defined as the bisection line of the two Mach waves. The flow on the upstream side should be uniform as Region 1, and the downstream side as Region 2.

A. Sasoh, *Compressible Fluid Dynamics and Shock Waves*,
https://doi.org/10.1007/978-981-15-0504-1_10

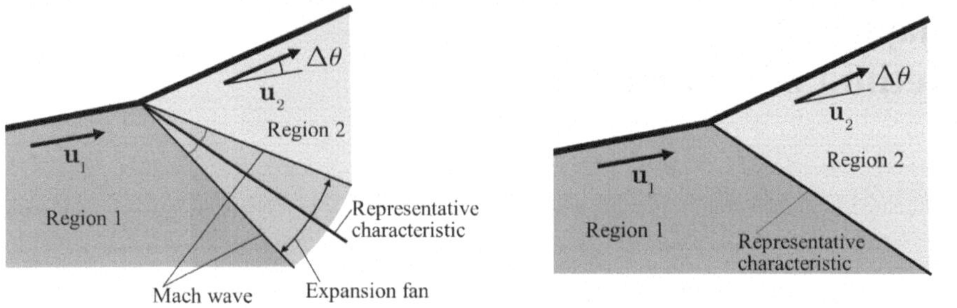

(a) Acceleration past an expansion fan **(b) Approximation by a characteristic.**

Fig. 10.1 Acceleration of supersonic flow past a corner

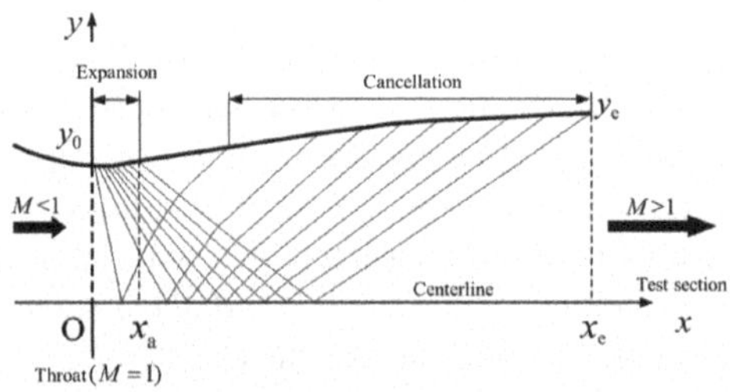

Fig. 10.2 Example of Laval nozzle resign ($\gamma = 1.4$, $M_e = 1.7$)

If the supersonic flow is deflected for compression, an oblique shock wave is generated, and the flow condition varies in a discontinuous manner. However, if the deflection angle, $|\Delta\theta|$, is small, the associated flow variation can be approximated as isentropic, and the oblique shock angle is approximated to that of the bisection line of the upstream and downstream Mach waves. In summary, no matter whether the flow deflection is expansive or compressive, the bisection line is regarded as the representative characteristic, across which the flow experiences a small discontinuous change. Applying this scheme to the whole flow field, it is divided into finite uniform regions. The finer the division, the better the accuracy of the flow design becomes.

Figure 10.2 shows a simple, two-dimensional Laval nozzle design employed in this section. The incoming subsonic flow is accelerated in the converging section, which then becomes sonic at the throat ($x = 0$). The flow is supersonically accelerated due to the characteristics generated on the nozzle wall in the expansion section. $y = 0$ corresponds to the centerline, which is the axis of symmetry. The characteristics are reflected on the centerline, such that the flow is deflected to the x-direction. The reflected characteristics are incident to the nozzle wall in the "cancelation" section, where the wall angles are set such that the respective incident characteristics do not cause a reflected wave, and so that the flow condition does not change in the downstream, meaning in the test section.

In the expansion section, the wall has a positive curvature, which is convex toward the downstream. With increasing the wall angle, expansion waves are generated, accelerating the expanding flow. The exit of the expansion section ($x = x_a$) is

Fig. 10.3 Relation among regional elements in contact

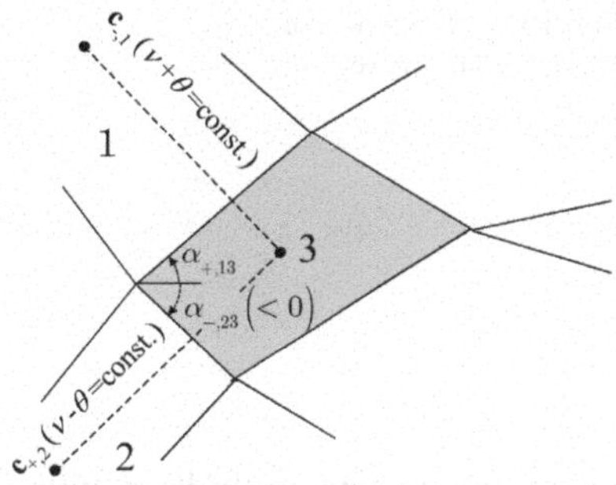

an inflection point; in the downstream the wall is either straight or has a negative curvature, which is concave toward the downstream. Thus, the expansion wave is not generated. In the expansion section of Fig. 10.2, the total deflection angle is equally divided to nine segments of an increment of $\Delta\theta$ and the expansive characteristic at each corner is drawn. Each characteristic is reflected on the centerline, and the reflected characteristics interact with other incident characteristics before arriving to the "cancelation" section on the wall. The sections between the expansion and cancelation are connected by a straight line.

Figure 10.3 shows the relation among regional elements that are in contact. The flow condition in the element "3" is determined by the interaction between the characteristic $\mathbf{c}_{-,1}$ from the element "1" and $\mathbf{c}_{+,2}$ from "2." As shown in Sect. 7.1, an invariant $\nu + \theta$ is conserved between "1" and "3," $\nu - \theta$ between "2" and "3", where ν and θ are Prandtl–Meyer function and a deflection angle, respectively.

$$\nu_3 + \theta_3 = \nu_1 + \theta_1 \tag{10.1}$$

$$\nu_3 - \theta_3 = \nu_2 - \theta_2 \tag{10.2}$$

Therefore,

$$\nu_3 = \frac{\nu_1 + \nu_2}{2} + \frac{\theta_1 - \theta_2}{2} \tag{10.3}$$

$$\theta_3 = \frac{\nu_1 - \nu_2}{2} + \frac{\theta_1 + \theta_2}{2}. \tag{10.4}$$

Let the angle of the "1"–"3" boundary from the x-axis be $\alpha_{+,13}$, and the angle of the "2"–"3" boundary be $\alpha_{-,23}$. They are obtained as follows:

$$\alpha_{+,13} = \frac{(\theta_1 + \beta_{\mathrm{M},1}) + (\theta_3 + \beta_{\mathrm{M},3})}{2} \tag{10.5}$$

$$\alpha_{-,23} = \frac{(\theta_2 - \beta_{\mathrm{M},2}) + (\theta_3 - \beta_{\mathrm{M},3})}{2}. \tag{10.6}$$

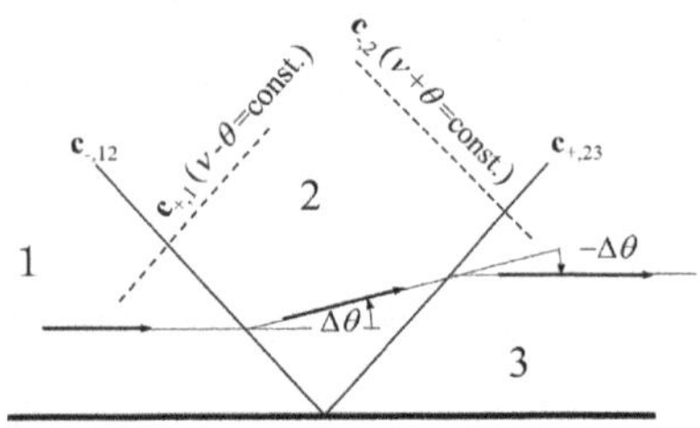

Fig. 10.4 Reflection of characteristic on the centerline

All angles have a positive value for the anticlockwise direction, and β_{M} is the Mach angle.

$$\beta_{\mathrm{M}} = \sin^{-1}\left(\frac{1}{M}\right). \tag{10.7}$$

Figure 10.4 shows the relation on the centerline. The flow in Region "1", before the reflection should be parallel to the centerline. The characteristic $\mathbf{c}_{-,12}$ is reflected on the centerline, and subsequently the characteristic $\mathbf{c}_{+,23}$ is generated as the reflected wave. In Region "3," after the reflection, the flow should be aligned to parallel to the centerline.

$$\theta_2 = \theta_1 + \Delta\theta = \Delta\theta(> 0) \tag{10.8}$$

$$\theta_3 = \theta_1 = 0. \tag{10.9}$$

Using the Riemann invariants,

$$\nu_2 - \theta_2 = \nu_1 - \theta_1 \tag{10.10}$$

$$\nu_3 + \theta_3 = \nu_2 + \theta_2. \tag{10.11}$$

Combining these equations,

$$\nu_2 = \nu_1 + \Delta\theta \tag{10.12}$$

$$\nu_3 = \nu_2 + \Delta\theta. \tag{10.13}$$

Therefore, the flow is accelerated past $\mathbf{c}_{-,12}$, and then again accelerated past $\mathbf{c}_{+,23}$. The corresponding deflection angles have equal magnitudes, but opposite signs.

Figure 10.5 shows the general case of the wave reflection on the upper nozzle wall. The characteristic $\mathbf{c}_{+,12}$ is reflected on the upper wall with a reflection angle $\Delta\theta(< 0)$, as the reflected wave, $\mathbf{c}_{-,23}$ is generated from the corner. Depending on $\Delta\theta$, it is compressive or expansive.

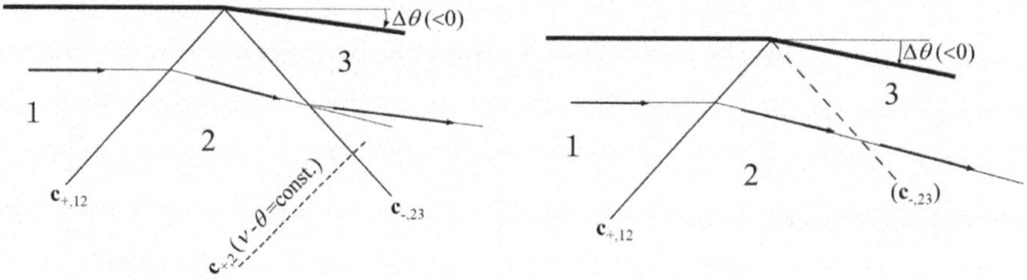

(a) With reflection, general case. **(b) Cancelled reflected wave.**

Fig. 10.5 Reflection on the upper wall

Fig. 10.6 Numbering of regional elements

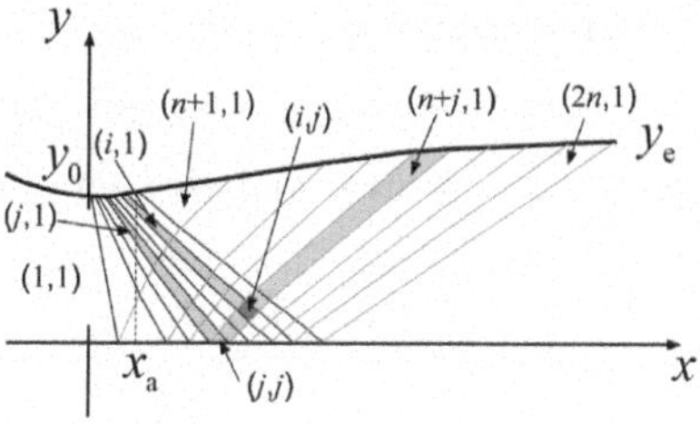

$$\theta_3 = \theta_1 + \Delta\theta \tag{10.14}$$

$$\nu_3 - \theta_3 = \nu_2 - \theta_2 \tag{10.15}$$

Therefore,

$$\nu_3 = \nu_2 + \theta_3 - \theta_2. \tag{10.16}$$

In order for the reflected wave to be canceled, $\Delta\theta$ should be set such that

$$\nu_3 = \nu_2 \tag{10.17}$$

$$\theta_3 = \theta_2. \tag{10.18}$$

From (10.14) and (10.18),

$$\Delta\theta = \theta_2 - \theta_1. \tag{10.19}$$

Following up the above procedure, the whole flow field is obtained.

Let us label a number (i, j) to the elements as shown in Fig. 10.6. The sonic flow at the throat is labeled (1, 1). Subsequently, along the upper wall, i is sequentially increased, j is sequentially increased after passing across the reflected waves. Let us specify the deflection angle $\Delta\theta(i, 1)$ such that

$$\theta(i,1) = \theta(1,1) + \sum_{k=1}^{i-1} \Delta\theta(k) = \sum_{k=1}^{i-1} \Delta\theta(k) \tag{10.20}$$

$$\Delta\theta(k) = \theta(k+1,1) - \theta(k,1). \tag{10.21}$$

Using the Riemann invariant,

$$\nu(i,1) - \theta(i,1) = \nu(1,1) - \theta(1,1) = \nu(1,1) = 0$$

$$\nu(i,1) = \theta(i,1) = \sum_{k=1}^{i-1} \Delta\theta(k). \tag{10.22}$$

In the same way,

$$\nu(j,j) + \theta(j,j) = \nu(j,1) + \theta(j,1) = 2\sum_{k=1}^{j-1} \Delta\theta(k)$$

$$\theta(j,j) = 0$$

$$\nu(j,j) = 2\sum_{k=1}^{j-1} \Delta\theta(k). \tag{10.23}$$

Using the above results,

$$\nu(i,j) + \theta(i,j) = \nu(i,1) + \theta(i,1) = 2\sum_{k=1}^{i-1} \Delta\theta(k)$$

$$\nu(i,j) - \theta(i,j) = \nu(j,j) - \theta(j,j) = 2\sum_{k=1}^{j-1} \Delta\theta(k).$$

Then,

$$\nu(i,j) = \sum_{k=1}^{i-1} \Delta\theta(k) + \sum_{k=1}^{j-1} \Delta\theta(k) = \theta(i,1) + \theta(j,1) \tag{10.24}$$

$$\theta(i,j) = \sum_{k=1}^{i-1} \Delta\theta(k) - \sum_{k=1}^{j-1} \Delta\theta(k) = \theta(i,1) - \theta(j,1). \tag{10.25}$$

Here, ν and θ were simply related to (i, j) by (10.24) and (10.25).

10.1.2 Design Procedure of Laval Nozzle

The design of the Laval nozzle of Fig. 10.6 is performed in the following procedure:

(1) The height, y_e, and Mach number, M_e, are given at the exit.
(2) Integrating (5.27) with the appropriate influence coefficient, an equation relating y_e and M_e is obtained.

$$\frac{dM^2}{M^2} = -\frac{(\gamma - 1)M^2 + 2}{1 - M^2}\frac{dA}{A}$$

$$\frac{y_e}{y_0} = \frac{1}{M_e}\left[\frac{(\gamma - 1)M_e^2 + 2}{\gamma + 1}\right]^{\frac{\gamma+1}{2(\gamma-1)}}. \tag{10.26}$$

(3) A smooth y curve is given in the subsonic section ($x < 0$). The choking condition is

$$\frac{dy}{dx} = 0 \quad (x = 0) \tag{10.27}$$

In order for the derivative in the flow variables to be continuous at the throat, the second derivative at the throat should likewise be continuous at $x = 0$..

(4) The y-variation in $0 \le x \le x_a$ is given under the following boundary conditions:
At $x = 0$,

$$y = y_0 \tag{10.28}$$

$$\frac{dy}{dx} = 0. \tag{10.29}$$

At $x = x_a$,

$$\frac{dy}{dx} = \tan\theta_{max} \tag{10.30}$$

$$\frac{d^2y}{dx^2} = 0. \tag{10.31}$$

To satisfy the above conditions, $y(x)$ should be cubic at least.

$$\frac{y - y_0}{x_a} = \tan\theta_{max}\left[\left(\frac{x}{x_a}\right)^2 - \frac{1}{3}\left(\frac{x}{x_a}\right)^3\right] \quad (0 \le x \le x_a). \tag{10.32}$$

(5) The number of divisions, n, is given, and Eq. (10.32) is drawn by an n polygonal line with $\theta(i, 1) \quad (i = 1, \ldots, n)$.

(6) $\nu(i, j)$ and $\theta(i, j)(i = 1, \ldots, n)$ are obtained using (10.24) and (10.25), respectively.
(7) $\nu(i, 1)$ and $\theta(i, 1)(i = n + 1, \ldots, 2n)$ are obtained using (10.19).
(8) Starting from $(0, y_0)$ in the element (1, 1), characteristics are sequentially obtained applying (10.5) and (10.6).

Here, we have designed the supersonic nozzle. The method presented is the simplest one, to ensure that readers easily understand the design principle. In real applications, more sophisticated schemes and/or know-hows should be introduced to improve the test flow quality. A possible modification is to assume a curved sonic line, and to set a partial expansion section to moderate the sudden cross-sectional area variation. One of the most important modifications is the boundary layer correction. In a real nozzle flow, a boundary layer is developed on the wall, decreasing the effective cross-sectional area. To obtain a desired test Mach number, the height of the nozzle should be increased by the boundary layer. If a desired test flow Mach number is high, the total enthalpy of the test gas becomes so high that the so-called *real gas effect* becomes significant, and consequently the assumption of a calorically perfect gas needs to be revised. In such complicated cases, the method of characteristics only provides limited information, and the design should be aided by computational fluid dynamics.

10.2 Wave Diagram of Shock Tube Operation

In unsteady, one-dimensional flow, one of two spatial coordinates in Sect. 10.1 is replaced by a time coordinate. Yet, a similar approach is possible. Even if shock waves appear, in many simple problems the method of characteristics is useful by using the solution of the Riemann problem. Here, we will analyze the behavior of expansion waves in the high-pressure channel in Fig. 9.8. This problem has an analytical solution presented in Sect. 9.4.

Consider, as shown in Fig. 10.7, the interaction between the characteristic $C_{+,L}$ originating in L and the characteristic $C_{-,R}$ in R. Using the results of Sect. 8.2,

$$u_I + \frac{2}{\gamma - 1} a_I = u_L + \frac{2}{\gamma - 1} a_L \tag{10.33}$$

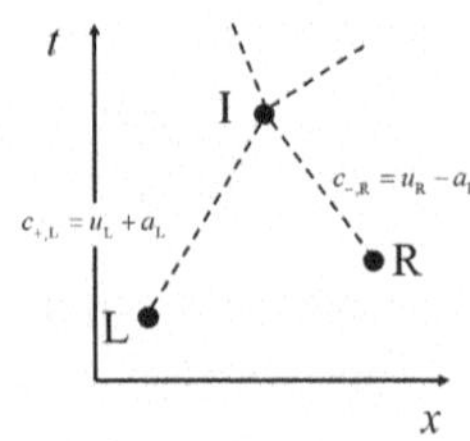

Fig. 10.7 Interaction between characteristics in an unsteady, one-dimensional flow

$$u_{\mathrm{I}} - \frac{2}{\gamma - 1} a_{\mathrm{I}} = u_{\mathrm{R}} - \frac{2}{\gamma - 1} a_{\mathrm{R}}. \tag{10.34}$$

Therefore,

$$u_{\mathrm{I}} = \frac{u_{\mathrm{L}} + u_{\mathrm{R}}}{2} + \frac{a_{\mathrm{L}} - a_{\mathrm{R}}}{\gamma - 1} \tag{10.35}$$

$$a_{\mathrm{I}} = \frac{\gamma - 1}{4}(u_{\mathrm{L}} - u_{\mathrm{R}}) + \frac{a_{\mathrm{L}} + a_{\mathrm{R}}}{2}. \tag{10.36}$$

Assuming that the total enthalpy is uniform in the entire flow field, p_{I} is obtained as

$$p_{\mathrm{I}} = p_{\mathrm{L}} \left(\frac{a_{\mathrm{I}}}{a_{\mathrm{L}}} \right)^{\frac{2\gamma}{\gamma-1}} = p_{\mathrm{R}} \left(\frac{a_{\mathrm{I}}}{a_{\mathrm{R}}} \right)^{\frac{2\gamma}{\gamma-1}}. \tag{10.37}$$

The above procedure should be sequentially applied to the incident and reflected waves.

As shown in Fig. 10.8a, the end wall of the high-pressure channel and the diaphragm are located at $x = x_{\mathrm{A}}$ and $x = x_{\mathrm{B}} = 0$, respectively. The diaphragm is instantaneously ruptured at time $t = 0$. As shown in Fig. 10.8b, the expansion fan incident to the end wall is divided into n expansion waves, which are numbered as $i = 1 \ldots n$. Here, $i = 1$ refers to the state "4," and $i = n$ to the state "3." In this example, the difference in the propagation velocity between adjacent expansion waves is set to a constant. The reflected waves are numbered by j based on the sequential order of reflections. $j = 0$ refers to the incident wave. The label (i, j) refers to the intersection between the i-th incident wave and j-th reflected wave.

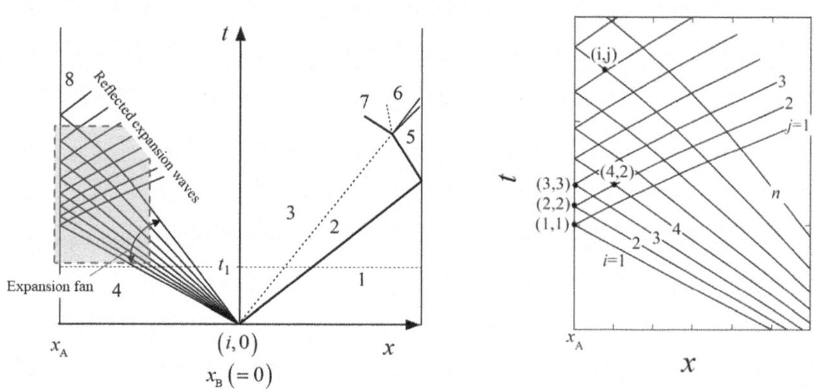

(a) Whole diagram

(b) Interaction among expansion waves, zoom into the gray zone in (a)

Fig. 10.8 Wave diagram of shock tube operation, the operation condition from Fig. 9.8

Because $u_4 = 0$, the propagation velocity of i-the incident wave is

$$c_{-,\mathrm{R}}(i,0) = -a_4 + (u_3 - a_3 + a_4)\frac{i-1}{n-1}. \tag{10.38}$$

The invariant and the propagation velocity of i-th incident wave before the interaction with a reflected wave is given by

$$J_-(i) = u(i,0) - \frac{2}{\gamma - 1}a(i,0) \tag{10.39}$$

$$c_{-,\mathrm{R}}(i,0) = u(i,0) - a(i,0) \tag{10.40}$$

$$u(i,0) + \frac{2}{\gamma-1}a(i,0) = \frac{2}{\gamma-1}a_4. \tag{10.41}$$

From (10.39) to (10.41),

$$a(i,0) = -\frac{\gamma-1}{\gamma+1}c_{-,\mathrm{R}}(i,0) + \frac{2}{\gamma+1}a_4 \tag{10.42}$$

$$u(i,0) = \frac{2}{\gamma+1}c_{-,\mathrm{R}}(i,0) + \frac{2}{\gamma+1}a_4 \tag{10.43}$$

$$J_-(i) = \frac{4}{\gamma+1}c_{-,\mathrm{R}}(i,0) + \frac{2(\gamma-3)}{(\gamma+1)(\gamma-1)}a_4. \tag{10.44}$$

Next, let us obtain the condition behind the reflected wave of the tail wave of the incident expansion fan (i, i). From (10.39), with

$$u(i,i) = 0 \tag{10.45}$$

$$J_-(i) = -\frac{2}{\gamma-1}a(i,i)$$

$$a(i,i) = -\frac{\gamma-1}{2}J_-(i). \tag{10.46}$$

The Riemann invariant and the propagation velocity of i-th reflected wave are

$$J_+(i) = \frac{2}{\gamma-1}a(i,i) \tag{10.47}$$

$$c_{+,\mathrm{L}}(i,i) = a(i,i). \tag{10.48}$$

The j-th reflected wave intersects with the i-th incident wave on the order of $i = j+1 \ldots n$. Let the intersection point between j-th reflected wave and i-th incident wave be P(i, j). Except for the end wall ($x = x_A$), the right-running, reflected wave past P($i-1$, j) in a Riemann invariant $J_+(j)$ and the left-running, incident wave past P(i, $j-1$) with a Riemann invariant $J_-(i)$ intersect at P(i, j).

$$J_+(j) = u(i, j) + \frac{2}{\gamma - 1} a(i, j) \tag{10.49}$$

$$J_-(i) = u(i, j) - \frac{2}{\gamma - 1} a(i, j). \tag{10.50}$$

Therefore,

$$a(i, j) = \frac{\gamma - 1}{4} \{J_+(j) - J_-(i)\} \tag{10.51}$$

$$u(i, j) = \frac{1}{2} \{J_+(j) + J_-(i)\}. \tag{10.52}$$

Following (10.51) and (10.52), solving with $i = j + 1 \ldots n$ sequentially for $j = 1 \ldots n$ provides the entire incident and reflected expansion fan.

$$c_{+,\mathrm{L}}(i-1, j) = u(i-1, j) + a(i-1, j) \tag{10.53}$$

$$c_{-,\mathrm{R}}(i, j-1) = u(i, j-1) - a(i, j-1) \tag{10.54}$$

$$x(i, j) = x(i, j-1) + c_{-,\mathrm{R}}(i, j-1)\{t(i, j) - t(i, j-1)\} \tag{10.55}$$

$$x(i, j) = x(i-1, j) + c_{+,\mathrm{L}}(i-1, j)\{t(i, j) - t(i-1, j)\} \tag{10.56}$$

$$t(i, j) = \frac{x(i, j-1) - x(i-1, j) + c_{+,\mathrm{L}}(i-1, j)t(i-1, j) - c_{-,\mathrm{R}}(i, j-1)t(i, j-1)}{c_{+,\mathrm{L}}(i-1, j) - c_{-,\mathrm{R}}(i, j-1)}. \tag{10.57}$$

Chapter 11
Generation and Utilization of Compressible Flows

By utilizing the characteristics of compressible flow, we can generate high-speed flow, high-pressure, and/or high-temperature states. The design of rocket engine nozzle and that of air intake for an aircraft engine need to be conducted based on the principle of compressible fluid dynamics. In this chapter, we will illustrate representative examples of such devices.

11.1 Nozzle and Orifice

In Chap. 5, we studied one-dimensional flows with varying cross-sectional areas. A *nozzle* is a device that accelerates the flow using this principle. The shape of the nozzle depends on the purpose and the flow Mach number range.

A *converging nozzle* is used to accelerate subsonic flow, as shown in Fig. 11.1a. It is used, for example, at the exit of a spray bottle. The same principle is applied when blowing out a candle flame.

A *Laval nozzle* (Fig. 11.1b) is composed of a converging part and a diverging part smoothly connected to each other. During the *choking condition*, the flow becomes sonic at its *throat* at which the cross-sectional area has a minimum value. In a rocket engine and a supersonic wind tunnel, the flow is accelerated to a supersonic speed. This shape is also used to set a constant flow rate. For this purpose, the device is termed a *critical nozzle* or a *sonic nozzle*.

An *orifice* (Fig. 11.1c) is a blockage in a flow with a circular hole that has a known cross-sectional area. Near the orifice, circulation zones are formed. The cross-sectional area of the flow past the orifice has an effective minimum value in the downstream that is smaller than the geometrical value.

A. Sasoh, *Compressible Fluid Dynamics and Shock Waves*,
https://doi.org/10.1007/978-981-15-0504-1_11

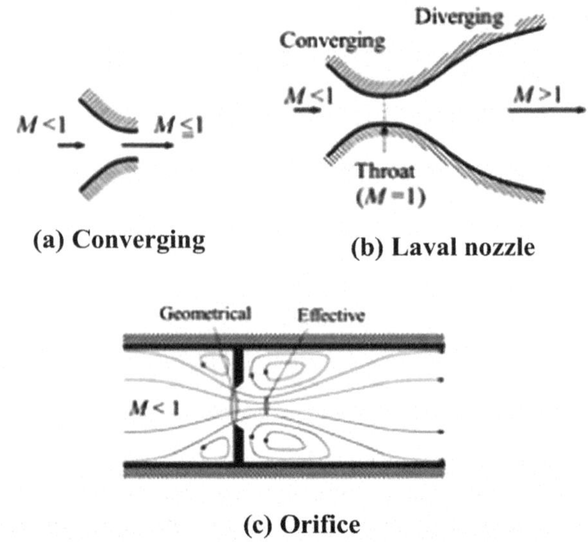

Fig. 11.1 Orifice and nozzles

The usage of a nozzle is categorized as follows:

(1) To obtain a constant flow rate (mass flow control valve, fuel injector, etc.)
(2) To generate high-speed flow (spray bottle, shower, blowing out a candle flame, whistle, etc.)
(3) To set a flow speed or Mach number (wind tunnel)
(4) To generate thrust (aircraft and rocket engines).

11.1.1 Isentropic Flow with Varying Cross Section

Assume here a steady, quasi-one-dimensional flow without heat transfer, body force, or shock waves. The flow variables are obtained using the equations in Chap. 5 if the pressure difference between the inlet and exit is sufficiently large. The relation between the flow Mach number M and the cross-sectional area A is given by

$$\frac{dM^2}{M^2} = -\frac{(\gamma - 1)M^2 + 2}{1 - M^2}\frac{dA}{A}. \tag{11.1}$$

Integrating (11.1),

$$\frac{\left(M^2 + \frac{2}{\gamma-1}\right)^{\frac{\gamma+1}{\gamma-1}}}{A^2 M^2} = \text{const.} \tag{11.2}$$

Attaching a subscript * to variables in the choking state, that is, $M_* = 1$ and $A = A_*$,

$$\frac{A}{A_*} = \frac{\left(\frac{\gamma-1}{\gamma+1}M^2 + \frac{2}{\gamma+1}\right)^{\frac{\gamma+1}{2(\gamma-1)}}}{M}. \tag{11.3}$$

Equation (11.3) gives M as an implicit function of A. In the same way,

$$\begin{aligned}\frac{du}{u} &= \frac{1}{(\gamma-1)M^2+2}\frac{dM^2}{M^2} = \left[-\frac{\frac{\gamma-1}{2}}{(\gamma-1)M^2+2} + \frac{\frac{1}{2}}{M^2}\right]dM^2 \\ \frac{u^2}{u_*^2} &= \frac{u^2}{a_*^2} = \frac{(\gamma+1)M^2}{(\gamma-1)M^2+2}\end{aligned} \tag{11.4}$$

$$\begin{aligned}\frac{dp}{p} &= \frac{\gamma M^2}{1-M^2}\frac{dA}{A} = -\frac{\gamma}{(\gamma-1)M^2+2}dM^2 \\ \frac{p}{p_*} &= \left(\frac{\gamma-1}{\gamma+1}M^2 + \frac{2}{\gamma+1}\right)^{-\frac{\gamma}{\gamma-1}}.\end{aligned} \tag{11.5}$$

The thermodynamics properties are expressed based on the *stagnation state* using a subscript t with $M_t = u_t = 0$.

$$\frac{p}{p_t} = \left(\frac{2}{(\gamma-1)M^2+2}\right)^{\frac{\gamma}{\gamma-1}} \tag{11.6}$$

where p_t is the *total pressure*. Likewise,

$$\begin{aligned}\frac{dT}{T} &= \frac{da^2}{a^2} = -\frac{(\gamma-1)dM^2}{(\gamma-1)M^2+2} \\ \frac{T}{T_t} &= \left(\frac{a}{a_t}\right)^2 = \frac{2}{(\gamma-1)M^2+2},\end{aligned} \tag{11.7}$$

where T_t is the *total temperature*.

$$\frac{\rho}{\rho_t} = \left(\frac{2}{(\gamma-1)M^2+2}\right)^{\frac{1}{\gamma-1}}. \tag{11.8}$$

As seen in Fig. 11.2, the flow variables that are normalized by the choked or stagnation value become a function only of M. The increment in the cross-sectional area, dA, has a negative value when $M < 1$ and a positive value when $M > 1$. A has a minimum at the choking point ($A = A_*$). With increasing M, the gas expands while p, T, and ρ monotonically decrease.

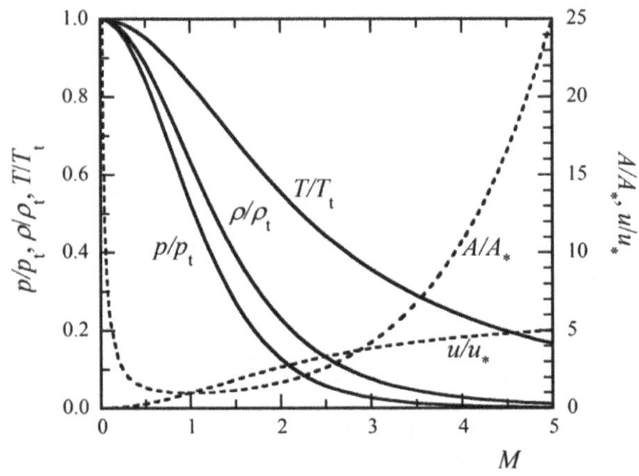

Fig. 11.2 Flow variables as a function of flow Mach number M

11.1.2 Mass Flow Rate

Let us consider a converging nozzle with $\mathrm{d}A = 0$ at the exit. The following applies to a choked flow in an orifice and Laval nozzle. The mass flow rate $\dot{m}$ is determined by the cross-sectional area at the exit, A_*, p_t, and the ambient pressure, p_a. Let us attach to variables at the exit. Using (11.7) and (11.8),

$$\dot{m} = \rho' u' A_* = \frac{\rho'}{\rho_t}\frac{u'}{a'}\frac{a'}{a_t}\rho_t a_t A_* = M' \left[\frac{2}{(\gamma-1)M'^2+2}\right]^{\frac{\gamma+1}{2(\gamma-1)}} \rho_t a_t A_*. \tag{11.9}$$

From (11.6),

$$M' = \sqrt{\frac{2}{\gamma-1}\left[\left(\frac{p'}{p_t}\right)^{-\frac{\gamma-1}{\gamma}} - 1\right]}. \tag{11.10}$$

From (11.6), (11.9), and (11.10),

$$\phi \equiv \frac{\dot{m}}{\rho_t a_t A_*} = \left(\frac{p'}{p_t}\right)^{\frac{\gamma+1}{2\gamma}} \sqrt{\frac{2}{\gamma-1}\left\{\left(\frac{p'}{p_t}\right)^{-\frac{\gamma-1}{\gamma}} - 1\right\}}. \tag{11.11}$$

Equation (11.11) gives the normalized mass flow rate ϕ as a function of p'/p_t. Nulling the differentiation of the rightmost hand of (11.11) with respect to p'/p_t, ϕ has a maximum of

$$\phi_c \equiv \left(\frac{2}{\gamma+1}\right)^{\frac{\gamma+1}{2(\gamma-1)}} \tag{11.12}$$

with

$$\frac{p'}{p_t} \equiv \left(\frac{p'}{p_t}\right)_c = \left(\frac{2}{\gamma+1}\right)^{\frac{\gamma}{\gamma-1}}. \tag{11.13}$$

Under this condition, (11.10) yields to

$$M' = 1$$

This is equivalent to the choking condition. This condition is termed the *critical state*, (11.12) is the *critical mass flow* rate, and (11.13) is the *critical pressure ratio*.[1] From (11.11) and (11.12),

$$\dot{m} = \left(\frac{2}{\gamma+1}\right)^{\frac{\gamma+1}{2(\gamma-1)}} \rho_t a_t A_* = \left(\frac{2}{\gamma+1}\right)^{\frac{\gamma+1}{2(\gamma-1)}} \frac{p_t}{RT_t}\sqrt{\gamma RT_t} A_* = \left(\frac{2}{\gamma+1}\right)^{\frac{\gamma+1}{2(\gamma-1)}} p_t A_* \sqrt{\frac{\gamma}{RT_t}}. \tag{11.14}$$

With a constant value of T_t, the critical mass flow rate is in proportion with the cross-sectional area at the throat and the total pressure. This is a useful relation to keep the flow rate as a constant.

If the flow is choked, the mass flow rate is kept constant even with further decreasing p_a because the downstream signals do not propagate past the choking point. In many flow devices, the mass flow rate is controlled using this principle. Substituting (11.6)–(11.8) with the choking condition $M = 1$,

$$\frac{p_*}{p_t} = \left(\frac{2}{\gamma+1}\right)^{\frac{\gamma}{\gamma-1}} \tag{11.15}$$

$$\frac{T_*}{T_t} = \left(\frac{a_*}{a_t}\right)^2 = \frac{2}{\gamma+1} \tag{11.16}$$

$$\frac{\rho_*}{\rho_t} = \left(\frac{2}{\gamma+1}\right)^{\frac{1}{\gamma-1}}, \tag{11.17}$$

where, for $\gamma = 1.4$, $p_*/p_t \simeq 0.528$. The flow is choked when $p_a \le p_*$. In other words, in order to obtain a constant mass flow rate using the choking condition, the stagnation pressure needs to be almost twice ($= 1/0.528 \simeq 1.89$ times) as high as the ambient pressure. With $p_a > p_*$, the flow is not choked; ϕ follows (11.11) with $p' = p_a$, see Fig. 11.3.

[1] The critical mass flow rate is also expressed by $\phi_c = C_m p_t/(\rho_t a_t)$, where C_m is termed the *mass flow coefficient*.

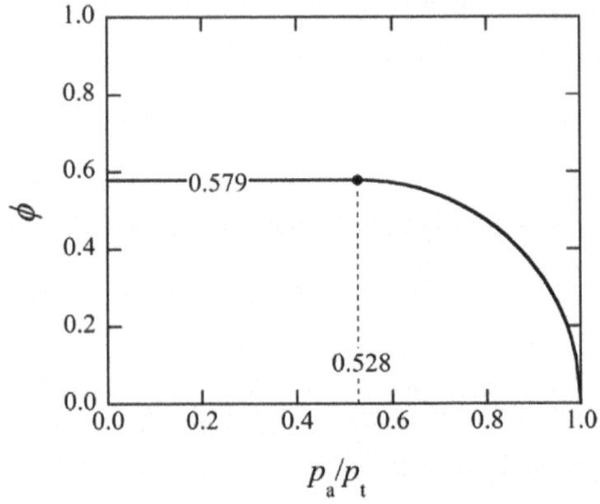

Fig. 11.3 ϕ versus p_a/p_t, $\gamma = 1.4$

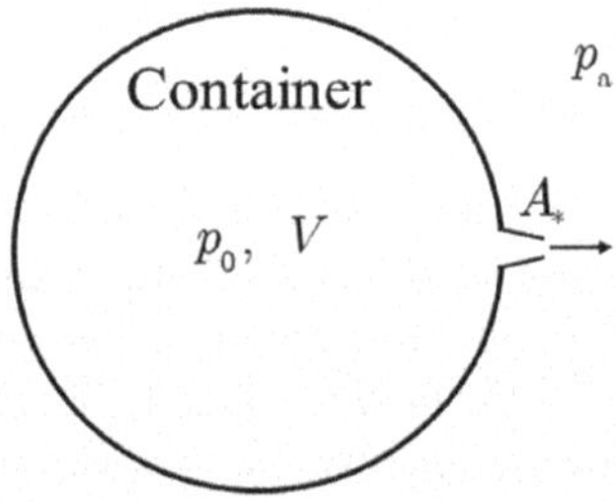

Fig. 11.4 Gas release from container

Column: Gas Release from a Container

Let us consider gas (initial pressure p_0, initial temperature T_0, and specific heat ratio γ) being released from a container with an inner volume of V through an orifice with a cross-sectional area of A_* to the atmosphere with p_a (Fig. 11.4). Let us obtain the period for the gas flowing out while keeping the critical condition.

Here, t is the time elapsed from the starting moment of the gas release. The critical mass flow rate is given by (11.14) and the equation of state for calorically perfect gas.

$$\dot{m} = -\frac{\mathrm{d}}{\mathrm{d}t}\left(\frac{p(t)V}{RT(t)}\right) = \left(\frac{2}{\gamma+1}\right)^{\frac{\gamma+1}{2(\gamma-1)}} pA_*\sqrt{\frac{\gamma}{RT}}. \tag{11.18}$$

For isentropic expansion, (2.91) applies.

$$\frac{p}{p_0} = \left(\frac{T}{T_0}\right)^{\frac{\gamma}{\gamma-1}}. \tag{11.19}$$

Combining (11.18) and (11.19),

$$\left(\frac{p}{p_0}\right)^{-\frac{\gamma+1}{2\gamma}} \mathrm{d}\left(\frac{p}{p_0}\right)^{\frac{1}{\gamma}} = \left(\frac{2}{\gamma+1}\right)^{\frac{\gamma+1}{2(\gamma-1)}} \frac{A_*}{V}\sqrt{\gamma R T_0}\mathrm{d}t. \tag{11.20}$$

The solution is

$$\frac{p}{p_0} = \left\{1 - \frac{\gamma-1}{2}\left(\frac{2}{\gamma+1}\right)^{\frac{\gamma+1}{2(\gamma-1)}} \frac{A_* a_0}{V} t\right\}^{\frac{\gamma-1}{2\gamma}}. \tag{11.21}$$

This equation gives the time variation of the pressure in the container. At $t = t_c$, the flow becomes the critical condition (11.13).

$$t_c = \frac{2}{\gamma-1}\left(\frac{2}{\gamma+1}\right)^{-\frac{\gamma+1}{2(\gamma-1)}} \left\{\left(\frac{p_0}{p_a}\right)^{\frac{\gamma-1}{2\gamma}} \left(\frac{2}{\gamma+1}\right)^{\frac{1}{2}} - 1\right\} \frac{V}{A_* a_0}. \tag{11.22}$$

11.1.3 Thrust

The gas charged in the chamber of Fig. 11.5 is released through the Laval nozzle. Let us obtain the thrust F. The chamber has a constant cross section of A_1 and is connected to the Laval nozzle. The gas on the upstream end is assumed to be under a stagnation condition, designated by the subscript t. The states at the nozzle entrance, throat, and exit are 1, *, and 2, respectively. The time variation of the pressure is neglected. Around the outer wall of the chamber and the nozzle, the gas is quiescent at an ambient pressure of p_a.

If we plug the exit of the nozzle, the chamber and the nozzle are exposed by a gas with p_a, and no net force is exerted no matter how the pressure is distributed on the inside. The thrust is the net force difference from this state and is obtained by

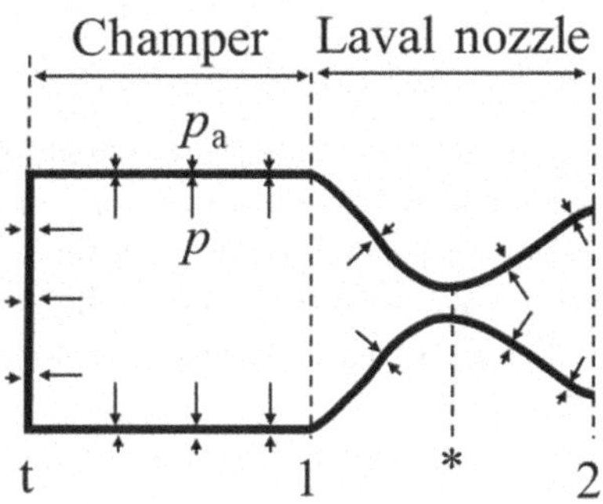

Fig. 11.5 Pressure distribution in a high-pressure chamber releasing gas jet

integrating the force in the right direction owing to the wall pressure on both sides of the wall. The exit of the real nozzle is not plugged by a wall; on its projection area on the left-side wall, a negative force of $p_a A_2$ is exerted on the outside. On the inside of the end wall, a positive force $p_t A_2$ is exerted. The thrust F is obtained by adding the force in the right direction inside of the side wall, $\int_1^2 pdA$, to the forces on the end wall and the exit.

$$\begin{aligned} F &= -\,p_a A_2 + p_t A_1 + \int_1^2 pdA \\ &= -\,p_a A_2 + p_t A_1 + [pA]_1^2 - \int_1^2 Adp \\ &= -\,p_a A_2 + p_t A_1 + p_2 A_2 - p_1 A_1 - \int_1^2 Adp. \end{aligned} \tag{11.23}$$

Neglecting the viscous force on the wall, the momentum conservation in the zone from t to 1 yields

$$p_t A_1 = p_1 A_1 + \rho_1 u_1^2 A_1 = p_1 A_1 + \dot{m} u_1 \qquad \because \dot{m} = \rho_1 u_1 A_1. \tag{11.24}$$

From (5.12),

$$dp = -\,\rho u du$$

$$\int_1^2 Adp = -\int_1^2 \rho u A du = -\dot{m}(u_2 - u_1). \tag{11.25}$$

Substituting (11.23) with (11.24) and (11.25),

$$F = \dot{m} u_2 + (p_2 - p_a) A_2. \tag{11.26}$$

Equation (11.26) gives F by using only the variables at the nozzle exit, implying that the thrust equals the momentum ejected in the unit time. The first term on the right-hand side is the *momentum thrust*, and the second term is the *pressure thrust*. Usually, a rocket engine is designed so that $p_2 < p_a$ on the ground; the pressure thrust has a negative value on the ground.

The thrust performance of a nozzle is evaluated using the thrust coefficient C_F, defined by

$$C_F \equiv \frac{F}{p_t A_*} = \frac{1}{p_t A_*}\{\rho_2 u_2^2 A_2 + (p_2 - p_a)A_2\} = \left\{\gamma M_2^2 \frac{p_2}{p_t} + \left(\frac{p_2}{p_t} - \frac{p_a}{p_t}\right)\right\}\frac{A_2}{A_*}. \tag{11.27}$$

The upper a rocket goes, the lower p_a becomes, and the larger C_F results owing to the increase in the pressure thrust. A_2/A_* is termed the *expansion ratio*. For an isentropic flow, the flow Mach number at the exit, M_2, and p_2/p_t are functions of the expansion ratio. From (11.10),

$$M_2^2 = \frac{2}{\gamma - 1}\left[\left(\frac{p_2}{p_t}\right)^{-\frac{\gamma-1}{\gamma}} - 1\right]. \tag{11.28}$$

Combining this with (11.3),

$$\frac{A_2}{A_*} = \frac{\left(\frac{2}{\gamma+1}\right)^{\frac{\gamma+1}{2(\gamma-1)}}\left(\frac{p_2}{p_t}\right)^{-\frac{\gamma+1}{2\gamma}}}{\sqrt{\frac{2}{\gamma-1}\left[\left(\frac{p_2}{p_t}\right)^{-\frac{\gamma-1}{\gamma}} - 1\right]}}. \tag{11.29}$$

Equation (11.29) gives p_2/p_t as an implicit function of A_2/A_*. Substituting (11.27) with (11.28) and (11.29), C_F is given as a function of A_2/A_*. Figure 11.6a shows C_F variation in the ultimate condition of $p_a = 0$. The larger the expansion ratio, the larger C_F becomes, although the pressure thrust decreases.

As seen in Fig. 11.6b, for a constant value of p_a/p_t there exists a value of A_2/A_* that maximizes C_F. Differentiating (11.27) with respect to p_2/p_t,

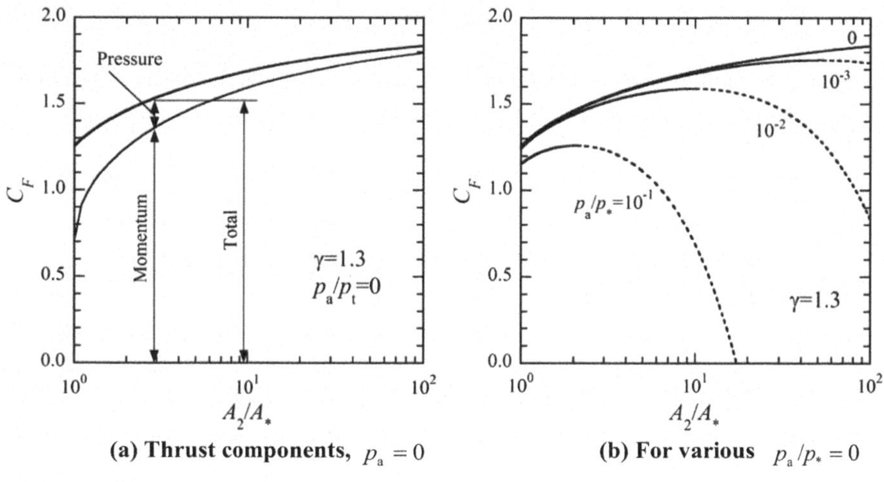

Fig. 11.6 C_F versus A_2/A_*, $\gamma = 1.3$

$$\frac{\partial C_F}{\partial\left(\frac{p_2}{p_t}\right)} = \frac{A_2}{A_*}\frac{\partial}{\partial\left(\frac{p_2}{p_t}\right)}\left\{\gamma M_2^2\frac{p_2}{p_t} + \left(\frac{p_2}{p_t} - \frac{p_a}{p_t}\right)\right\} + \left\{\gamma M_2^2\frac{p_2}{p_t} + \left(\frac{p_2}{p_t} - \frac{p_a}{p_t}\right)\right\}\frac{\partial}{\partial\left(\frac{p_2}{p_t}\right)}\frac{A_2}{A_*}$$

$$= \frac{A_2}{A_*}\left(\gamma M_2^2 + 1\right) + \frac{A_2}{A_*}\gamma\frac{p_2}{p_t}\frac{\partial M_2^2}{\partial\left(\frac{p_2}{p_t}\right)} + \left\{\gamma M_2^2\frac{p_2}{p_t} + \left(\frac{p_2}{p_t} - \frac{p_a}{p_t}\right)\right\}\frac{\partial\left(\frac{A_2}{A_*}\right)}{\partial\left(\frac{p_2}{p_t}\right)}.$$

From (11.6),

$$\frac{\partial M_2^2}{\partial\left(\frac{p_2}{p_t}\right)} = -\frac{(\gamma - 1)M_2^2 + 2}{\gamma\left(\frac{p_2}{p_t}\right)}.$$

Then, using (11.5),

$$\frac{\partial\left(\frac{A_2}{A_*}\right)}{\partial\left(\frac{p_2}{p_t}\right)} = \frac{1 - M_2^2}{\gamma M_2^2}\frac{\frac{A_2}{A_*}}{\frac{p_2}{p_t}}.$$

Therefore,

$$\frac{\partial C_F}{\partial\left(\frac{p_2}{p_t}\right)} = \frac{A_2}{A_*}\left(\gamma M_2^2 + 1\right) - \frac{(\gamma - 1)M_2^2 + 2}{\gamma\left(\frac{p_2}{p_t}\right)}\frac{A_2}{A_*}\gamma\frac{p_2}{p_t} + \left\{\gamma M_2^2\frac{p_2}{p_t} + \left(\frac{p_2}{p_t} - \frac{p_a}{p_t}\right)\right\}\frac{1 - M_2^2}{\gamma M_2^2}\frac{\frac{A_2}{A_*}}{\frac{p_2}{p_t}}.$$

$$= \left(1 - \frac{p_a}{p_2}\right)\frac{1 - M_2^2}{\gamma M_2^2}\frac{A_2}{A_*}$$

In order to maximize C_F for $M_2 > 1$,

$$p_2 = p_a \tag{11.30}$$

C_F is at a maximum if the pressure at the nozzle exit equals the ambient pressure. This operation corresponds to *optimum expansion*, as seen in (11.27), the pressure thrust vanishes.

11.1.4 Nozzle Flow Patterns with Various Nozzle Pressure Ratios

So far, we assumed an isentropic flow in the nozzle and neglected the influence of the ambient pressure on the nozzle flow. However, under most operation conditions, the pressure at the nozzle exit differs from the ambient pressure. In order to compensate

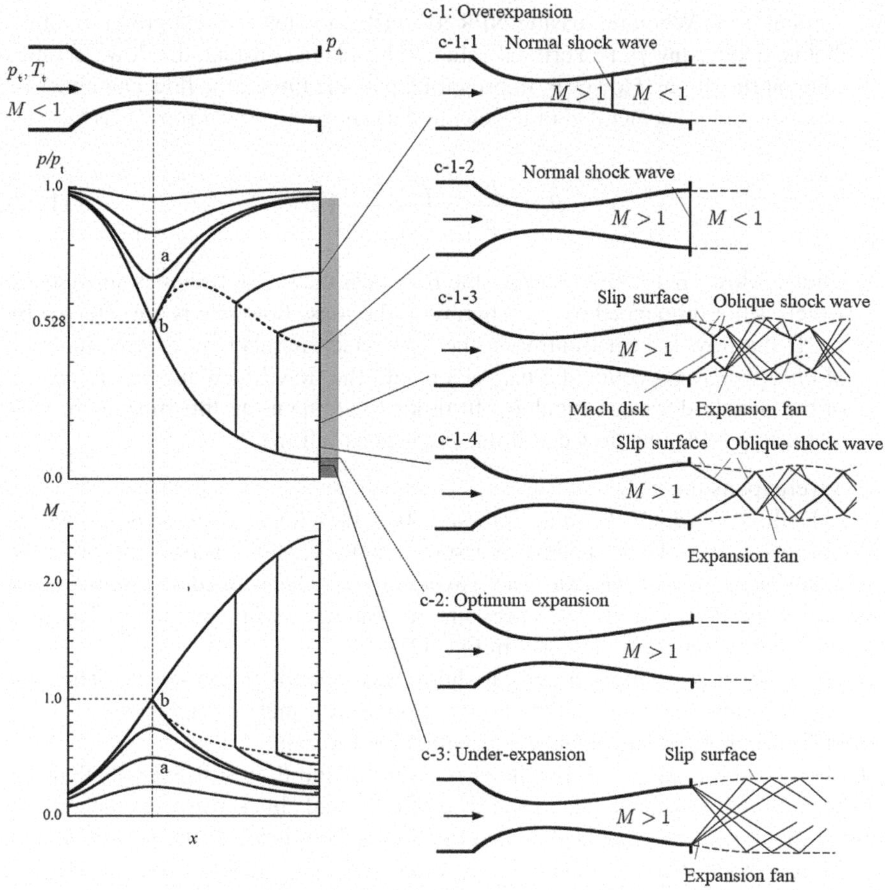

Fig. 11.7 Nozzle flows under various ambient pressures, $\gamma = 1.4$, expansion ratio of 2.4

this difference, shock waves and/or expansion waves appear in and/or outside the nozzle.[2]

We define the *nozzle pressure ratio* (NPR) as

$$\mathrm{NPR} \equiv \frac{p_\mathrm{t}}{p_\mathrm{a}} (> 1). \tag{11.31}$$

Using NPR, the flow patterns are categorized as shown in Fig. 11.7.

(a) Subsonic flow in entire region: With NPR close to unity, the subsonic flow is accelerated in the converging section. Past the throat, the flow decelerates in the diverging section.

[2]In real nozzle flows, complicated phenomena (for example, interactions with the boundary layer) and flow separation can occur.

(b) Critical flow: When increasing NPR to a critical value (critical pressure ratio), that is, decreasing p_a to a critical value with p_t kept constant, the flow becomes sonic at the throat. However, in principle, past the throat, the flow can either be supersonically accelerated or decelerated as subsonic flow. From (11.15),

$$\beta_c \equiv \frac{p_*}{p_t} = \left(\frac{2}{\gamma+1}\right)^{\frac{\gamma}{\gamma-1}}. \tag{11.32}$$

(c) Choked flow: With NPR $> 1/\beta_c$, the flow past the throat becomes supersonic, which is not influenced by p_a. Therefore, the mass flow rate is not affected by p_a. If the flow is kept isentropic, the flow Mach number becomes a function of only the cross-sectional area. As a result, the flow Mach number at the exit of the nozzle does not equal p_a. In order to compensate this mismatch, with increasing NPR, the flow condition changes as follows:

(c-1) Overexpansion
A nozzle flow in which the static pressure becomes lower than p_a is termed an *overexpansion*. In order to achieve pressure matching, shock waves appear in the inside and/or outside of the exit. The wave pattern is determined so that the static pressure of the flow and the back pressure are balanced with each other. The flow pattern changes from c-1-1 to c-1-4 in Fig. 11.7.

(c-1-1) When p_a is high, a normal shock wave stands in the nozzle. The flow becomes subsonic behind it so that the static pressure equals p_a at the exit.

(c-1-2) The normal shock wave is generated at the exit.

(c-1-3) At the nozzle exit, the static pressure of the flow still is lower than p_a, yet not so much as to have the appearance of a normal shock wave. At the exit, an oblique shock wave is generated so that the flow is bent inward. On the center axis, the oblique shock wave is reflected as a *Mach reflection*; a *Mach disk*, which is a normal shock wave, appears.

(c-1-4) At the nozzle exit, an oblique shock wave appears. On the center axis, there is a regular reflection in which the oblique shock wave appears as a reflected wave. Then, the reflected wave experiences reflection on the free jet boundary and slip surface. Expansion waves appear as reflected waves. These processes are repeated until the flow becomes subsonic.

(c-2) Optimum expansion
At the nozzle exit, the flow static pressure equals p_a. If the passage at the nozzle exit is parallel to the flow, no wave appears because the flow is not bent. A slip surface appears against the ambient gas.

(c-3) Underexpansion
Even with further decreasing p_a, the flow in the diverging section is kept supersonic and unchanged. At the nozzle exit, the flow static pressure is still higher than p_a (*underexpansion*); expansion waves are generated with the nozzle flow bent outward across them. The expansion waves are reflected as compression waves on the outer boundary where the flow static pressure is matched with p_a.

11.2 Supersonic Diffuser

A *diffuser* is a device that decelerates flow. With deceleration, the flow static pressure is increased. In an aircraft engine, air is introduced to its intake and is decelerated by a diffuser. In a wind tunnel, the test flow is decelerated in a diffuser so that it is damped or recirculated with low noise and pressure loss. A *supersonic diffuser* should decelerate a supersonic flow to subsonic with a small pressure loss. Figure 11.8 illustrates the operation modes of a supersonic diffuser.

In a real flow, it is impossible to isentropically decelerate the supersonic flow as shown in Fig. 11.8a. Instead, as shown in Fig. 11.8b, the supersonic flow is decelerated once in the converging section, supersonically passes the throat, and is accelerated again in the diverging section. In the diverging section, a normal shock wave is generated so that the flow is decelerated to a subsonic speed. The flow static pressure is matched at the exit condition. This flow pattern is robust against flow fluctuations.

However, if the flow matching is not accomplished, the flow reaches an *unstart* situation, as shown in Fig. 11.8c. This leads to large losses in the flow rate and total pressure and an increase in the drag.

Here, we deal with the supersonic diffuser performance based on quasi-one-dimensional flow without taking the effects of the boundary layer, flow separation, etc., into consideration.

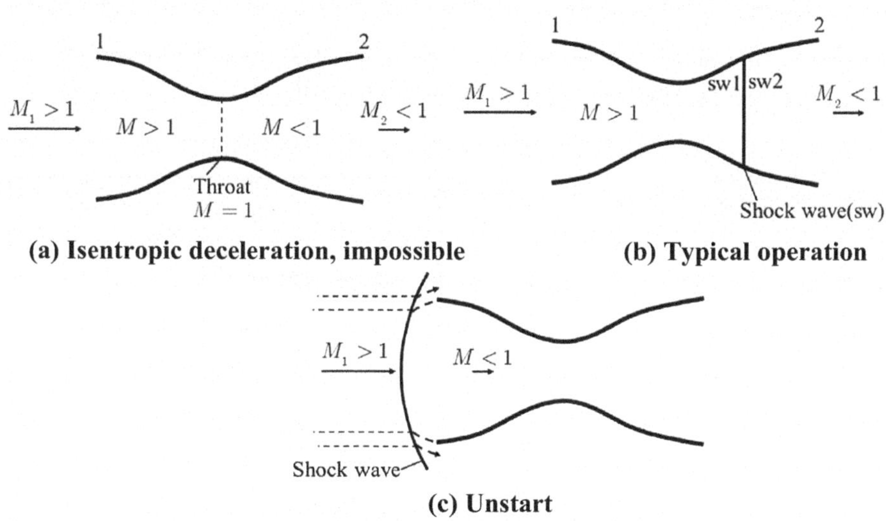

Fig. 11.8 One-dimensional operation model of supersonic diffuser

11.2.1 Quasi-One-Dimensional Operation

Let us analyze the operation of a supersonic diffuser based on quasi-one-dimensional flow. In real operation, the entropy of the supersonic flow is increased during its deceleration with the appearance of shock waves. However, it is assumed here that the flow is isentropic as long as the flow is kept supersonic. From (11.2), the relation between the flow Mach number M and the cross-sectional area A is

$$\frac{A}{A_1} = \frac{M_1}{M}\left\{\frac{(\gamma - 1)M^2 + 2}{(\gamma - 1)M_1^2 + 2}\right\}^{\frac{\gamma+1}{2(\gamma-1)}} \tag{11.33}$$

Figure 11.9 shows the relation among M, p, and A in supersonic flow. The flow enters the diffuser at the entrance with $A = A_1$. Then, the flow is decelerated in the converging section, and A decreases. The diffuser works only when the flow does not reach the sonic condition ($M = 1$), in other words, only if the flow is not choked. Past the throat, the flow is accelerated in the diverging section with A increased, as shown in Fig. 11.10a. Figure 11.10b shows a critical operation in which the flow becomes choked at the throat. The condition in which the flow becomes sonic in the converging section (see Fig. 11.10c) is impossible. Therefore, the necessary condition for supersonic diffuser operation is that the flow is not choked in the converging section.

$$A_{\text{throat}} \geq A_* \tag{11.34}$$

where A_* is the value of A for the choked condition. Hereafter, the subscript * will designate the choked condition. Substituting (11.34) with (11.33),

Fig. 11.9 Isentropic operation in supersonic regime $\gamma = 1.4$

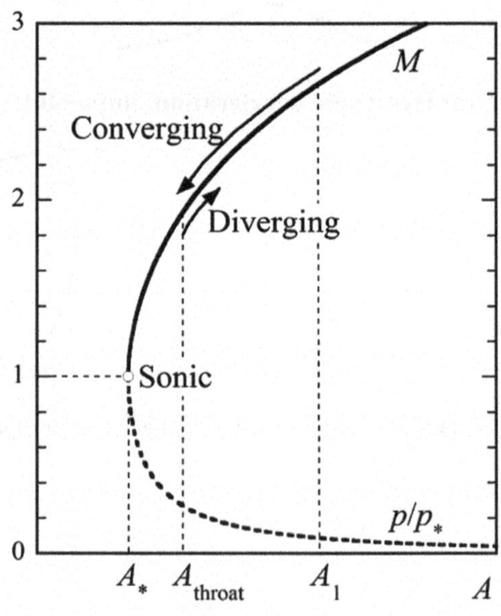

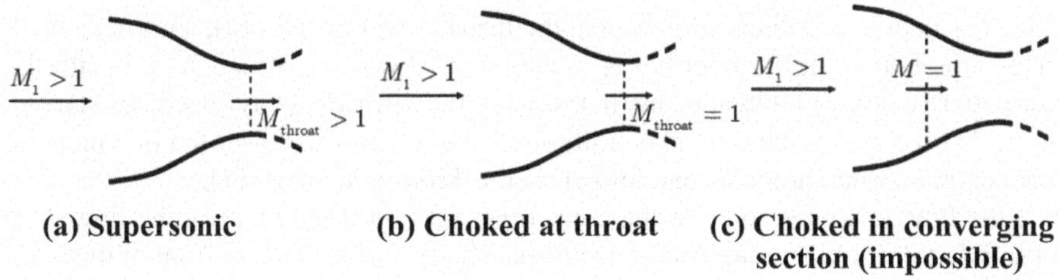

Fig. 11.10 Isentropic flow patterns in converging section

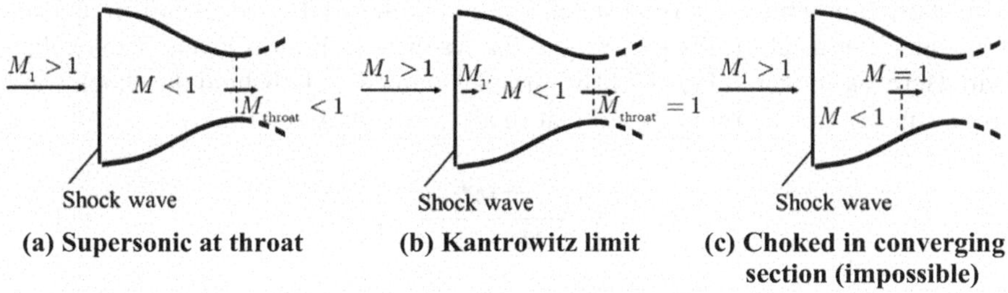

Fig. 11.11 Flow patterns with normal shock wave at inlet

$$\frac{A_{\text{throat}}}{A_1} = \frac{M_1}{M_{\text{throat}}} \left\{ \frac{(\gamma - 1)M_{\text{throat}}^2 + 2}{(\gamma - 1)M_1^2 + 2} \right\}^{\frac{\gamma+1}{2(\gamma-1)}}$$

$$\geq \frac{A_*}{A_1} = \frac{M_1}{1} \left\{ \frac{(\gamma - 1) \cdot 1^2 + 2}{(\gamma - 1)M_1^2 + 2} \right\}^{\frac{\gamma+1}{2(\gamma-1)}} = M_1 \left\{ \frac{\gamma + 1}{(\gamma - 1)M_1^2 + 2} \right\}^{\frac{\gamma+1}{2(\gamma-1)}}.$$

$$\therefore \frac{A_{\text{throat}}}{A_1} \geq M_1 \left\{ \frac{\gamma + 1}{(\gamma - 1)M_1^2 + 2} \right\}^{\frac{\gamma+1}{2(\gamma-1)}} \tag{11.35}$$

The left-hand side of (11.35) is termed the *contraction ratio*. Equation (11.35), which corresponds to the *isentropic unchoked condition*, is a necessary condition for a supersonic diffuser to operate properly. The equality corresponds to a choked condition at the throat, for example, when $M_1 = 3$ $A_{\text{throat}}/A_1 \geq 0.236$.

From (11.6) (see Fig. 11.9), the pressure is given by

$$\frac{p}{p_1} = \left[\frac{(\gamma - 1)M_1^2 + 2}{(\gamma - 1)M^2 + 2} \right]^{\frac{\gamma}{\gamma-1}}. \tag{11.36}$$

Next, let us consider flow patterns with a normal shock wave at the inlet (see Fig. 11.11). Here, we do not consider the flow spillage at the inlet. The flow behind the shock wave is subsonic and is thus accelerated in the converging section. In the present quasi-one-dimensional flow, the flow Mach number M is a function of only A.

If M is kept lower than unity even at the throat (see Fig. 11.11a), a normal shock wave cannot stand at the inlet but is "shallowed" down the throat. As a result, the flow pattern of Fig. 11.11a appears. In this case, the diffuser always becomes *started*. Figure 11.11b shows the critical condition for the diffuser to be started in which the flow becomes sonic at the throat, known as the *Kantrowitz limit* [1].

If the flow becomes sonic in the converging section (Fig. 11.11c), the flow does not satisfy the choking condition. The normal shock wave stands in front of the inlet, and part of the incoming flow that should be introduced into the diffuser spills out, thereby resulting in unstart (see Fig. 11.8c).

The Kantrowitz limit corresponds to a sufficient condition for a supersonic diffuser to be started, in which a normal shock wave is swallowed down even if it appears at the inlet. For a calorically perfect gas, the Kantrowitz limit is given in an explicit form. Here, as shown in Fig. 11.11b, variables immediately behind the shock wave have a subscript of 1'. From the normal shock wave condition (4.60),

$$M_{1'} = \left\{ \frac{(\gamma - 1)M_1^2 + 2}{2\gamma M_1^2 - \gamma + 1} \right\}^{1/2}. \tag{11.37}$$

Substituting (11.37) with (11.3) and the choking condition,

$$\frac{A_1}{A_{\text{throat}}} = \frac{\left(\frac{\gamma-1}{\gamma+1} M_{1'}^2 + \frac{2}{\gamma+1}\right)^{\frac{\gamma+1}{2(\gamma-1)}}}{M_{1'}}.$$

Therefore, the condition for the shock wave being swallowed down is

$$\frac{A_{\text{throat}}}{A_1} \geq \frac{\{(\gamma - 1)M_1^2 + 2\}^{1/2} \{2\gamma M_1^2 - (\gamma - 1)\}^{\frac{1}{\gamma-1}}}{(\gamma + 1)^{\frac{\gamma+1}{2(\gamma-1)}} M_1^{\frac{\gamma+1}{\gamma-1}}}. \tag{11.38}$$

The equality corresponds to the Kantrowitz limit. Based on (11.35) and (11.38), three operation domains exist:

(1) Always *started*, wherein both (11.35) and (11.38) are satisfied.
(2) *Dual-solution domain*, wherein (11.35) is satisfied but (11.38) is not. In this domain, whether the diffuser is started or in unstart mode depends on the flow history.
(3) Always unstart, wherein neither equation is satisfied.

These conditions are displayed using Mach number-flow passage area coordinates (Fig. 11.12). In the dual-solution domain, whether the diffuser is in a started or unstart state depends on the flow history and disturbances experienced by the flow. For example, if the flow enters this domain from the low-Mach-number side, that is, the unstart side, the operation to some extent stays in unstart even in the dual-solution domain. Conversely, if the flow enters from the higher Mach-number side, that is, the start domain, to some extent, the diffuser is maintained in a started state in the dual-solution domain.

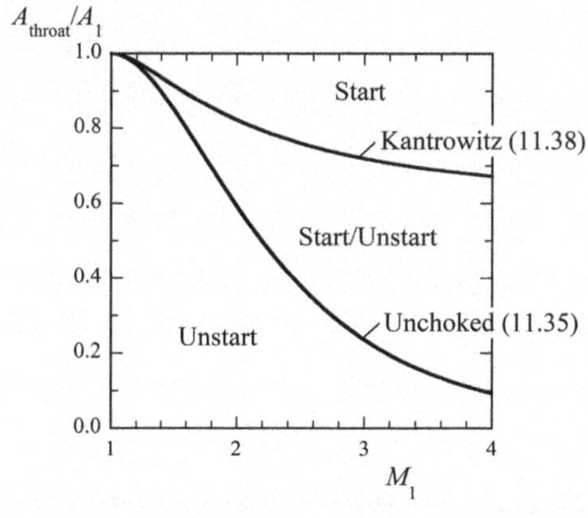

Fig. 11.12 Supersonic diffuser operation domains on $M_1 - A_t/A_1$ coordinates, $\gamma = 1.4$

In order for the supersonic flow to be decelerated to a subsonic speed, a normal shock wave needs to stand as is in Fig. 11.8b. Behind the shock wave, the entropy is increased. The higher the Mach number in front of the shock wave, the greater the increase in entropy. In order for the entropy increment to be suppressed, the location of the shock wave should be close to the throat so that the flow Mach number is close to unity. However, the shock wave cannot stay at the throat.

As will be shown later, the location of the shock wave is determined by the pressure at the exit, p_2. Here, the flow passage area at the shock wave is designated by A_{sw}. The flow variables in front of and behind the shock wave have subscripts sw1 and sw2, respectively. For the isentropic flow up to the shock wave, by using (11.2), M_{sw1} is obtained as the following implicit form:

$$\frac{A_{sw}}{A_1} = \frac{M_1}{M_{sw1}} \left\{ \frac{(\gamma - 1)M_{sw1}^2 + 2}{(\gamma - 1)M_1^2 + 2} \right\}^{\frac{\gamma+1}{2(\gamma-1)}}.$$

The pressure is obtained as

$$\frac{p_{sw1}}{p_1} = \left\{ \frac{(\gamma - 1)M_1^2 + 2}{(\gamma - 1)M_{sw1}^2 + 2} \right\}^{\frac{\gamma}{\gamma-1}}. \tag{11.39}$$

From normal shock wave relations (4.47) and (4.60),

$$\frac{p_{sw2}}{p_{sw1}} = 1 + \frac{2\gamma}{\gamma + 1}(M_{sw1}^2 - 1) \tag{11.40}$$

$$M_{sw2} = \left\{ \frac{(\gamma - 1)M_{sw1}^2 + 2}{2\gamma M_{sw1}^2 - \gamma + 1} \right\}^{1/2}. \tag{11.41}$$

The flow Mach number and the pressure at the exit are obtained as

$$\frac{A_2}{A_{\mathrm{sw}}} = \frac{M_{\mathrm{sw2}}\left\{(\gamma - 1)M_2^2 + 2\right\}^{\frac{\gamma+1}{2(\gamma-1)}}}{M_2\left\{(\gamma - 1)M_{\mathrm{sw2}}^2 + 2\right\}^{\frac{\gamma+1}{2(\gamma-1)}}} \tag{11.42}$$

$$\frac{p_2}{p_{\mathrm{sw2}}} = \left\{\frac{(\gamma - 1)M_{\mathrm{sw2}}^2 + 2}{(\gamma - 1)M_2^2 + 2}\right\}^{\frac{\gamma}{\gamma-1}}. \tag{11.43}$$

Note that in (11.42), the subsonic solution should be chosen. With the exit pressure p_2 being given, the location of the shock wave is determined so that (11.43) is satisfied.

11.2.2 Multidimensional Effects

The diffuser operation analysis so far was based on a quasi-one-dimensional flow. In reality, however, the flow condition is not only determined by the cross-sectional area ratio. Figure 11.13 shows three configurations with an equal throat-to-inlet flow passage area ratio but with different configurations. In Fig. 11.13a, an axisymmetric center body is set on the center axis of a circular cylinder duct. An oblique shock wave stands at the leading edge of the center body. It repeatedly reflects between the duct and the center body walls. Among the three configurations shown in Fig. 11.13, the shock wave is weakest; the pressure loss is smallest.

In Fig. 11.13b, the configuration is two dimensional, and the shock wave is stronger than that in Fig. 11.13a. Hence, the pressure loss is greater. In Fig. 11.13c,

(a) Circular cylinder with axisymmetric center body

(b) Two-dimensional view

(c) Axisymmetrical converging symmetric center body

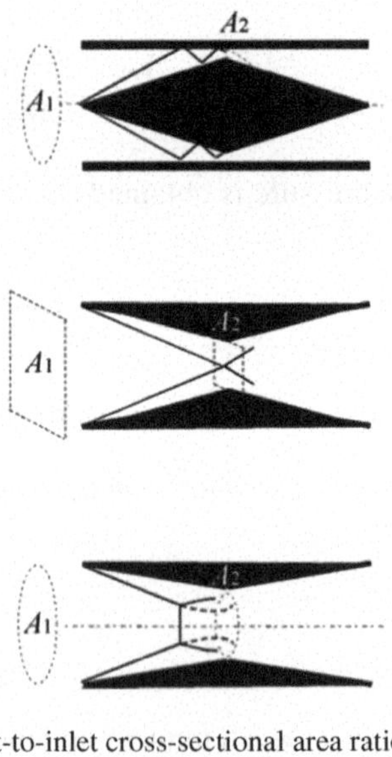

Fig. 11.13 Three diffuser configurations with equal throat-to-inlet cross-sectional area ratios

the flow passage is converged in an axisymmetric manner. The shock wave generated at the inlet is strongest and experiences Mach reflection. On the axis, a normal shock wave (the so-called Mach stem) appears, accompanied by a subsonic flow behind it. The pressure loss is largest among the three configurations, readily leading to unstart.

In a supersonic combustion ram (SCRAM) jet engine, supersonic flow is introduced in an intake. The flow is then decelerated in a diffuser, mixed with fuel, and burned with its Mach number kept higher than unity. The smaller the pressure loss, the better the diffuser performance with a widened operational regime. For this purpose, the configuration of Fig. 11.13a is favorable, although the structure to support the center body is necessary. The configurations of Fig. 11.13b, c have structural simplicity but have aerodynamics drawbacks.

11.2.3 Pseudo-Shock

So far, we have not dealt with the boundary layer in a duct flow, while a real flow often is strongly influenced by the boundary layer. *The shock wave-boundary layer interaction (SWBLI)* is an important problem in high-speed flows. The flow speed in a boundary layer is lower than of the outside. If a shock wave exists over the boundary layer, an inverse pressure gradient is formed in the boundary layer. In other words, the high-pressure region behind the shock wave penetrates the boundary layer. If SWBLI becomes significant in duct flows, a *pseudo-shock* is formed, and thus the supersonic flow is decelerated to subsonic under a condition different from that by a single shock wave. A pseudo-shock is generated in the intake of a supersonic engine and strongly influences the engine performance.

A pseudo-shock (Fig. 11.14) is composed of a *shock train* in which a boundary layer and shock waves repeat their interaction, and a *mixing region* [2]. In the shock train, Mach/regular reflections appear over the boundary layer. Even a subsonic flow behind a Mach stem in the Mach reflection can be accelerated through expansion. By

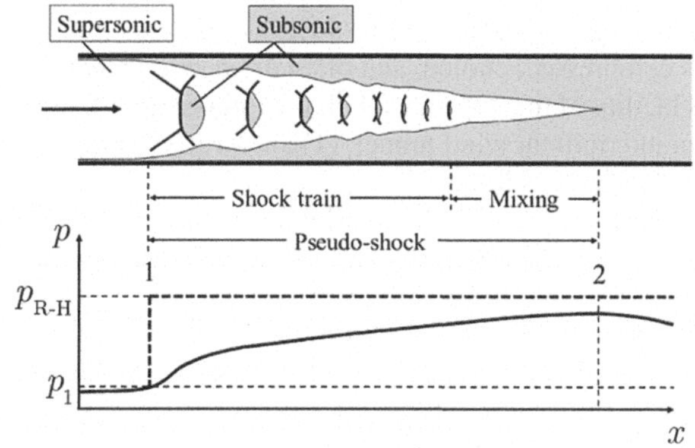

Fig. 11.14 Transition from supersonic to subsonic past a pseudo-shock

repeating the reflections, the flow is gradually decelerated with the subsonic portions increased.

In the mixing region, the flow is supersonic around the center but subsonic near the wall. Even without shock waves, the flow is decelerated by compression waves, thus eventually becoming subsonic all through.

With regard to the pseudo-shock, the supersonic flow at the entrance (label 1) becomes subsonic at the exit (label 2) over a distance that is greater than the tube diameter. At the exit, the static pressure is increased; however, the pressure is lower than that behind a shock wave for the same entrance Mach number. In the downstream of the pseudo-shock wave, the flow is accelerated while the static pressure is decreased owing to a friction force against the wall.

11.3 Supersonic Test Facilities

Special facilities are necessary for supersonic flow testing. Although their operation principle is simple, they are expensive with respect to size and cost. There are two methodologies: the generation of supersonic flow or the launching of a model at a supersonic speed. In this section, representative facilities will be introduced.

11.3.1 Supersonic Wind Tunnel

In order to generate steady-state supersonic flow, a Laval nozzle is used (Fig. 11.15a). Gas under a high-pressure stagnation condition is introduced to a converging section, is choked at the throat, and is accelerated supersonically in a diverging section before the *test section*, where the flow is diagnosed and/or the force components are measured. After the test section, the flow is decelerated to subsonic speed, and exhausted or recirculated. There are three categories of tunnel:

In a *continuous, closed-circuit tunnel* (Fig. 11.15b), steady-state operation is possible with (1) and (2) in Fig. 11.15a connected to form a closed cycle. Flow from (2) is compressed, cooled, and often dried, and is then introduced to (1) again. Usually, a facility of this type usually has a large size and necessitates a high running cost. In an intermittent wind tunnel, (1) and/or (2) in Fig. 11.15a is connected to a chamber with a finite volume.

A facility connected to a high-pressure gas reservoir at (1) is a *blow-down tunnel* (Fig. 11.15c), and that with a vacuum chamber at (2) is an *in-draft tunnel* (Fig. 11.15d). In the latter, the air sucked in the Laval nozzle contains moisture that may be frozen through the expansion. In order to eliminate this influence, in some in-draft tunnels, dry air is packed in a bag that is connected to the intake of the nozzle (1). In intermittent wind tunnels, the test time is limited by the volume of the airbag or the vacuum chamber.

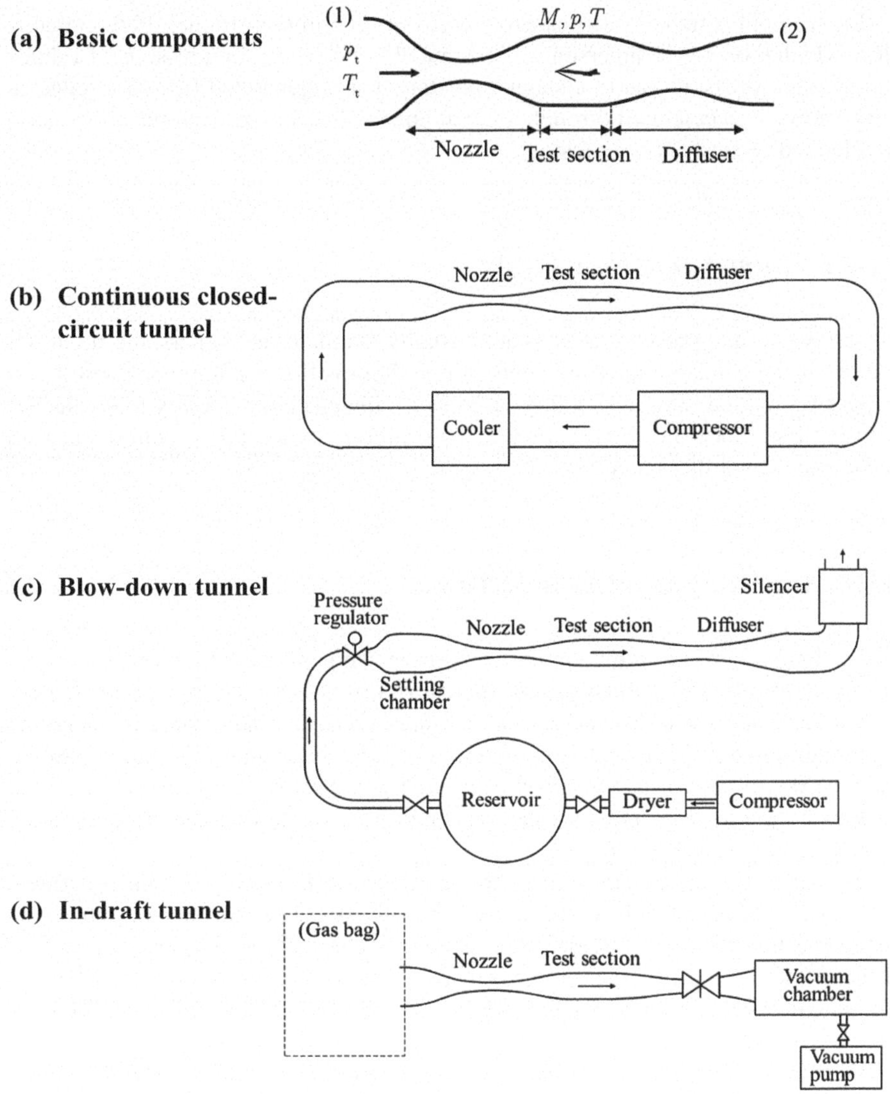

Fig. 11.15 Supersonic wind tunnels using Laval nozzle

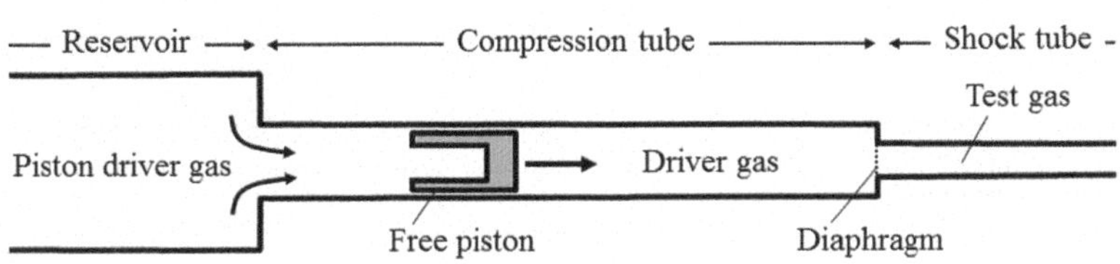

Fig. 11.16 Free-piston driver connected to shock tube

In an impulse tunnel, also categorized as an intermittent tunnel, high-enthalpy flow, which is of course supersonic, is generated by unsteady compression. In a shock tunnel, high-pressure gas in a stagnation state that is generated behind a reflected shock wave is accelerated through a Laval nozzle. In an expansion tube, the gas is accelerated by unsteady expansion.

11.3.2 Supersonic Free Flight

Experiments of supersonic flow should usually simulate real supersonic flight. Of course, using available measurements of real flight is the best approach, but this is expensive and has slim availability. *A ballistic range* is a test facility that launches a scale flight model in a test section. If the similarity exists, this is convenient with respect to availability and cost.

11.4 Unsteady Operation Driver

In an impulsive wind tunnel, an unsteady drive can generate a high-pressure/high-temperature test flow, which cannot be realized in steady-state operation. A *free-piston driver* boosts a driver gas to a high-pressure/temperature state by using the inertia of a piston. This is widely used in shock tubes, shock tunnels, expansion tubes, ballistic ranges, etc. It is composed of a (high-pressure) reservoir, free piston, and *compression tube* (Fig. 11.16), which is connected to a shock wave or acceleration tube in a ballistic range, etc.

The high-pressure gas, usually air, is charged in the reservoir. Once the separation at the end of the reservoir is removed, the free piston is accelerated in the compression tube. The free-piston compresses the driver gas charged in the compression tube using its inertia. Usually, the compression processes are almost isentropic. When the pressure of the driver gas reaches a preset value, the diaphragm is ruptured. Thus, in the case of Fig. 11.16, a shock wave propagates in the shock tube.

In order to increase the pressure and temperature of the driver gas, the inertia and then the speed and mass of the free piston need to be increased. However, if the

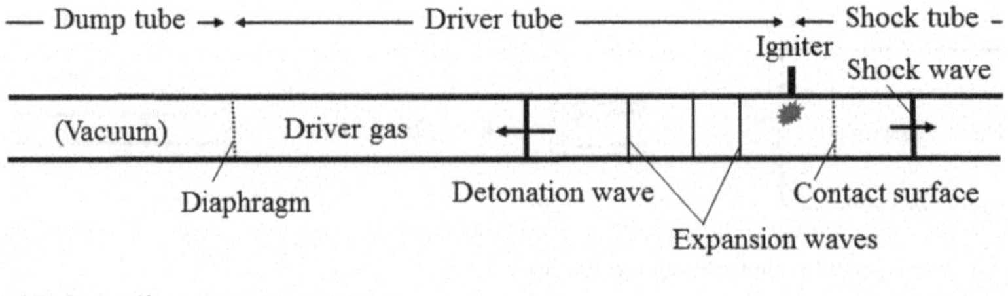

Fig. 11.17 Retarding detonation driver

initial pressure of the test gas in the shock tube is too low, the free piston will collide against the end wall of the compression tube, causing serious damage.

On the other hand, if the initial pressure in the compression tube is too high, the free piston stops moving before a sufficiently high pressure is obtained. Moreover, it is important to maintain the high-pressure state of the driver gas in a sufficiently long duration time. In the *tuned operation* by Ito et al. [3], the free-piston motion is designed so that the duration for the driver pressure to be kept constant is maximized, and the free-piston lands softly before the end wall.

A free-piston driver is a powerful tool to obtain high-enthalpy flow. However, it is difficult to obtain a long test time. Yu [4] invented a *retarding detonation driver* (Fig. 11.17). It does not use a free piston but a high-pressure/temperature state behind a detonation wave, thereby increasing the duration time of a constant pressure. The driver tube is initially filled with a detonable mixture. A detonation is ignited at the diaphragm to the shock tube, not at the other end. A detonation wave thus initiated propagates toward the other end, in the direction opposite the shock wave.

When the detonation wave arrives at the other diaphragm of the dump tube, the diaphragm is ruptured, and the incident detonation wave is attenuated in the dump tank. During the period in which the detonation wave propagates in the driver tube, a shock wave is driven and propagates in the shock tube. The test gas between the shock wave and the tail of the expansion fan is shock-compressed to a constant pressure. Compared to a free-piston driver, a retarding detonation driver can maintain a constant pressure of the test gas during a much longer period, although its attainable total enthalpy is lower.

11.5 Shock Tunnel

When a vehicle (re)enters the atmosphere of Earth or other planets, the flight Mach number exceeds 20 or is even higher. In front of the vehicle, a strong shock wave is generated. Behind the shock wave, a *shock layer* at a high temperature is formed. The heat shield from the shock layer is critical in designing the (re)entry vehicle.

In addition, when designing an air-breathing engine at a flight Mach number of 5 or higher, the heat transfer and aerodynamic forces need to be accurately estimated. In such flows, the largest part of the total enthalpy is the kinetic energy. In order to characterize the flow, the total enthalpy is used more often than the Mach number. For example, a total enthalpy in Mach-10 flight at an altitude of 20 km equals about 4.6 MJ/kg, which is comparable to the internal energy of typical explosives.[3] With an entry speed of 7.9 km/s at an altitude of 200 km, the total enthalpy becomes 31 MJ/kg.

A *shock tunnel* is an impulsive wind tunnel composed of an unsteady driver, shock tube, and a Laval nozzle at the end. Initially, the shock tube and the nozzle are separated by a diaphragm (Fig. 11.18a). After the incident shock wave is reflected at

[3]The energy released in the detonation of trinitrotoluene (TNT) equals about 4.2 MJ/kg.

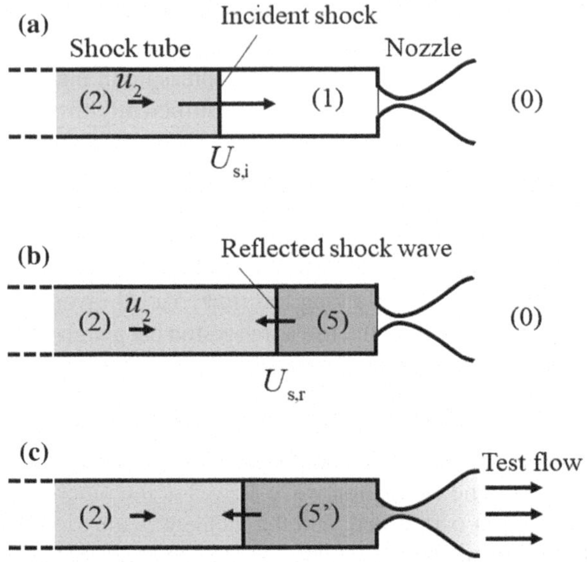

Fig. 11.18 Operation principle of shock tunnel

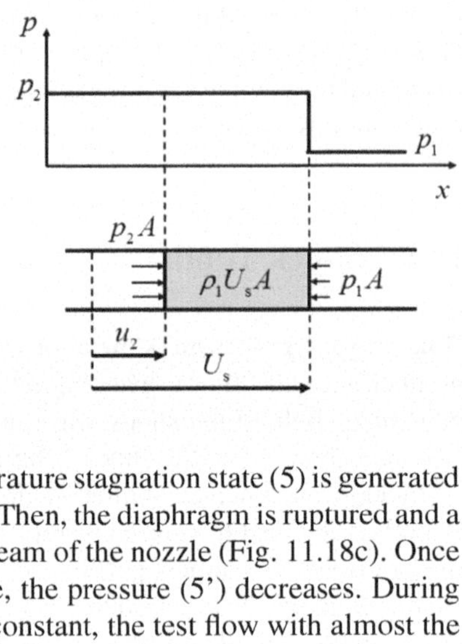

Fig. 11.19 Normal shock wave propagating to right, and forces exerted on fluid element

the shock tube end wall, a high-pressure/temperature stagnation state (5) is generated behind the reflected shock wave (Fig. 11.18b). Then, the diaphragm is ruptured and a high-enthalpy flow is generated in the downstream of the nozzle (Fig. 11.18c). Once the gas (5) starts to flow out of the shock tube, the pressure (5') decreases. During a period in which the pressure is kept almost constant, the test flow with almost the same condition continues.

Let us understand the principle of the entropy increase owing to the incident shock wave. Figure 11.19 illustrates the total enthalpy gain mechanisms behind the shock wave. Initially, the gray gas segment (which has a null velocity) is compressed by the incident shock wave. The mass that is compressed in a unit time equals $\rho_1 U_{s,i} A$ ($U_{s,i}$, propagation speed of the incident shock wave; A, cross-sectional area of the shock tube). While the left-hand end of the segment is compressed with a force of $p_2 A$ at a velocity of u_2, the right-hand end does not experience any work. Therefore, the segment gains an energy of $p_2 A u_2$.

$$\rho_1 U_s A \Delta\left(e + \frac{1}{2}u^2\right) = p_2 A u_2 > 0. \tag{11.44}$$

Next, let us obtain the increment in h_t of the test flow with a constant specific heat ratio of γ. According to our custom, the quiescent state in front of the incident shock wave, the state behind the incident shock wave, and that behind the reflected wave are labeled as 1, 2, and 5, respectively (see Fig. 11.18). The shock Mach number of the reflected shock wave is $M_{s,r}$. Note here that the flow speed u refers to the laboratory frame. From

$$\frac{p_2}{p_1} = 1 + \frac{2\gamma}{\gamma+1}(M_{s,i}^2 - 1) \tag{11.45}$$

$$\frac{\rho_2}{\rho_1} = \frac{(\gamma+1)M_{s,i}^2}{(\gamma-1)M_{s,i}^2 + 2} \tag{11.46}$$

$$u_2 = \frac{2a_1}{\gamma+1}\left(M_{s,i} - \frac{1}{M_{s,i}}\right), \tag{11.47}$$

$$\frac{h_{t,2}}{h_1} = \frac{\frac{\gamma}{\gamma-1}\frac{p_2}{\rho_2} + \frac{1}{2}u_2^2}{h_1} = \frac{2(\gamma-1)M_{s,i}^2 + 3 - \gamma}{\gamma+1}. \tag{11.48}$$

Let us consider the post-shock state of the reflected shock wave (5). From (9.49),

$$M_{s,r} = \sqrt{\frac{2\gamma M_{s,i}^2 - (\gamma-1)}{(\gamma-1)M_{s,i}^2 + 2}}. \tag{11.49}$$

Because $u_5 = 0$,

$$\frac{h_{t,5}}{h_2} = \frac{\frac{p_5}{\rho_5}}{\frac{p_2}{\rho_2}} = \frac{1 + \frac{2\gamma}{\gamma+1}(M_{s,r}^2 - 1)}{\frac{(\gamma+1)M_{s,r}^2}{(\gamma-1)M_{s,r}^2+2}} = \frac{(2\gamma M_{s,r}^2 - \gamma + 1)\{(\gamma-1)M_{s,r}^2 + 2\}}{(\gamma+1)^2 M_{s,r}^2} \tag{11.50}$$

$$\frac{h_{t,5}}{h_1} = \frac{h_{t,5}}{h_2}\frac{h_2}{h_1} = \frac{\{(3\gamma-1)M_{s,i}^2 - 2(\gamma-1)\}\{2(\gamma-1)M_{s,i}^2 - \gamma + 3\}}{(\gamma+1)^2 M_{s,i}^2}. \tag{11.51}$$

As seen in Fig. 11.20, h_t is increased by the incident and reflected shock waves. The higher the $M_{s,i}$, the larger the increments become. For example, when $M_{s,i} = 10$, the increment becomes 34 and 45 times larger than the initial value. However, in real flows, this ratio becomes lower because the degree of freedom is increased at such high temperatures, and thus γ is decreased.

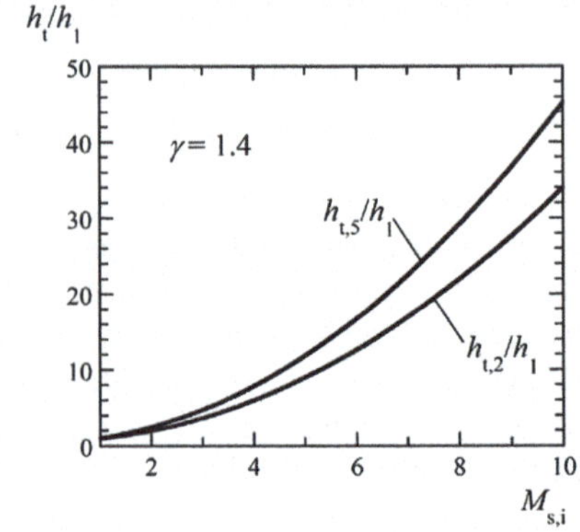

Fig. 11.20 Variation of total enthalpy owing to shock wave, $\gamma = 1.4$

From (11.48) and (11.50), the upper limit in the total enthalpy increment behind the reflected shock wave is

$$\left.\frac{h_{t,5}}{h_{t,2}}\right|_{M_{s,i}\to\infty} = \frac{3\gamma - 1}{\gamma + 1}. \tag{11.52}$$

When $\gamma = 1.4$, this ratio equals about 1.3, implying that the total enthalpy is not increased significantly behind the reflected shock wave.

If we neglect the heat loss to the wall, the total enthalpy of the test flow is determined uniquely by the shock Mach number of the incident shock wave. In order to increase this, a free-piston driver is used in many high-enthalpy facilities. A *free-piston shock tunnel* (see Fig. 11.21), that is, a shock tunnel with a free-piston driver, is also known as a *Stalker tube* [5].

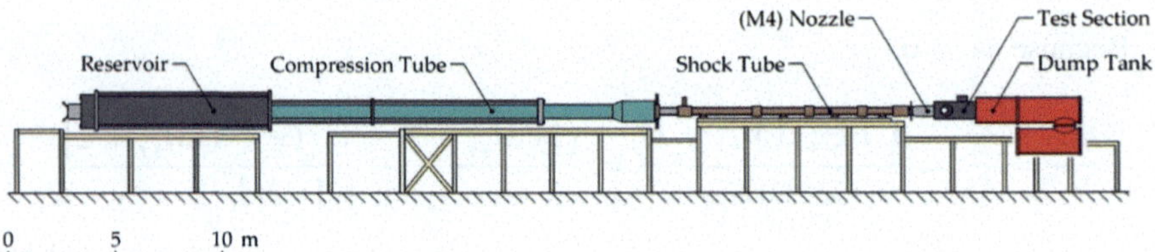

(a) T4 at University of Queensland (courtesy of University of Queensland)

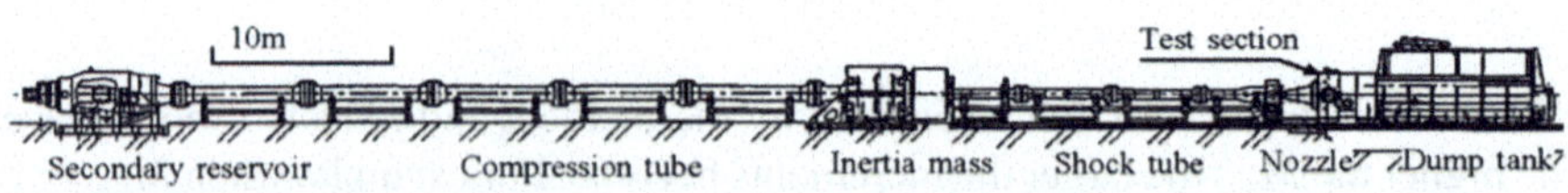

(b) High-Enthalpy Shock Tunnel (HIEST) at Japan Aerospace Exploration Agency (JAXA) (courtesy of JAXA)

Fig. 11.21 Free-piston-driven shock tunnels

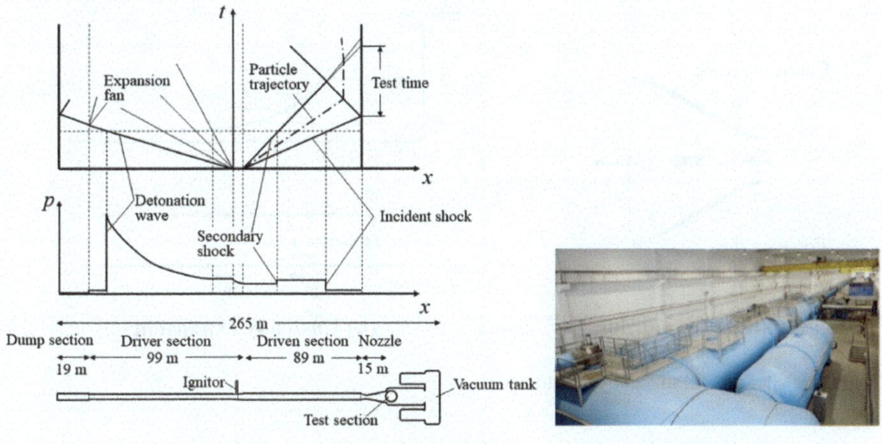

(a) Wave diagram and instant pressure distribution **(b) Photograph**

Fig. 11.22 Retarding detonation-driven shock tunnel JF12 (HYPER DRAGON), Institute of Mechanics, China Academy of Science. Courtesy of China Academy of Science

Jiang et al. developed a 265-m-long shock tunnel in which a retarding detonation driver is employed (Fig. 11.22). The researchers achieved a test time longer than 200 ms in a test section with a nozzle diameter of 2.5 m [6, 7].

With either driver, in order to obtain a long test time, it is favorable that pressure waves are not reflected on the contact surface between the test and driver gases toward the end wall of the shock tube. This condition is referred to as the *tailored condition*, which is obtained by solving the relevant Riemann problem.

11.6 Expansion Tube

An *expansion tube* is a device in which the flow behind a normal shock wave is further accelerated through unsteady expansion without stagnation. Let us compare here the unsteady expansion from a steady expansion through a nozzle. In the steady expansion through a nozzle (Fig. 11.23a), the flow passage area increases in the supersonic acceleration section. The resultant force on the flow element in the control volume is directed downstream.

However, no work is done on the element owing to this force because the wall does not move, so the total enthalpy is not increased. In an isentropic flow, the flow Mach number M, static pressure p, and total pressure p_t are related by (11.10).

$$M = \left[\frac{2}{\gamma - 1}\left\{\left(\frac{p}{p_t}\right)^{-\frac{\gamma-1}{\gamma}} - 1\right\}\right]^{\frac{1}{2}}. \tag{11.53}$$

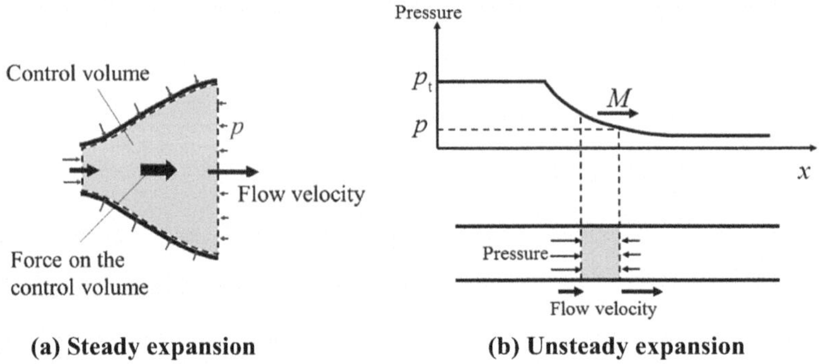

Fig. 11.23 Steady and unsteady expansions

The force balance in an unsteady expansion of a driver gas in a shock tube is shown in Fig. 11.23b. In the flow element (gray section), the force owing to the pressure on the high-pressure (left) side is larger than that on the right side; it is accelerated to the right. A work is done to the element from the high-pressure gas on the left, and a work is done from the element to the low-pressure gas on the right. The work imbalance per area equals to the difference in the product of the pressure and flow velocity between the left and the right. The pressure decreases and the velocity increases from the left to the right. If the product is larger on the left than on the right, then the total enthalpy of the flow element is increased.

Let us formulate the gain in the total enthalpy by unsteady expansion. The substantial derivative of total enthalpy is

$$\frac{\mathrm{D}h_t}{\mathrm{D}t} = \frac{\mathrm{D}h}{\mathrm{D}t} + \frac{\mathrm{D}}{\mathrm{D}t}\left(\frac{u^2}{2}\right). \tag{11.54}$$

Because we assume isentropic flow,

$$T\frac{\mathrm{D}s}{\mathrm{D}t} = \frac{\mathrm{D}h}{\mathrm{D}t} - \frac{1}{\rho}\frac{\mathrm{D}p}{\mathrm{D}t} = 0. \tag{11.55}$$

Taking the inner product of Euler's Eq. (3.18) with $\mathbf{f} = \mathbf{0}$ and $\mathbf{u}$,

$$\rho\left(\frac{\mathrm{D}\mathbf{u}}{\mathrm{D}t}\right)\cdot\mathbf{u} = -(\nabla\mathbf{p})\cdot\mathbf{u}. \tag{11.56}$$

Combining (11.54)–(11.56),

$$\frac{\mathrm{D}h_t}{\mathrm{D}t} = \frac{1}{\rho}\frac{\mathrm{D}p}{\mathrm{D}t} - \frac{1}{\rho}(\nabla\mathbf{p})\cdot\mathbf{u} = \frac{1}{\rho}\left\{\frac{\partial p}{\partial t} + (\mathbf{u}\cdot\nabla)p - (\nabla\mathbf{p})\cdot\mathbf{u}\right\} = \frac{1}{\rho}\frac{\partial p}{\partial t}. \tag{11.57}$$

Equation (11.57) implies that the total enthalpy is increased by unsteady expansion during which the pressure measured at a fixed point increases, see Fig. 11.23b.

Let a quiescent, high-pressure state (with subscript t) and a state after unsteady expansion (without a subscript) correspond to the states L and L*, respectively, in the Riemann problem of Chap. 9. From (9.11),

$$\frac{u}{a_t} = \frac{u}{a}\frac{a}{a_t} = M\frac{a}{a_t} = \frac{2}{\gamma - 1}\left\{1 - \left(\frac{p}{p_t}\right)^{\frac{\gamma-1}{2\gamma}}\right\}. \tag{11.58}$$

$$M = \frac{u}{a} \tag{11.59}$$

From (2.91),

$$\frac{a}{a_t} = \left(\frac{p}{p_t}\right)^{\frac{\gamma-1}{2\gamma}}. \tag{11.60}$$

Substituting (11.58) with (11.60),

$$M = \frac{2}{\gamma - 1}\left\{\left(\frac{p}{p_t}\right)^{-\frac{\gamma-1}{2\gamma}} - 1\right\}. \tag{11.61}$$

Equation (11.61) is the relation between the flow Mach number and the pressure ratio in the unsteady expansion.

As seen in Fig. 11.24, if the flow element is accelerated to a sufficiently low pressure, the unsteady expansion (11.61) yields a higher flow Mach number than the steady expansion (11.53). During steady expansion, the total enthalpy is kept constant. In unsteady expansion, the total enthalpy can be locally increased, although the total energy in the system remains unchanged.

An *expansion tube* (Fig. 11.25) has an *acceleration tube* attached to the end of a

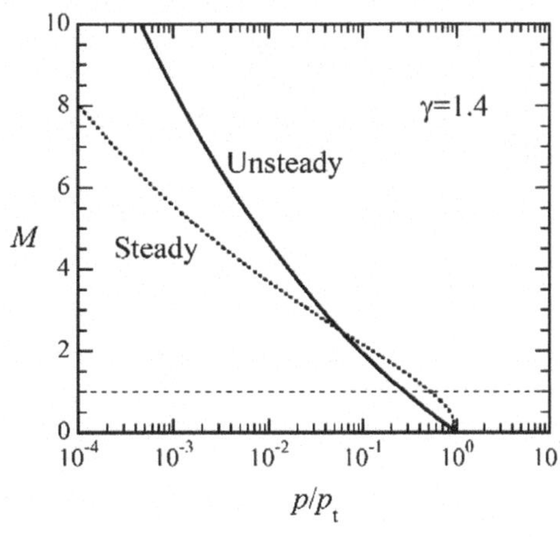

Fig. 11.24 p/p_t versus M. Steady expansion with $M < 1$ is possible in converging duct

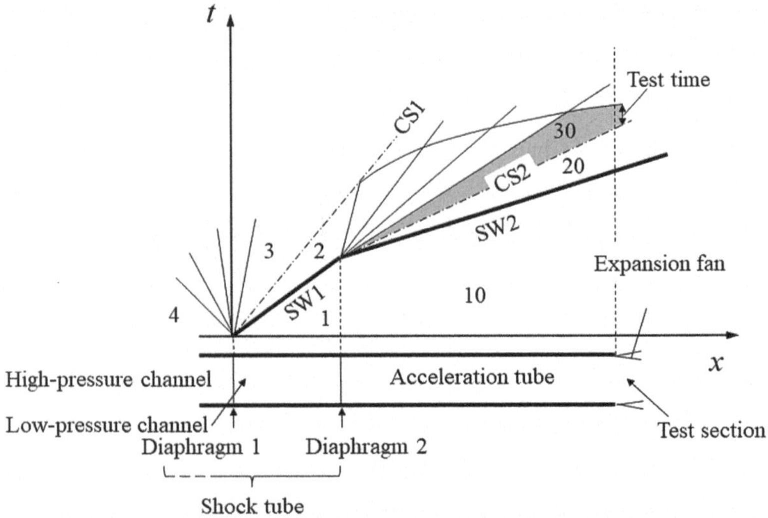

Fig. 11.25 Expansion tube, components, and operation wave diagram

low-pressure channel of a shock tube. The initial pressure in the acceleration tube is even lower than that in the low-pressure channel. A flow element that is shock-compressed in the low-pressure channel experiences unsteady expansion, thereby being accelerated with its total enthalpy increased in the acceleration tube. The test section is at the exit of the acceleration tube. The test flow state 30 (Fig. 11.25) is kept intact in the region enclosed by the expansion fan during the test time. Since the flow element is not stagnated, it does not experience an extremely high-temperature process that is accompanied by a nonequilibrium condition such as electronic excitation, dissociation, or ionization during combustible mixture ignition in the acceleration processes.

However, the test time is considerably short and should be kept intact against the expansion waves and reflected waves from the contact surface to the driver gas. The wave diagram of the associated processes (Fig. 11.25) is rather complicated. Depending on the operation condition and the location of the test section, the test time might not be obtained.

Let us analyze the unsteady acceleration of flow element 2 in Fig. 11.25. The total enthalpy is given by

$$h_{t,2} = h_2 + \frac{1}{2}u_2^2 = h_2\left(1 + \frac{\gamma - 1}{2}M_2^2\right). \tag{11.62}$$

When this flow element experiences unsteady expansion, from (8.44),

$$u + \frac{2a}{\gamma - 1} = u_2 + \frac{2a_2}{\gamma - 1}. \tag{11.63}$$

Labeling the stagnation state of 2 with the subscript "t,2", the corresponding isentropic relation is

$$a = a_2 \left(\frac{p}{p_2}\right)^{\frac{\gamma-1}{2\gamma}} = a_{t,2}\left(\frac{p}{p_{t,2}}\right)^{\frac{\gamma-1}{2\gamma}}.$$

Therefore,

$$\frac{h_t}{h_{t,2}} = \frac{h + \frac{1}{2}u^2}{h_{t,2}} = \frac{\frac{a^2}{\gamma-1} + \frac{1}{2}u^2}{\frac{a_{t,2}^2}{\gamma-1}} = \frac{1}{\gamma - 1}\left[(\gamma + 1)\left(\frac{p}{p_{t,2}}\right)^{\frac{\gamma-1}{\gamma}} - 4\left(\frac{p}{p_{t,2}}\right)^{\frac{\gamma-1}{2\gamma}} + 2\right]. \tag{11.64}$$

From (11.61),

$$M = \frac{2}{\gamma - 1}\left\{\left(\frac{p}{p_{t,2}}\right)^{-\frac{\gamma-1}{2\gamma}} - 1\right\}. \tag{11.65}$$

Differentiating (11.64) with (11.65),

$$\frac{dh_t}{h_{t,2}} = \frac{\gamma - 1}{\gamma}\left(\frac{p}{p_{t,2}}\right)^{-\frac{1}{\gamma}}(1 - M)\frac{dp}{p_{t,2}}. \tag{11.66}$$

h_t is increased during supersonic expansion in which $M > 1$ and $dp < 0$. As seen in Fig. 11.26, at a relatively low pressure $\left(p/p_{t,2} > 0.058\right)$ or low Mach number $(M < 2.5)$, the total enthalpy decreases. Yet, when accelerating to lower pressure, it increases.

From (11.60), in the ultimate condition of null pressure,

$$\left.\frac{h_t}{h_{t,2}}\right|_{\frac{p}{p_{t,2}} \to 0} = \frac{2}{\gamma - 1}. \tag{11.67}$$

This corresponds to the total enthalpy at the escape speed (9.23). When $\gamma = 1.4$, (11.67) equals 5. Considering that the total enthalpy behind a reflected shock wave is increased by a factor of 1.3 at most, the unsteady expansion is very effective for locally increasing the total enthalpy.

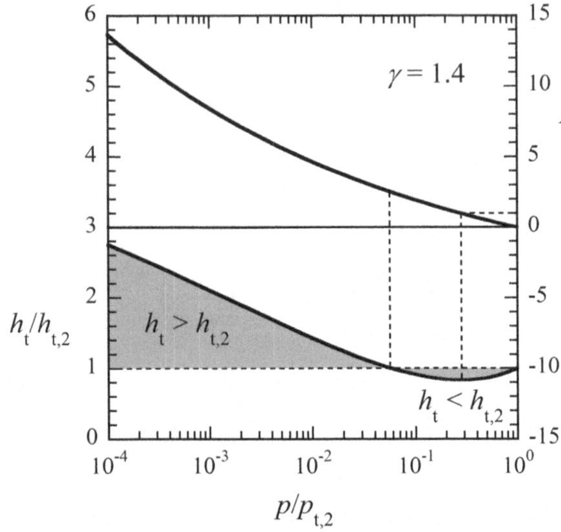

Fig. 11.26 Variation of total enthalpy and Mach number through unsteady expansion

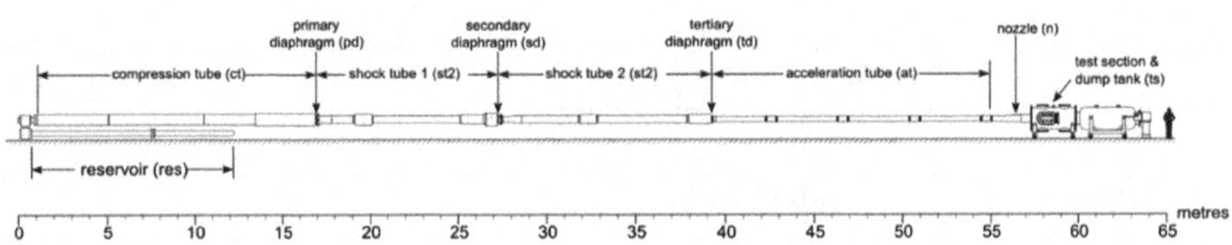

Fig. 11.27 Expansion tube X-3 at University of Queensland. Courtesy of University of Queensland

The X-3, a large expansion tube at the University of Queensland (Fig. 11.27), uses a free-piston driver as the shock tunnel (Stalker tube) in the same place.

11.7 Ballistic Range

A ballistic range (Fig. 11.28) is a high-speed launcher in which a flight model, often termed a *projectile*, is accelerated using high-pressure gas. This realizes a supersonic flight experiment in the laboratory, thus vastly saving on the costs of full-scale flight. The range is also utilized in hypervelocity impact experiments. The flight model is accelerated by a driver gas at high pressure. The higher the speed of the flight model, the lower the driver gas pressure behind the flight model.

Before launch operation, the gas in front of the flight model is quiescent with a static pressure in the test section. The higher the acceleration of the flight model,

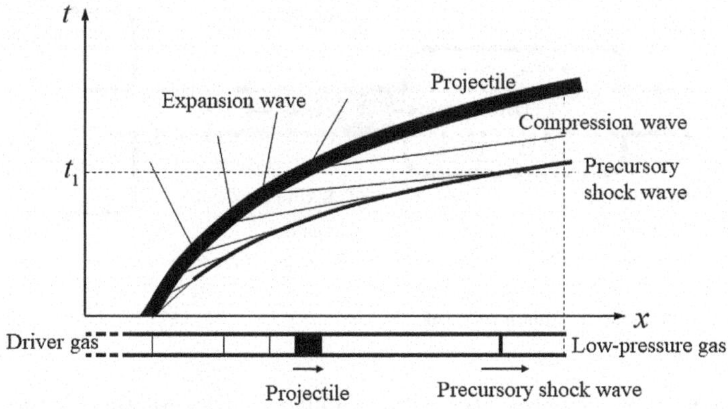

Fig. 11.28 Ballistic range, operation principle, and operation wave diagram

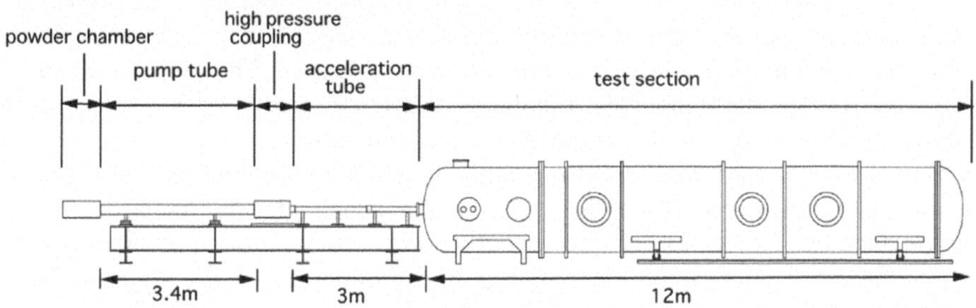

Fig. 11.29 Two-stage, light-gas gun at Institute of Fluid Science (IFS), Tohoku University. Courtesy of IFS

the greater the compression of the test gas in front of the flight model. Eventually, a *precursor shock wave* appears and propagates ahead of the flight model.

Figure 11.28 shows a wave diagram of the projectile (or flight model), the precursor shock wave, and associated pressure waves. The lower picture indicates their locations at $t = t_1$. If the driver section has a finite length, then the expansion waves behind the projectile can be reflected from the (leftmost) end wall. As long as the reflected waves do not catch up with the projectile, the condition immediately behind the projectile is obtained from a Riemann invariant.

High-speed launch experiments are important not only in fluid dynamics but also for investigating hypervelocity impact phenomena. For example, in low Earth orbits, space debris moves at an orbital speed on the order of 8 km/s. During a mutual collision, the relative speed can even double. In order to obtain a high launch speed, the speed of sound of the driver gas needs to be high. For this purpose, light gas (hydrogen and helium) is often used.

In order to further increase the speed of sound, the driver gas is sometimes heated. Another method to increase the speed of sound is to utilize a free-piston driver. A two-stage, light-gas gun (Fig. 11.29) is a typical hypervelocity ballistic range. So far, we

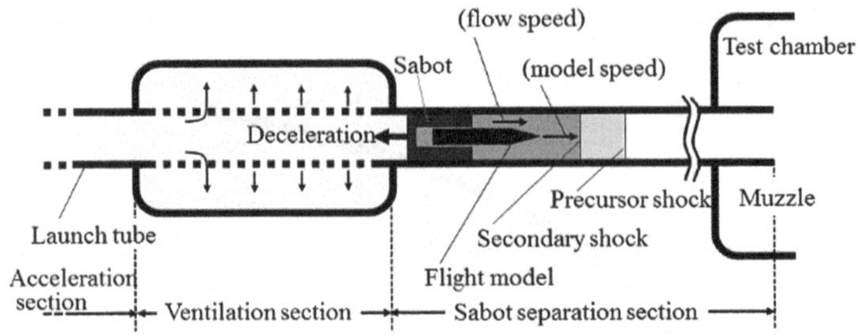

Fig. 11.30 In-tube catapult launch method

have dealt with only the *aeroballistic range*. Other devices utilizing electromagnetic forces (rail gun, coil gun, etc.) can be used.

Usually, the cross section of the acceleration tube in a ballistic range has a circular bore. Many or most flight models do not fit this shape. In order to fit the projection area to the tube cross section, a *sabot* of the same cross-sectional shape as the bore, and contacts the flight model during the acceleration is used. The sabot needs to be separated from the flight model before the test section. Usually, the sabot separation is done after the *muzzle*, which is the exit of the launch tube.

The *in-tube catapult launch* [8] is a unique method to separate the sabot before the muzzle. In this method (see Fig. 11.30), in the launch tube after the acceleration section there exist a *ventilation section* and a *sabot separation section*, thereby separating the sabot before the muzzle. In the acceleration section, the flight model is accelerated together with the sabot. In the ventilation section, the gas in the launch tube is exhausted through many ventilation holes so that the precursor shock wave is attenuated and the pressure behind the sabot is also decreased.

When the flight model and the sabot enter the sabot separation section, which has the same cross-sectional shape as the acceleration section, the pressure in front of the sabot and behind the precursor shock wave should be higher than that behind the sabot, thereby decelerating the sabot. However, the pressure around the flight model is kept uniform because the speed of the post-shock flow behind the precursor shock wave equals that of the flight model, and the flight model is drag-free. Owing to the difference in drag between the sabot and the flight model, the sabot is separated from the flight model in the sabot separation section. However, this separation should be completed near the muzzle in order to suppress the yaw and pitch motions of the light model to be as small as possible.

Figure 11.31 shows an aeroballistic range with an in-tube catapult launch scheme and an example of a free-flight experiment. Helium was used as the driver gas. Before the launch operation, the driver gas was separated from the launch tube using a free piston. Once the gas behind the free piston was released, the separation is broken with the free-piston pushed back by the driver gas. Then, the flight model held in the sabot starts to be accelerated.

The launch tube has a rectangular bore (Fig. 11.31b) so that a flight model with a large aspect ratio such as a winged body (Fig. 11.31c) can be launched. In the

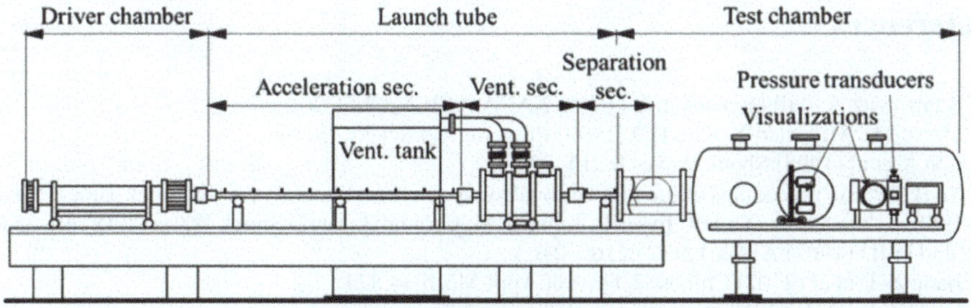

(a) Entire system

(b) Bore

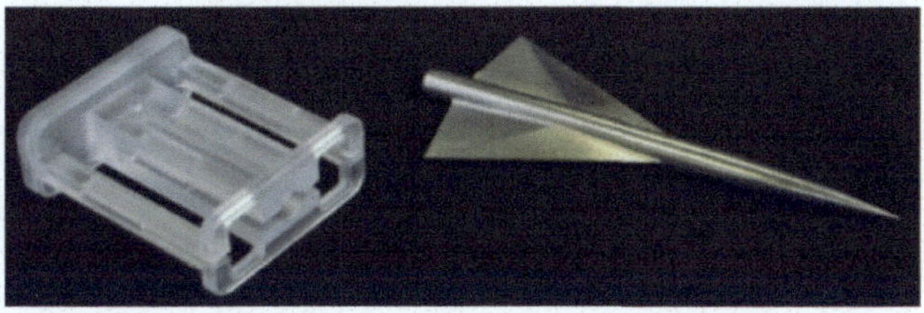

(c) Example of sabot and flight model

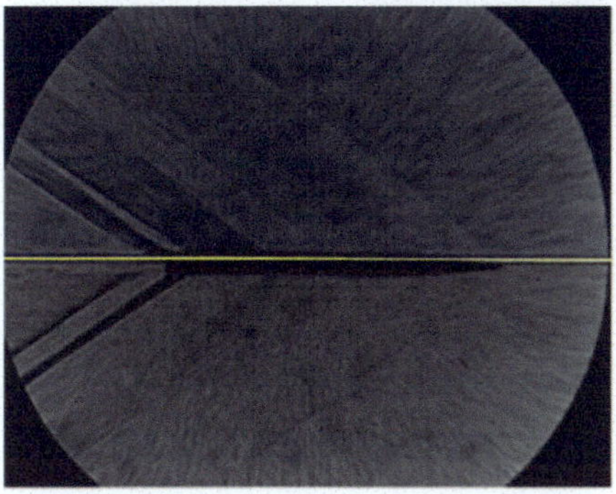

(d) Schlieren image, Mach number 1.7, cient under over flight model of (c)

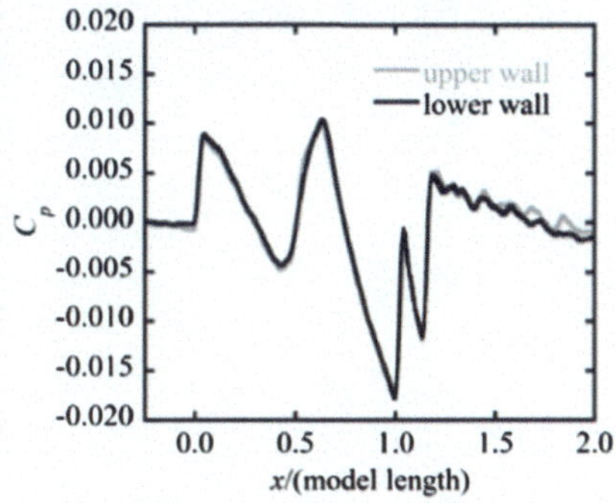

(e) Time variation of pressure coefficient unsame condition as (d)

Fig. 11.31 Rectangular-bore ballistic range with shot example (Nagoya University)

experiment, the flow field around the model was visualized by a high-speed Schlieren arrangement (Fig. 11.31d) and the pressure history was measured on a flat plate at a separation distance of 1.5 model lengths (shown as the variation in the pressure coefficient, C_p, in Fig. 11.31e).

References

1. Kantrowitz A, duP Donaldson C (1945) NACA ACR No. L5D20
2. Matsuo K, Miyazato Y, Kim HD (1999) Prog Aerosp Sci 35:33–100
3. Itoh K et al (1998) Shock Waves 8:215–233
4. Yu H (1988) Proceedings of the fifth national symposium on Shock Waves & Shock Tubes, Nov, 1988, Xi'an, China./Yu HR, Esser B, Lenartz M, Grönig H (1992) Shock Waves 2(4):245–254
5. Stalker RJ (1967) AIAA J 5(12):2160–2165
6. Jiang Z–L et al (2102) Chinese J Theoret Appl Mech 44:824
7. Jiang Z–L, Yu H (2017) Natl Sci Rev 4(3):290–296
8. Sasoh A et al (2015) AIAA J 53:2781–2784

The manufacturer's authorised representative in the EU is Springer Nature Customer Service Centre GmbH, Europaplatz 3, 69115 Heidelberg, Germany. If you have any concerns regarding our products, please contact ProductSafety@springernature.com

Printed and bound by CPI Group (UK) Ltd, Croydon, CR0 4YY
25/09/2026
02222901-0003